CIRRUS LOGIC - COLORADO
INTERLOCKEN BUSINESS PARK
100 Technology Drive, Suite 300
Broomfield, Colorado 80021
Telephone (303) 466-5228
FAX (303) 466-5482

If you wish to receive updates to this book, please copy this form, complete it and send it to the above address.

Please add my name to your permanent mailing list for updates to the book: PRACTICAL ERROR CORRECTION DESIGN FOR ENGINEERS (Revised Second Edition)

Name ______________________________

Title ______________________________

Organization ______________________________

Address ______________________________

____

Phone ______________________________

Other Areas of Interest ______________________________

____

PRACTICAL ERROR CORRECTION DESIGN FOR ENGINEERS

REVISED SECOND EDITION

PRACTICAL ERROR CORRECTION DESIGN FOR ENGINEERS

REVISED SECOND EDITION

Neal Glover and Trent Dudley

Published By:
Cirrus Logic - Colorado
Broomfield, Colorado 80020
(303) 466-5228

Second Edition Revision 1.1

Published by:

CIRRUS LOGIC - COLORADO
INTERLOCKEN BUSINESS PARK
100 Technology Drive, Suite 300
Broomfield, Colorado 80021
Phone (303) 466-5228 FAX (303) 466-5482

Second Printing 1990.
ISBN #0-927239-00-0

To my children,

Rhonda, Karen, Sean, and Robert

Neal Glover

To the memory of my parents,

Robert and Constance

Trent Dudley

PREFACE

The study of error-correcting codes is now more than forty years old. There are several excellent texts on the subject, but they were written mainly by coding theorists and are based on a rigorous mathematical approach. This book is written from a more intuitive, practical viewpoint. It is intended for practicing engineers who must specify, architect, and design error-correcting code hardware and software. It is an outgrowth of a series of seminars presented during 1981 and 1982 on practical error-correction design.

An engineer who must design an error-control system to meet data recoverability, data accuracy, and performance goals must become familiar with the characteristics and capabilities of different types of EDAC codes as well as their implementation alternatives, including tradeoffs between hardware and software complexity, speed/space/ cost, etc. Our goal is to provide this information in a concise manner from a practical engineering viewpoint. Numerous examples are used throughout to develop familiarity and confidence in the methods presented. Most proofs and complex derivations have been omitted; these may be found in theoretical texts on error correction coding.

We would like to thank our friends for their assistance and advice. The engineers attending DST's seminars also deserve thanks for their suggestions.

Neal Glover
Trent Dudley

Broomfield, Colorado
August 1988

ABOUT CIRRUS LOGIC - COLORADO

Cirrus Logic - Colorado was originally founded in 1979 as Data System Technology (DST) and was sold to Cirrus Logic, Inc., of Milpitas, California, on January 18, 1990. Cirrus Logic - Colorado provides error detection and correction (EDAC) products and services to the electronics industries. We specializes in the practical implementation of EDAC, recording and data compression codes to enhance the reliability and efficiency of data storage and transmission in computer and communications systems, and all aspects of error tolerance, including framing, synchronization, data formats, and error management.

Cirrus Logic - Colorado also develops innovative VLSI products that perform complex peripheral control functions in high-performance personal computers, workstations and other office automation products. The company develops advanced standard and semi-standard VLSI controllers for data communications, graphics and mass storage.

Cirrus Logic - Colorado was a pioneer in the development and implementation of computer-generated codes to improve data accuracy. These codes have become widely used in magnetic disk systems over the past few years and are now defacto standards for 5¼ inch Winchester drives. Cirrus Logic - Colorado developed the first low-cost high-performance Reed-Solomon code integrated circuits; the codes implemented therein have become worldwide standards for the optical storage industry. EDAC codes produced by Cirrus Logic - Colorado have become so associated with high data integrity that many users include them in their lists of requirements when selecting storage subsystems.

Cirrus Logic - Colorado licenses EDAC software and discrete and integrated circuit designs for various EDAC codes, offers books and technical reports on EDAC and recording codes, and conducts seminars on error tolerance and data integrity as well as EDAC, recording, and data compression codes.

PRODUCTS

- Error tolerant controller designs for magnetic and optical storage.
- Turnkey integrated circuit development.
- Low-cost, high-performance EDAC integrated circuit designs.
- Discrete and integrated circuit designs for high-performance Reed-Solomon codes, product codes, and computer-generated codes.
- Universal encoder/decoder designs for Reed-Solomon codes including bit-serial, time slice, and function sharing designs.
- Multiple-burst EDAC designs for high-end storage devices with high-speed parallel interfaces, supporting record lengths beyond 100,000 bytes.

- EDAC designs supporting QIC tape formats.
- Software written for a number of processors to support integrated circuits implementing Cirrus Logic - Colorado's EDAC technology.
- *Practical Error Correction Design for Engineers,* a book on EDAC written especially for engineers.
- Cirrus Logic - Colorado develops polynomials for use in storage products.

CONSULTING SERVICES

Consulting services are offered in the following areas:

- Semiconductor memories and large cache memories
- Magnetic disk devices
- Magnetic tape devices
- Optical storage devices using read-only, write-once, and erasable media
- Smart cards
- Communications

Consulting services offered include:

- Code selection
- Design of discrete hardware and integrated circuits
- Design of software
- Advice in the selection of synchronization, header, and defect management strategies
- Complex MTBF computations
- Analysis of existing codes and/or designs
- Establishing EDAC requirements from defect data
- Assistance in system integration of integrated circuits implementing Cirrus Logic's EDAC technology.

PROLOGUE
THE COMING REVOLUTION IN ERROR CORRECTION TECHNOLOGY

By: Neal Glover

Presented at ENDL's 1988 Disk/Test Conference

INTRODUCTION

The changes that are occurring today in error detection and correction, error tolerance, and failure tolerance are indeed revolutionary. Two major factors are driving the revolution: need and capability. The need arises from more stringent error and failure tolerance requirements due to changes in capacity, through-put, and storage technology. The capability is developing due to continuing increases in VLSI density and decreases in VLSI cost, along with more sophisticated error-correction techniques. This preface discusses the changes in requirements, technology, and techniques that are presently occurring and those that are expected to occur over the next few years.

Some features of today's error-tolerant systems would have been hard to imagine a few years ago.

Some optical storage systems now available are so error tolerant that user data is correctly recovered even if there exists a defect situation so gross that the sector mark, header and sync mark areas of a sector are totally obliterated along with dozens of data bytes.

Magnetic disk drive array systems under development today are so tolerant to errors and failures that simultaneous head crashes on two magnetic disk drives would neither take the system down nor cause any loss of data. Some of these systems will also be able to detect and correct many errors that today go undetected, such as transient errors in unprotected data paths and buffers and even software errors that result in the transfer of the wrong sector. Some magnetic disk drive array systems specify mean time between data loss (MTBDL) in the hundreds of thousands of hours.

The contrast with prior-generation computer systems is stark. Before entering development I spent some time on a team maintaining a large computer at a plant in California that developed nuclear reactors. I will never forget an occasion when the head of computer operations pounded his fist on a desk and firmly stated that if we saw a mushroom cloud over Vallecito it would be the fault of our computer. The mainframe's core memory was prone to intermittent errors. The only checking in the entire computer was parity on tape. Punch card decks were read to tape twice and compared.

By the mid-seventies, the computer industry had come a long way in improving data integrity. I had become an advisory engineer in storage-subsystem development, and in 1975 I was again to encounter a very unhappy operations manager when a micro-code bug, which I must claim responsibility for, intermittently caused undetected erroneous data to be transferred in a computer system at an automobile manufacturing plant. Needless to say, the consequences were disastrous. This experience taught me the importance of exhaustive firmware verification testing and has influenced my desire to incorporate data-integrity features in Cirrus Logic's designs that are intended to detect and in some cases even correct for firmware errors as well as hardware errors.

Changes in hardware and software data-integrity protection methods are occurring today at a truly revolutionary rate and soon the weaknesses we knew of in the past and those that we live with today will be history forever.

THE CHANGING REQUIREMENTS

Requirements for error and failure tolerance increase with capacity and through-put, and changing storage technology. Over the years, many storage systems have specified their non-recoverable read error rate as 1.E-12 events per bit. In many cases this is no longer acceptable. As more sophisticated applications require ever faster access to ever larger amounts of information, system integrators will demand that storage system manufacturers meet much higher data-integrity standards.

As an example of how capacity influences error tolerance requirements, consider a hypothetical write-once optical storage device employing removable 5 gigabyte cartridges. Twenty-five such cartridges would hold 1.E+12 bits, so a non-recoverable read error rate of 1.E-12 would imply the existence of a non-recoverable read error on about one in twenty-five cartridges. Is this acceptable? Would one non-recoverable read error in every 250 platters be acceptable?

As an example of how through-put influences error tolerance requirements, consider a magnetic disk array subsystem which is designed to transfer data simultaneously from all drives and has no redundant drives. The through-put of ten 10-megabit-per-second magnetic disk drives operating with a ten percent read duty cycle would be 8.64E+11 bits per day. A 1.E-12 non-recoverable read error rate would imply one non-recoverable read error every eleven days. Is this acceptable? Would one non-recoverable read error per year be acceptable?

For new storage technologies, it is often not practical to achieve the low media defect event rates which we have been accustomed to handling in magnetic storage. New techniques have been and must continue to be developed and implemented to accommodate higher defect rates and different defect characteristics.

THE CHANGING TECHNOLOGY

VLSI density continues to increase, allowing us to incorporate logic on a single integrated circuit today that a few years ago would have required several separate boards. This allows us to implement very complex data-integrity functions within a single IC. Cirrus Logic's low-cost, high-performance, Reed-Solomon code IC's for optical storage devices are a good example. As VLSI densities increase, such functions will occupy a small fraction of the silicon area of a multi-function IC. The ability to place very complex functions on a single IC and further to integrate multiple complex functions on a single IC opens the door for greater data integrity. Our ability to achieve greater data integrity at reasonable cost is clearly one of the forces behind the revolution in error and failure tolerant technology.

Even with the development of cheaper, higher density VLSI technology, it is often more economical to split the implementation of high-performance EDAC systems between hardware and software. Using advanced software algorithms and buffer management techniques, nearly "on-the-fly" correction performance can be achieved at lower cost than using an all-hardware approach.

CHANGES IN ERROR CORRECTION

For single-burst correction, Cirrus Logic - Colorado still recommends computer-generated codes. Most new designs employing computer-generated codes are using binary polynomials of degree 48, 56, and 64. In many cases, implementations of the higher degree polynomials include hardware to assist in performing on-the-fly correction.

Economic and technical factors are driving the industry to accommodate higher defect rates to which single-burst error-correction codes are not suited. Consequently, Reed-Solomon codes, a class of powerful codes which allow efficient correction of multiple bursts, are currently being designed into a wide variety of storage products including magnetic tape, magnetic disk, and optical disk. Reed-Solomon codes were discovered more than twenty-five years ago but only recently have improved encoding and decoding algorithms, along with decreased VLSI costs, made them economical to implement. Using software decoding techniques running on standard processors, Cirrus Logic - Colorado now routinely achieves correction times for Reed-Solomon codes that were difficult to achieve with bit-slice designs just a few years ago.

IBM has announced a new version of its 3380 magnetic disk drive that employs multiple-burst error detection and correction, using Reed-Solomon codes, to achieve track densities significantly higher than realizable with previous technology. Single-burst error correction can handle modest defect densities, but defect densities increase exponentially with track density. On-the-fly, multiple-burst error correction and error-tolerant synchronization are required to handle these higher defect densities. On earlier models of the 3380, IBM corrected a single burst in a record of up to several thousand bytes. Using IBM's 3380K error-correction code, under the right circumstances it would be possible to correct hundreds of bursts in a record. A unique feature of the 3380K code is that it can be implemented to perform on-the-fly correction with a data delay that is roughly 100 bytes.

The impact of this IBM announcement, coupled with the general push toward higher track densities, the success of high-performance error detection and correction on optical storage devices, and the availability of low-cost, high-performance EDAC IC's, will stimulate the use of high-performance EDAC codes on a wide range of magnetic disk products. Cirrus Logic - Colorado itself is currently implementing double-burst correcting, Reed-Solomon codes on a wide range of magnetic disk products, ranging from low-end designs which process one bit per clock edge to high-end designs which process sixteen bits per clock edge.

CHANGES IN ERROR DETECTION

When an error goes undetected, erroneous data is transferred to the user as if it were error free. The transfer of undetected erroneous data can be one of the most catastrophic failures of a data storage system. Some causes of undetected erroneous data transfer are listed below.

- Miscorrection by an error-correction code.
- Misdetection by an error-detection or error-correction code.
- Synchronization failure in an implementation without synchronization framing error protection.
- Intermittent failure in an unprotected data path on write or read.
- Intermittent failure in an unprotected RAM buffer on write or read.
- A software error resulting in the transfer of the wrong sector.
- Failed hardware, such as a failed error latch that never flags an error.

It is important to understand that no error-correction code is perfect; all are subject to miscorrection when an error event occurs that exceeds the code's guarantees. However, it is also important to understand that the miscorrection probability for a code can be reduced to any arbitrarily low level simply by adding enough redundancy. As VLSI costs go down, more redundancy is being added to error-detection and error-correction codes to achieve greater detectability of error events exceeding code guarantees. New single-burst error-correction code designs use polynomials of degree 48, 56, and 64 to accomplish the same correctability achieved with degree 32 codes several years ago, but with significantly improved detectability. If correctability is kept the same, detectability is improved more than nine orders of magnitude in moving from a degree 32 code to a degree 64 code.

Error-detection codes are not perfect either; they are subject to misdetection. Like miscorrection, misdetection can be reduced to any arbitrarily low level by adding enough redundancy. Unfortunately, the industry has not, in general, increased the level of detectability of implemented error-detection codes significantly in the last twenty-five years. Two degree 16 polynomials, CRC-16 and CRC-CCITT, have been in wide use for many years. For many storage device applications, there are degree 16 polynomials with superior detection capability, and moreover, the requirements of many applications would be better met by error-detection polynomials of degree 32 or greater.

In the last few years, the industry has been doing a better job of avoiding pattern sensitivities of error-detection and error-correction codes. Cirrus Logic - Colorado avoids using the Fire code because of its pattern sensitivity, and we use 32-bit auxiliary error detection codes in conjunction with our Reed-Solomon codes in order to overcome their interleave pattern sensitivity.

Auxiliary error-detection codes that are used in conjunction with ECC codes to enhance detectability have special requirements. The error-detection code check cannot be made until after correction is complete. It is undesirable to run corrected data through error-detection hardware after performing correction due to the delay involved. It is also not feasible to perform the error-detection code check as data is transferred to the host after correction, since some standard interfaces have no provision for a device to flag an uncorrectable sector after the transfer of data has been completed. To meet these requirements, some error-detection codes developed over the last few years are specially constructed so that their residues can be adjusted as correction occurs. When correction is complete, the residue should have been adjusted to zero. Cirrus Logic - Colorado has been using such error-detection codes since 1982, and such a code is included within Cirrus Logic - Colorado Reed-Solomon code IC's for optical storage. IBM's 3380K also uses such an auxiliary error-detection code.

As the requirements for data integrity have increased, Cirrus Logic - Colorado has tightened its recommendations accordingly. One of the areas needing more attention in the industry is synchronization framing error protection. To accomplish this protection, Cirrus Logic - Colorado now recommends either the initialization of EDAC shift registers to a specially selected pattern or the inversion of a specially selected set of EDAC redundancy bits.

The magnetic disk drive array segment of the industry is making significant gains in detectability. Some manufacturers are adding two redundant drives to strings of ten data drives in order to handle the simultaneous failure of any two drives without losing data. The mean time between data loss (MTBDL) for such a system computed from the MTBF for individual drives may be in the millions of hours. In order for these systems to meet such a high MTBDL, all sources of errors and transient failures that could dominate and limit MTBDL must be identified, and means for detection and correction of such errors and failures must be developed. For these systems, Cirrus Logic - Colorado recommends that a four-byte error-detection code be appended and checked at the host adapter. We also recommend that the logical block number and logical drive number be included in this check. This allows the detection with high probability of a wide variety of errors and transient failures, including the transfer of a wrong sector or transfer of a sector from the wrong drive.

CHANGES IN TRACK-FORMAT ERROR TOLERANCE

In many of today's single-burst-correcting EDAC designs, tolerance to errors is limited by the ability to handle errors in the track format rather than by the capability of the data-field EDAC code. In upgrading such designs, it is pointless to change from single-burst to multiple-burst error correction without also improving track-format error tolerance. In the future, all magnetic disk products will use error-tolerant synchronization and header strategies.

The optical storage industry has already proved the feasibility of handling error rates as high as 1.E-4 through track-format error tolerance as well as powerful data-field EDAC codes. Optical track-format error tolerance has been achieved using multiple headers, error-tolerant sync marks, and periodic resynchronization within data fields. Some systems now available are so error tolerant that user data is correctly recovered even if there exists a defect situation so gross that the sector mark, header, and sync mark areas of a sector are totally obliterated along with dozens of data bytes.

CHANGES IN DEFECT MANAGEMENT

As track densities increase in magnetic recording, and as erasable optical technology becomes more common, many companies will implement defect skipping to handle higher defect densities without significantly affecting performance. This technique is not applicable to write-once optical applications, where sector retirement and reassignment will be used. Such techniques also work well within dynamic defect management strategies. Combining the two will allow the full power of the EDAC code to be used for margin against new defects. Dynamic defect management will become more common, especially for write-once and erasable optical technologies subject to relatively high new defect rates and defect growth.

As more complex and intelligent device interfaces and controllers are implemented, more responsibility for defect management will be shifted from the host to the device controller.

CHANGES IN SELF-CHECKING

As data integrity requirements increase, it becomes very important to detect transient hardware failures. New designs for component IC's for controller implementations are carrying parity through the data paths of the part when possible, rather than just checking and regenerating parity. Cirrus Logic - Colorado sees this as a step forward, but we also look beyond, to the day when all data paths are protected by CRC as well.

It is especially important to detect transient failures in EDAC hardware. Some companies have implemented parity-predict circuitry to continuously monitor their EDAC shift registers for proper operation.

When possible, Cirrus Logic - Colorado has incorporated circuitry to divide codewords on write by a factor of the code generator polynomial and check for zero remainder. This function is performed as close to the recording head as possible.

Cirrus Logic - Colorado's 8520 IC uses dynamic cells for the major EDAC shift registers. To detect transient failures in the shift registers themselves, we incorporated a feature whereby the parity of all bits going into a shift register is compared with the parity of all bits coming out of the shift register.

CHANGES IN VERIFICATION AND TESTING

The traditional diagnostic technique for storage-device EDAC circuitry uses write long and read long. For write-once optical media, this technique has two problems. Since these are high error rate devices, real errors may be encountered along with simulated errors. Also, each write long operation uses up write-once media. Cirrus Logic - Colorado incorporates a special diagnostic mode in its EDAC IC's that allows the EDAC hardware to be tested without writing to or reading from the media.

The introduction of complex, high-performance hardware and software algorithms for error correction and track-format error tolerance introduce new verification and testing challenges. Cirrus Logic - Colorado verifies its error-correction software for optical storage devices against millions of test cases. To verify the track-format error tolerance of optical storage devices, Cirrus Logic - Colorado recommends a track format simulator that allows all forms of errors to be simulated, including slipped PLL cycles. Cirrus Logic - Colorado plans to market such a track simulator in the future. Cirrus Logic - Colorado also recommends programmable buggers to allow all forms of errors to be simulated during the performance of a wide range of operational tasks on real devices.

CHALLENGES FOR THE FUTURE

Many of the factors shaping the future of error correction and error tolerance have already been discussed. One of the most significant will be carry-through error detection that will be generated and checked for each sector at the host adapter. The redundancy for this overall check will include the logical block number and the logical drive number and will cover the entire path from the host adapter to the media and back. A logical next step will be for hosts to provide an option for carrying all or part of the overall check code redundancy through host memory when data is being moved from one device to another. Looking further into the future, we may also see the redundancy for the overall check maintained in host memory for those sectors that are to be updated. In this case, an updatable error-detection code will be used and the error-detection redundancy will be adjusted for each change made to the contents of the sector.

An area that needs more attention is verification that we will be able to properly read back all the data that we write. To avoid adversely impacting performance, we must be able to accomplish this without following each write with a verify read. At the closest possible point to the head we need to verify that the written user write data and associated redundancy constitute a valid codeword. A good forward step in this direction would be to decode the write encoded RLL bits back to data bits and to divide this data stream by the code generator polynomial or compare it to the write data stream going into the encoder.

CONTENTS

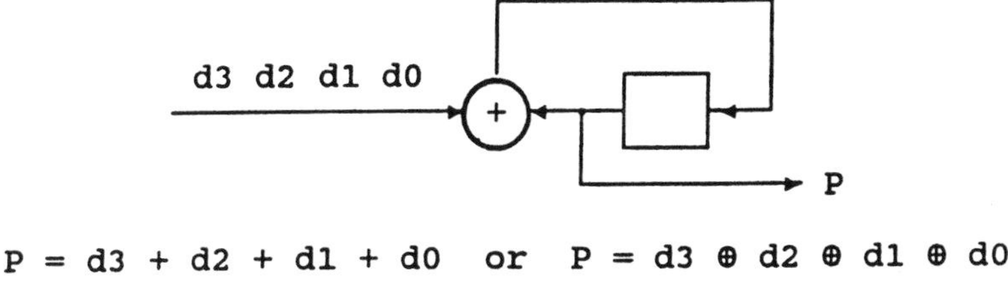

CHAPTER 1 - INTRODUCTION

1.1 INTRODUCTION TO ERROR CORRECTION

1.1.1 *A REVIEW OF PARITY*

A byte, word, vector, or data stream is said to have odd parity if the number of '1's it contains is odd. Otherwise, the byte, word, vector, or data stream is said to have even parity. Parity may be determined with combinational or sequential logic.

The parity of two bits may be determined with an EXCLUSIVE-OR (XOR) gate. The circled '+' symbol is understood to represent XOR throughout this book.

d1 Odd -OR- d1 + Odd
d0 d0

Parity across a nibble may be determined with a parity tree.

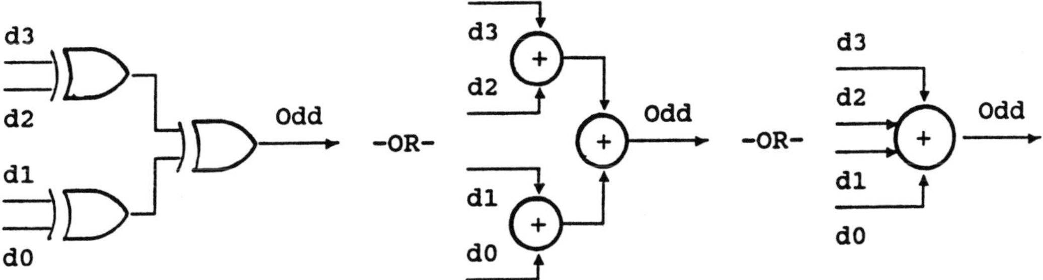

Parity of a bit stream may be determined by a single shift register stage and one XOR gate. The shift register is assumed to be initialized to zero. The highest numbered bit is always transmitted and received first.

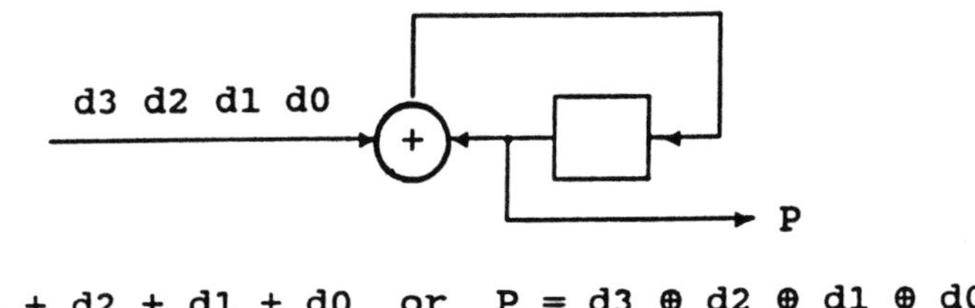

$$P = d3 + d2 + d1 + d0 \quad \text{or} \quad P = d3 \oplus d2 \oplus d1 \oplus d0$$

The circuit below determines parity across groups of data stream bits.

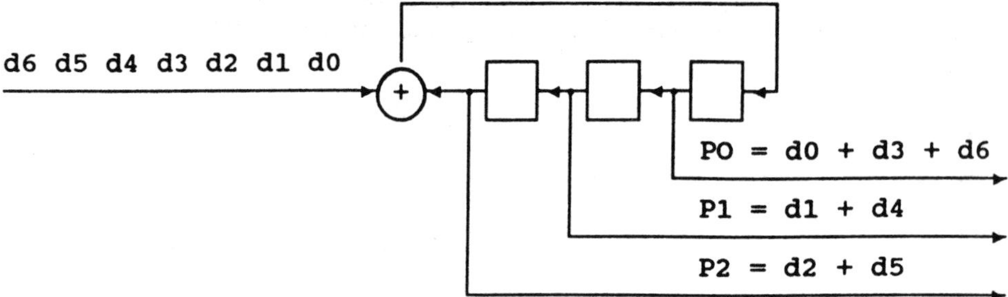

Note that each bit is included in only one parity check.

The circuit below will also determine parity across groups of data stream bits.

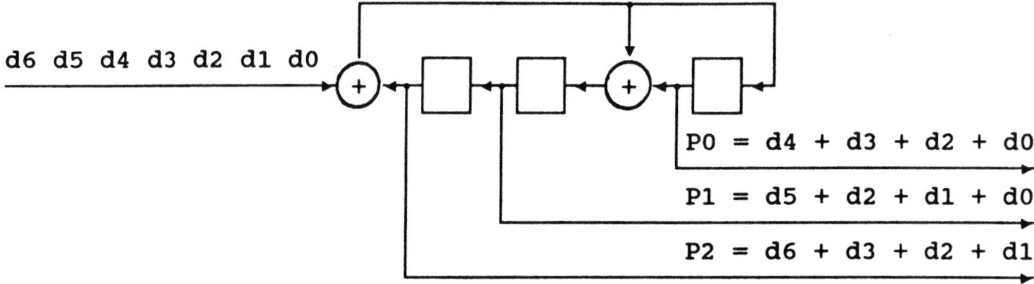

The contribution of each data bit to the final shift register state is shown below. Each data bit affects a unique combination of parity checks.

Data Bit	Contribution P2 P1 P0
d6	100
d5	010
d4	001
d3	101
d2	111
d1	110
d0	011

The contributions to the final shift register state made by several strings of data bits are shown below.

String		Contribution P2 P1 P0
d6,d4	=>	101
d3,d2,d0	=>	001
d4,d0	=>	010

The contribution to the final shift register state by each string is the XOR sum of contributions from individual bits of the string, because the circuit is linear. For a linear function f:

$$f(x+y) = f(x)+f(y)$$

The parity function P is linear, and therefore

$$P(x+y) = P(x)+P(y)$$

Circuits of this type are the basis of many error-correction systems.

1.1.2 *A FIRST LOOK AT ERROR CORRECTION*

This discussion presents an introduction to single-bit error correction using a code that is intuitive and simple. Consider the two-dimensional parity-check code defined below.

Check-Bit Generation	Syndrome Generation
P0 = d0 + d4 + d8 + d12	S0 = d0 + d4 + d8 + d12 + P0
P1 = d1 + d5 + d9 + d13	S1 = d1 + d5 + d9 + d13 + P1
P2 = d2 + d6 + d10 + d14	S2 = d2 + d6 + d10 + d14 + P2
P3 = d3 + d7 + d11 + d15	S3 = d3 + d7 + d11 + d15 + P3
P4 = d12 + d13 + d14 + d15	S4 = d12 + d13 + d14 + d15 + P4
P5 = d8 + d9 + d10 + d11	S5 = d8 + d9 + d10 + d11 + P5
P6 = d4 + d5 + d6 + d7	S6 = d4 + d5 + d6 + d7 + P6
P7 = d0 + d1 + d2 + d3	S7 = d0 + d1 + d2 + d3 + P7

				Row Checks
d0	d1	d2	d3	P7
d4	d5	d6	d7	P6
d8	d9	d10	d11	P5
d12	d13	d14	d15	P4
P0	P1	P2	P3	

Column Checks

One of the eight required check/syndrome circuits is shown below.

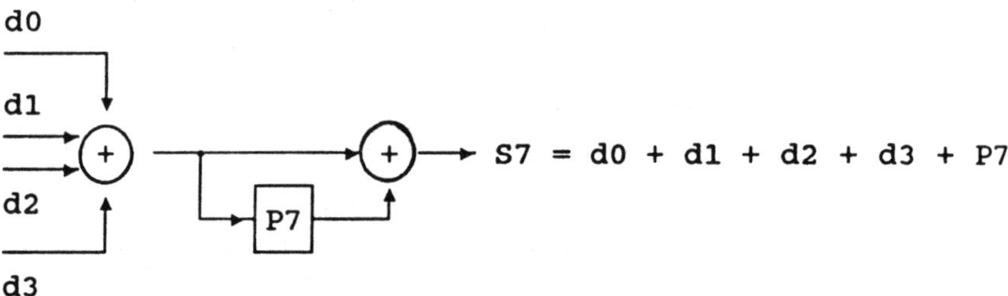

On write, each row check bit is selected to make the parity of its row even. Each column check bit is selected to make the parity of its column even. The data bits and the parity bits together are called a codeword.

On read, row syndrome bits are generated by checking parity across each row, including the row check bit. Column syndrome bits are generated in a similar fashion. Syndrome means symptom of error. For this code, syndrome bits can be viewed as the XOR differences between read checks and write checks. If there is no error, all syndrome bits are zero.

When a single-bit error occurs, one row and one column will have inverted syndrome bits (odd parity). The bit in error is at the intersection of this row and column.

The circuit above shows the logic necessary for generating the write-check bit and the syndrome bit for one row. For parallel decoding, this logic is required for each row and column. Also, 16 AND gates are required for detecting the intersections of inverted row and column syndrome bits. In addition, 16 XOR gates are required for inverting data bits. The correction circuit for one particular data bit is shown below.

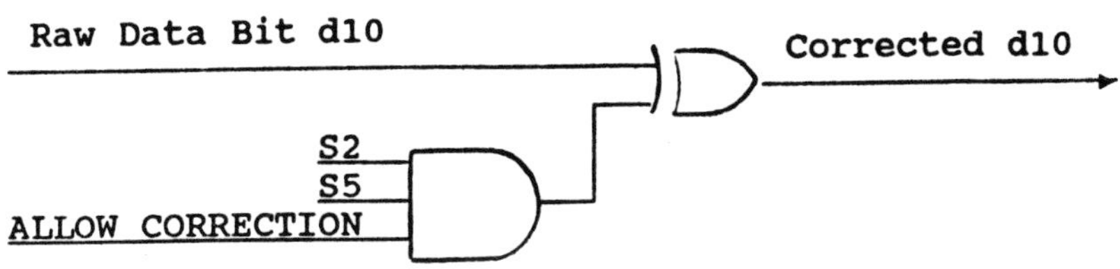

Two data bits in error will cause either two rows, two columns, or both to have inverted syndrome bits (odd parity). This condition can be trapped to give the code the capability to detect double-bit errors in data.

All single check-bit errors are detected, but not all double check-bit errors. One row and one column check bit in error will result in miscorrection (false correction). If an overall check bit across data is added, the code is capable of detecting all double-bit errors in data and check bits. This includes the case where one data bit and one parity bit are in error. The overall check bit can be generated by forming parity across all row or all column check bits. With the overall check bit added, all double-bit errors are detectable but uncorrectable.

Miscorrection occurs when three bits are in error on three corners of a rectangle. For example:

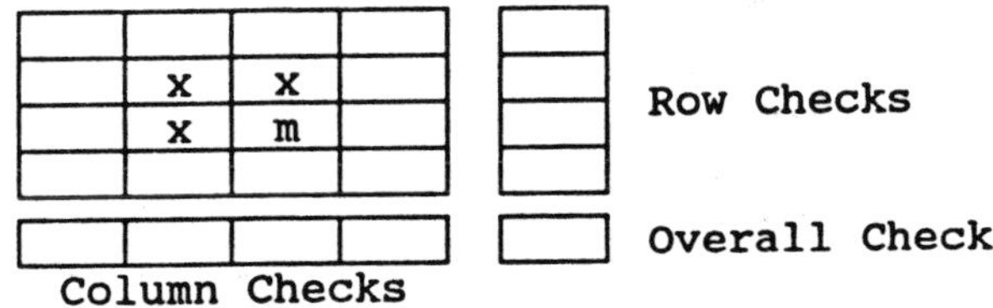

The three errors which are illustrated above cause the decoder to respond as if there were a single-bit error at location m. Miscorrection does not result for all combinations of three bits in error, only for those where there are errors on three corners of a rectangle.

Miscorrection probability for three-bit errors is the ratio of three-bit error combinations that result in miscorrection to all possible three bit-error combinations.

Misdetection (error condition not detected at all) occurs when four-bits are in error on the corners of a rectangle. For example:

	x	x			Row Checks
	x	x			
					Overall Check

Column Checks

This error condition leaves all syndrome bits equal to zero.

Misdetection does not result for all combinations of four bits in error, only those where there are errors on four corners of a rectangle. Misdetection probability for four-bit errors is the ratio of four-bit error combinations that result in misdetection to all possible four-bit error combinations.

This discussion introduced the following error-correction concepts:

- Check bits
- Syndromes
- Codeword
- Correctable error
- Detectable error
- Miscorrection
- Misdetection
- Miscorrection probability
- Misdetection probability

PROBLEMS

1. Write the parity check equations for the circuit below.

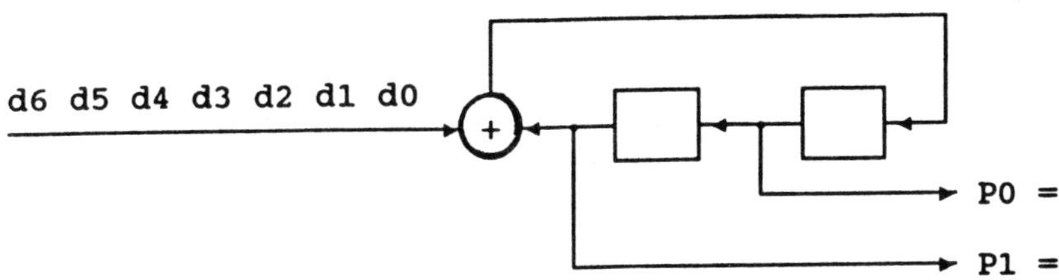

2. Write the parity check equations for the circuit below.

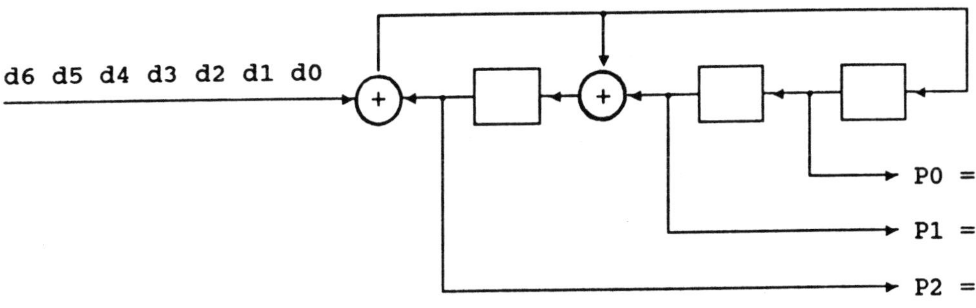

3. Generate a chart showing the contribution of each data bit to the final shift register state for the circuits shown above.

   If the data stream is zeros except for d3 and d1, what is the final shift register state?

1.2 MATHEMATICAL FOUNDATIONS

1.2.1 *SOME DEFINITIONS, THEOREMS AND ALGORITHMS FOR INTEGERS*

Definition 1.2.1. When we say an integer a divides an integer b we mean a divides b with zero remainder. "a divides b" is written as $a|b$. "a does not divide b" is written as $a\nmid b$.

Examples: $3|6$, $3\nmid 4$, $2\nmid 1$

Definition 1.2.2. An integer a is called prime if a is greater than 1 and there are no divisors of a that are less than a but greater than 1. If an integer a greater than 1 is not prime, then it is called composite.

Examples: 2, 3, and 5 are prime

4, 6, and 8 are composite

Definition 1.2.3. The greatest common divisor (GCD) of a set of integers $\{a_1,a_2,\cdots,a_n\}$ is the largest positive integer that divides each of $a_1,a_2,\cdots,a_n$. The greatest common divisor may be written as $GCD(a_1,a_2,\cdots,a_n)$.

Algorithm 1.2.1. To find $GCD(a_1,a_2,\cdots,a_n)$, express each integer as the product of prime factors. Form the product of their common factors. For repeated factors, include in the product the highest power that is a factor of all the given integers. The GCD is the absolute value of the product. If there are no common factors, the GCD is one.

Examples:

$GCD(3,9,15) = GCD(3,3^2,3*5) = 3$

$GCD(-165,231) = GCD(-3*5*11,3*7*11) = 33$

$GCD(105,165) = GCD(3*5*7,3*5*11) = 15$

$GCD(45,63,297) = GCD(3^2*5,3^2*7,3^3*11) = 9$

The GCD can also be found using Euclid's Algorithm.

Definition 1.2.4. The least common multiple (LCM) of a set of integers $\{a_1,a_2,\cdots,a_n\}$ is the smallest positive integer that is divisible by each of $a_1,a_2,\cdots,a_n$. The least common multiple may be written $LCM(a_1,a_2,\cdots,a_n)$.

Algorithm 1.2.2. To find $LCM(a_1,a_2,\cdots,a_n)$, express each integer as a product of prime factors. Form the product of primes that are a factor of any of the given integers. Common factors between two or more integers are included in the product only once. For repeated factors, include in the product the highest power that occurs in any of the prime factorizations. The LCM is the absolute value of the product.

Examples:

$LCM(6,15,21) = LCM(2*3,3*5,3*7) = 210$

$LCM(30,42,66) = LCM(2*3*5,2*3*7,2*3*11) = 2310$

$LCM(-15,21,11) = LCM(-3*5,3*7,11) = 1155$

$LCM(45,63,297) = LCM(3^2*5,3^2*7,3^3*11) = 10395$

Theorem 1.2.1. Every integer a>1 can be expressed as the product of primes, (with at least one factor).

Examples: 3 =3
6 =2*3
15 =3*5

Definition 1.2.5. Integers a and b are relatively prime if their greatest common divisor is 1.

Examples: 3, 7
3, 4
15, 77

Theorem 1.2.2. Let integers a, b, and c be relatively prime in pairs, then a*b*c divides d if, and only if, each of a, b, and c divide d.

Examples: $3|15$, $5|15$, $7\nmid 15$, therefore, $(3*5*7)\nmid 15$

$3|210$, $5|210$, $7|210$, therefore, $(3*5*7)|210$

Theorem 1.2.3. Let an integer a be prime, then a divides b*c*d if, and only if, a divides b or c or d.

Examples: $3|6$, therefore, $3|(6*5*7)$

$3\nmid 5$, $3\nmid 7$, $3\nmid 11$, therefore, $3\nmid 385$

Definition 1.2.6. Let x be any real number. The integer function of x, written as INT(x), is the greatest integer less than or equal to x.

Examples: INT(1/2) = 0
INT(5/3) = 1
INT(-1/2) = -1

Definition 1.2.7. Let x and y be any real numbers. x modulo y, written as x MOD y, is defined as follows:

$$x \text{ MOD } y = x - y*\text{INT}(x/y)$$

Examples: 5 MOD 3 = 2
9 MOD 3 = 0
-5 MOD 7 = 2

1.2.2 *SOME DEFINITIONS, THEOREMS AND ALGORITHMS FOR POLYNOMIALS*

Definition 1.2.8. A polynomial is said to be monic if the coefficient of the term with the highest degree is 1.

Definition 1.2.9. The greatest common divisor of two polynomials is the monic polynomial of greatest degree which divides both.

Definition 1.2.10. The least common multiple of a(x) and b(x) is some c(x) divisible by each of a(x) and b(x), which itself divides any other polynomial that is divisible by each of a(x) and b(x).

Definition 1.2.11. If the greatest common divisor of two polynomials is 1, they are said to be relatively prime.

Definition 1.2.12. A polynomial of degree n is said to be irreducible if it is not divisible by any polynomial of degree greater than 0 but less than n.

Theorem 1.2.4. Let a(x), b(x), and c(x) be relatively prime in pairs, then $a(x) \cdot b(x) \cdot c(x)$ divides d(x) if, and only if, a(x) and b(x) and c(x) divide d(x).

Theorem 1.2.5. Let a(x) be irreducible, then a(x) divides $b(x) \cdot c(x) \cdot d(x)$ if, and only if, a(x) divides b(x) or c(x) or d(x).

Definition 1.2.13. A function is said to be linear if the properties stated below hold:

a. Linearity: $f(a \cdot x) = a \cdot f(x)$

b. Superposition: $f(x+y) = f(x)+f(y)$

1.2.3 *THE CHINESE REMAINDER METHOD*

There are times when integer arithmetic in a modular notation is preferred to a fixed radix notation. The integers are represented by residues modulo a set of relatively prime moduli.

Example: Assume integers are represented by residues modulo the moduli 3 and 5.

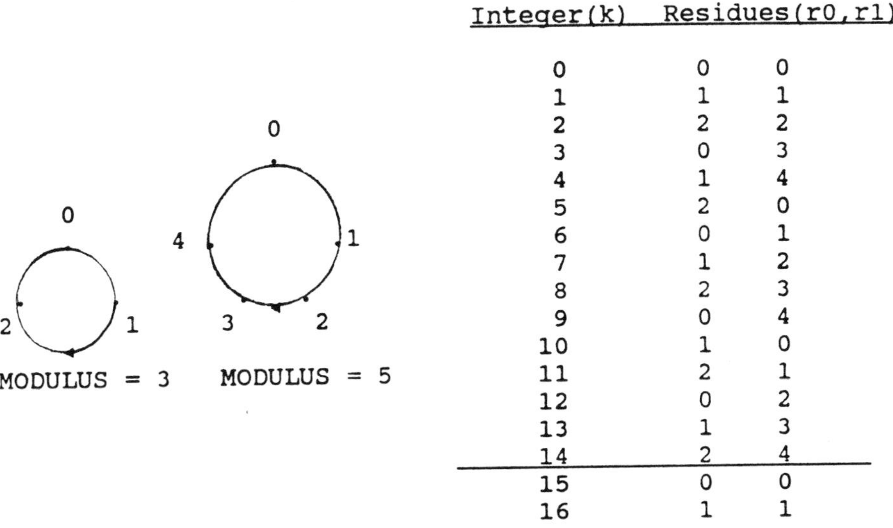

Integer(k)	Residues(r0,r1)	
0	0	0
1	1	1
2	2	2
3	0	3
4	1	4
5	2	0
6	0	1
7	1	2
8	2	3
9	0	4
10	1	0
11	2	1
12	0	2
13	1	3
14	2	4
15	0	0
16	1	1

Notice that the integer k has a unique representation in residues from k=0 through k=14. The integer k=15 has the same representation as k=0. In this case, the total number of integers that have unique representation is 15. In general, the total number of integers n having unique representation is given by the equation:

$$n = \mathrm{LCM}(e_0, e_1, \cdots)$$

where the e_i are moduli.

There are also times when an integer d must be determined if its residues modulo a set of moduli are given. This can be accomplished with the Chinese Remainder Method. This method is based on the Chinese Remainder Theorem. See any number theory text.

METHOD

e_i = Moduli (The e_i must be relatively prime in pairs)

n = LCM $(e_0, e_1, \cdots)$

$m_i = n/e_i$

A_i = Constants such that $(A_i * m_i)$ MOD $e_i = 1$

r_i = Residues

d = desired integer = $(A_0*m_0*r_0 + A_1*m_1*r_1 + \cdots)$ MOD n

EXAMPLE

$e_i = 3,5$ $(e_0=3,\ e_1=5)$

n = LCM(3,5) = 15

$m_0 = n/e_0 = 15/3 = 5$

$m_1 = n/e_1 = 15/5 = 3$

A_0*5 MOD 3 = 1, therefore $A_0 = 2$

A_1*3 MOD 5 = 1, therefore $A_1 = 2$

$d = (10*r_0 + 6*r_1)$ MOD 15

— This calculation is performed at development time.

If $r_0, r_1 = 2,3$ then $d = 8$

If $r_0, r_1 = 1,3$ then $d = 13$

— This calculation is performed at execution time.

A PROCEDURE FOR PERFORMING THE CHINESE REMAINDER METHOD WITHOUT USING MULTIPLICATION

Frequently, the Chinese Remainder Method must be solved on a processor that does not have a multiply instruction. A procedure using only addition and compare instructions is described below.

The integer d is to be determined where d is the least integer such that:

d MOD $e_0 = r_0$ and simultaneously d MOD $e_1 = r_1$

or equivalently,

$$\frac{d}{e_0} = n_0 + \frac{r_0}{e_0} \quad \text{and simultaneously} \quad \frac{d}{e_1} = n_1 + \frac{r_1}{e_1}$$

Rearranging gives

$$d = n_0{*}e_0 + r_0 \text{ and simultaneously } d = n_1{*}e_1 + r_1$$

or,

$$d = n_0{*}e_0 + r_0 = n_1{*}e_1 + r_1$$

Multiplication can be expressed as repeated addition. Therefore,

$$d = r_0 + \underbrace{e_0 + e_0 + \cdots}_{n_0 \text{ times}} = r_1 + \underbrace{e_1 + e_1 + \cdots}_{n_1 \text{ times}}$$

A procedure for finding d based on the relationship above is detailed in the following flowchart.

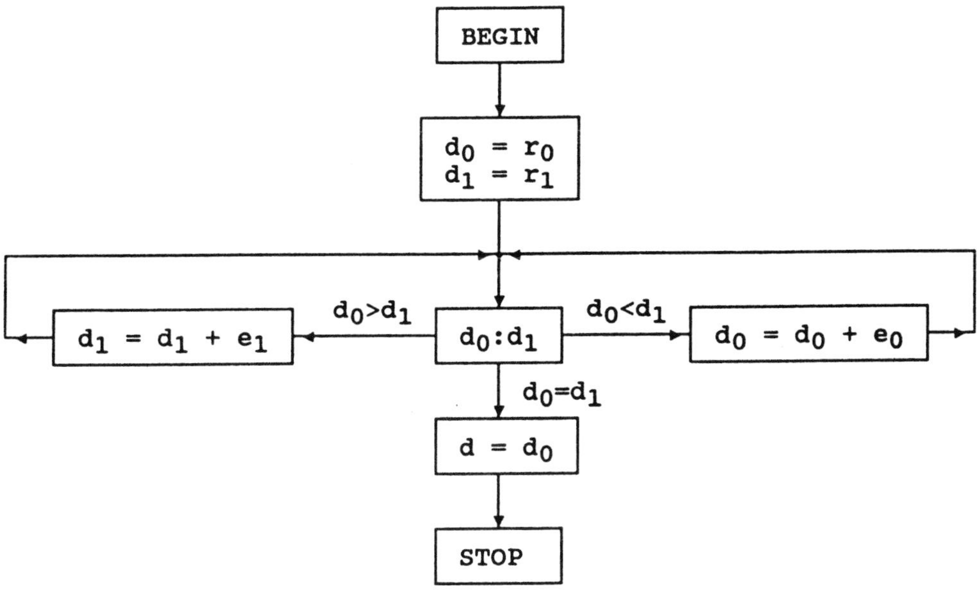

1.2.4 *MULTIPLICATION BY SHIFTING, ADDING, AND SUBTRACTING*

Many 8-bit processors do not have a multiply instruction. This discussion describes techniques to minimize the complexity of multiplying a variable by a constant, when these processors are used. These techniques provide another alternative for accomplishing the multiplications required in performing the Chinese Remainder Method.

On an 8-bit processor any shift that is a multiple of 8 bits can be accomplished with register moves. Therefore, multiplying by a power of 2 that is a multiple of 8 can be accomplished by register moves. Any string of ones in a binary value can be represented by the power of 2 that is just greater than the highest power of 2 in the string minus the lowest power of 2 in the string. These results can be used to minimize the complexity of multiplying a variable by a constant using register moves, shifts, adds and subtracts.

<u>Examples</u>: In all examples, x is less than 256. The results are shown in a form where register moves and shifts are identifiable.

$$
\begin{aligned}
y &= 255*x \\
&= (2^8-1)*x \\
&= 2^8*x-x
\end{aligned}
$$

$$
\begin{aligned}
y &= 257*x \\
&= (2^8+1)*x \\
&= 2^8*x+x
\end{aligned}
$$

$$
\begin{aligned}
y &= 992*x \\
&= (2^9+2^8+2^7+2^6+2^5)*x \\
&= (2^{10}-2^5)*x \\
&= 2^{10}*x-2^5*x
\end{aligned}
$$

$$
\begin{aligned}
y &= 32131*x \\
&= (2^{14}+2^{13}+2^{12}+2^{11}+2^{10}+2^8+2^7+2^1+2^0)*x \\
&= (2^{15}-2^9-2^7+2^1+2^0)*x \\
&= 2^{15}*x-2^9*x-2^7*x+2^1*x+2^0*x \\
&= 2^8*(2^7*x)-(2^7*x)-2^8*(2^1*x)+(2^1*x)+x
\end{aligned}
$$

In the last example, only two unique shift operations are required even though the original constant contains nine powers of 2. This particular example is from the Chinese Remainder Method when moduli 255 and 127 are used.

PROBLEMS

1. Find the GCD of 70 and 15.

2. Find the GCD of 70 and 11.

3. Find the LCM of 30 and 42.

4. Find the LCM of 33 and 10.

5. Express 210 as a product of primes.

6. Are 70 and 15 relatively prime?

7. Are 70 and 11 relatively prime?

8. Determine a

   a = INT(7/3) =
   a = -INT(1/3) =
   a = INT(-1/3) =
   a = 10 MOD 3 =
   a = -3 MOD 15 =
   a = 254 MOD 255 =

9. Is $2 \cdot x^2 + 1$ a monic polynomial?

10. Write the residues modulo the moduli 5 and 7 of the integer 8.

11. The residues for several integers modulo 5 and 7 are listed below. Compute the A_i of the Chinese Remainder Method. Then use the Chinese Remainder Method to determine the integers.

    a MOD 5 = 4, a MOD 7 = 6, a = ?
    a MOD 5 = 3, a MOD 7 = 5, a = ?
    a MOD 5 = 0, a MOD 7 = 4, a = ?

    What is the total number of unique integers that can be represented by residues modulo 5 and 7?

12. Define a fast division algorithm for dividing by 255 on an 8-bit processor that does not have a divide instruction. The dividend must be less than 65536.

13. What is the total number of unique integers that can be represented by residues modulo 4 and 11?

1.3 POLYNOMIALS AND SHIFT REGISTER SEQUENCES

1.3.1 *INTRODUCTION TO POLYNOMIALS*

It is convenient to consider the symbols of a binary data stream to be coefficients of a polynomial in a variable x, with the powers of x serving as positional indicators. These polynomials can be treated according to the laws of ordinary algebra with one exception: coefficients are to be added modulo-2 (EXCLUSIVE-OR sum). The '+' operator will be used to represent both ordinary addition and modulo-2 addition; when used to represent modulo-2 addition, it will usually be separated from its operands by a preceding and a following space.

As with polynomials of ordinary algebra, these polynomials have properties of associativity, distributivity, and commutativity. These polynomials also factor into prime or irreducible factors in only one way, just as do those of ordinary algebra.

For now, the value of coefficients will be either '1' or '0' depending on the value of the corresponding data bit. Such polynomials are said to have binary coefficients or to have coefficients from the field of two elements. Later, polynomials with coefficients other than '1' and '0' will be discussed. When transmitting and receiving polynomials, the highest order symbol is always transmitted or received first.

MULTIPLICATION OF POLYNOMIALS

Multiplication is just like ordinary multiplication of polynomials, except the addition of coefficients is accomplished with the XOR operation (modulo-2 addition).

Example #1:

$$\begin{array}{r} x^3 \\ \cdot\ x^3 + x + 1 \\ \hline x^6 + x^4 + x^3 \end{array} \qquad \text{-or-} \qquad \begin{array}{l} 1000 \\ \cdot\ 1011 \\ \hline 1000 \\ \ \ 1000 \\ \ \ \ 1000 \\ \hline 1011000 \end{array}$$

Example #2:

$$\begin{array}{r} x + 1 \\ \cdot\ x^3 + x + 1 \\ \hline x^4 + x^3 \qquad\qquad \\ x^2 + x \qquad \\ x + 1 \\ \hline x^4 + x^3 + x^2 + 1 \end{array} \qquad \text{-or-} \qquad \begin{array}{l} 11 \\ \cdot\ 1011 \\ \hline 11 \\ \ \ 11 \\ \ \ \ 11 \\ \hline 11101 \end{array}$$

In example #2, unlike in ordinary polynomial multiplication, the two x terms cancel.

DIVISION OF POLYNOMIALS

Division is just like ordinary division of polynomials, except the addition of coefficients is accomplished with the XOR operation (modulo-2 addition).

Example #1:

$$\begin{array}{r|l} & x^2 + 1 \\ \hline x^3 + x + 1 & x^5 + 1 \\ & \underline{x^5 + x^3 + x^2} \\ & x^3 + x^2 + 1 \\ & \underline{x^3 + x + 1} \\ & x^2 + x \end{array}$$

-OR-

```
          101
      ________
1011 | 100001
       1011
       ----
         1101
         1011
         ----
         0110
```

Example #2:

$$\begin{array}{r|l} & x^2 + 1 \\ \hline x^3 + x + 1 & x^5 + x^2 + 1 \\ & \underline{x^5 + x^3 + x^2} \\ & x^3 + 1 \\ & \underline{x^3 + x + 1} \\ & x \end{array}$$

-OR-

```
          101
      ________
1011 | 100101
       1011
       ----
         1001
         1011
         ----
         0010
```

1.3.2 *INTRODUCTION TO SHIFT REGISTERS*

A linear sequential circuit (LSC) is constructed with three building blocks. Any connection is permissible as long as a single output arrow of one block is mated to a single input arrow of another block.

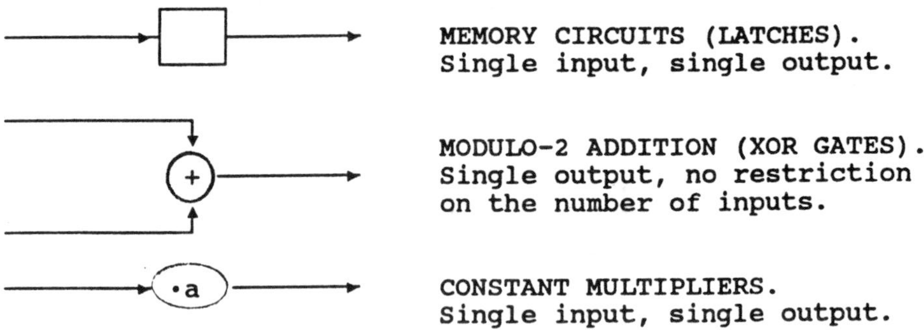

Latches are clocked by a synchronous clock. The output of a latch at any point in time is the binary value that appeared on its input one time unit earlier.

The output of a modulo-2 adder at any point in time is the modulo-2 sum of the inputs at that time.

For now, a constant multiplier '•a' will be either '•1' or '•0'. If such a constant multiplier is '•1', a connection exists. No connection exists for a constant multiplier of '•0'.

AN EXAMPLE OF AN LSC

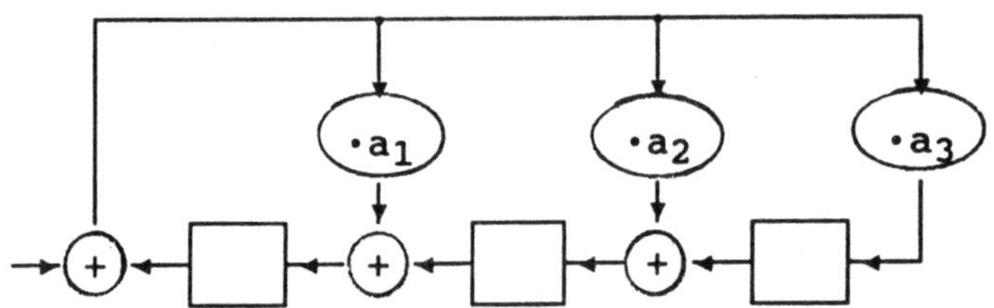

A linear sequential circuit of the above form is also called a linear feedback shift register (LFSR), a linear shift register (LSR) or simply a shift register (S/R).

AN EQUIVALENT CIRCUIT WHERE $a_1 = 0$, $a_2 = 1$, $a_3 = 1$

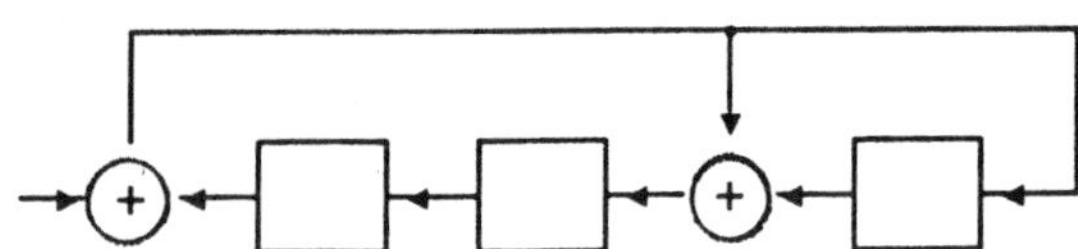

SHIFT REGISTER IMPLEMENTATION OF MULTIPLICATION

Polynomial multiplication can be implemented with a linear shift register.

The circuit below will multiply any input bit stream (input polynomial) by (x + 1). The product appears on the output line. The number of shifts required is equal to the sum of the degrees of the input polynomial and the multiplier polynomial plus one.

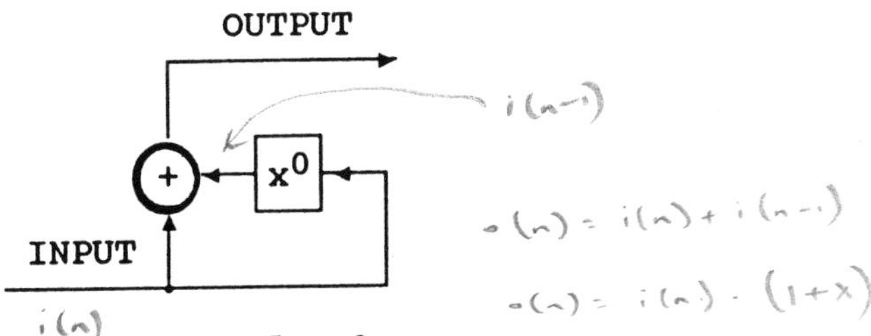

Example #1: Assume the input polynomial to be $(x^5 + x^3 + 1)$.

Input Bit	Shift Reg State	Output Bit
1 (x^5)	1	1 (x^6)
0	0	1 (x^5)
1 (x^3)	1	1 (x^4)
0	0	1 (x^3)
0	0	0
1 (1)	1	1 (x)
0	0	1 (1)

Example #2: Assume the input polynomial to be x^3.

Input Bit	Shift Reg State	Output Bit
1 (x^3)	1	1 (x^4)
0	0	1 (x^3)
0	0	0
0	0	0
0	0	0

NOTE: The shift register state is shown after the indicated input bit is clocked.

The circuits below will multiply any input bit stream (input polynomial) by $(x^3 + x + 1)$.

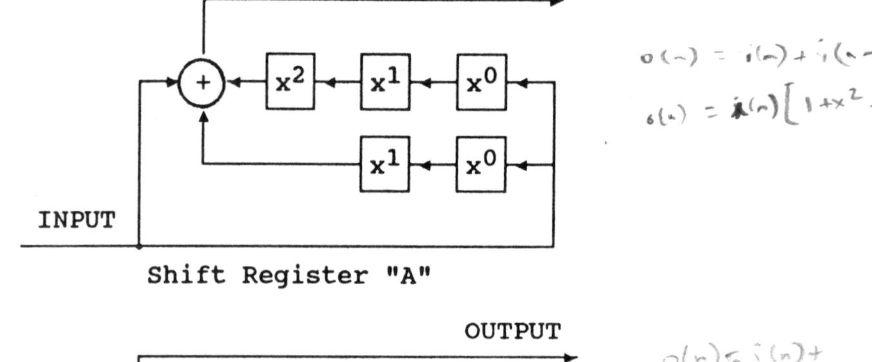

Shift Register "A"

OUTPUT

INPUT

Shift Register "B"

Example #1: Assume the input polynomial to be x^3.

Input Bit	Shift Req 'B' State	Output Bit
$1(x^3)$	011	$1(x^6)$
0	110	0
0	100	$1(x^4)$
0	000	$1(x^3)$
0	000	0
0	000	0
0	000	0

Example #2: Assume the input polynomial to be $(x + 1)$.

Input Bit	Shift Req 'B' State	Output Bit
$1(x)$	011	$1\ (x^4)$
$1(1)$	101	$1\ (x^3)$
0	010	$1\ (x^2)$
0	100	0
0	000	1 (1)

NOTE: The shift register state is shown after the indicated input bit is clocked.

A GENERAL MULTIPLICATION CIRCUIT

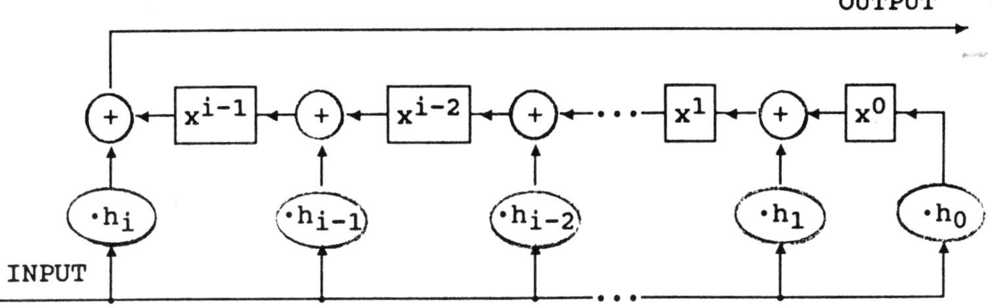

The circuit shown above multiplies any input polynomial D(x) by a fixed polynomial P(x). The product appears on the output line.

$$P(x) = h_i \cdot x^i + h_{i-1} \cdot x^{i-1} + h_{i-2} \cdot x^{i-2} + \ldots + h_1 \cdot x + h_0$$

The number of shifts required is equal to the sum of the degrees of the input polynomial and multiplier polynomial, plus one.

MULTIPLY CIRCUIT EXAMPLES

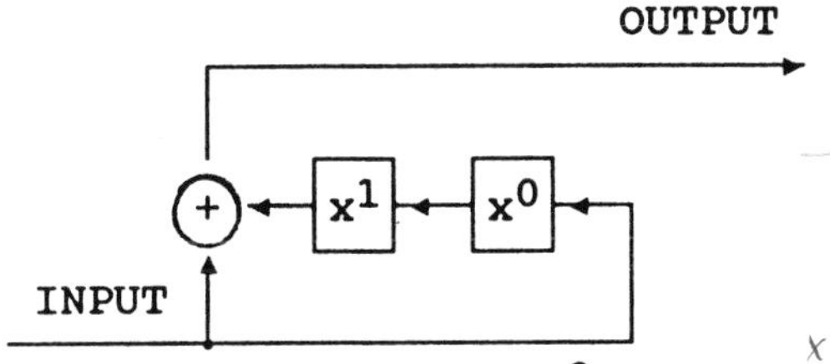

OUTPUT

x^1 x^0

INPUT

Multiply by $x^2 + 1$

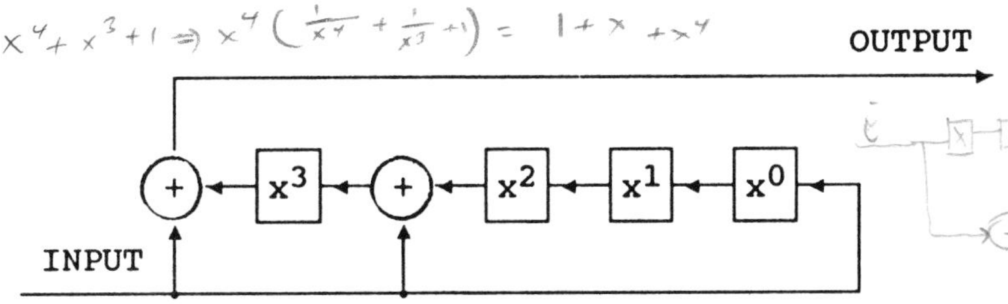

OUTPUT

x^3 x^2 x^1 x^0

INPUT

Multiply by $x^4 + x^3 + 1$

OUTPUT

x^4 x^3 x^2 x^1 x^0

INPUT

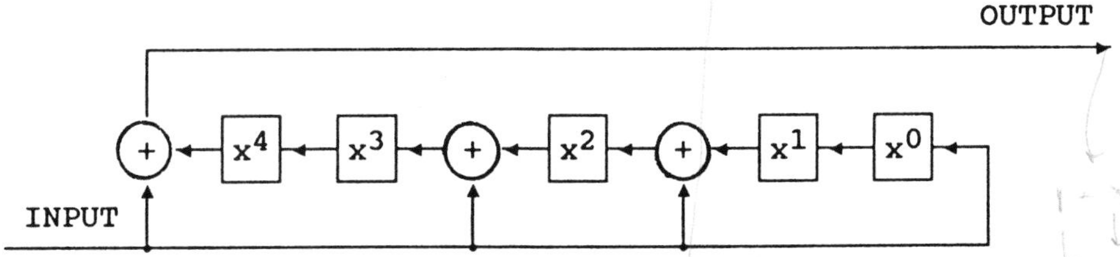

Multiply by $x^5 + x^3 + x^2 + 1$

SHIFT REGISTER IMPLEMENTATION OF DIVISION

Polynomial division can be implemented with an LSR.

The circuit below will divide any input bit stream by (x + 1). One shift is required for each input bit. The quotient appears on the output line. The final state of the LSR represents the remainder.

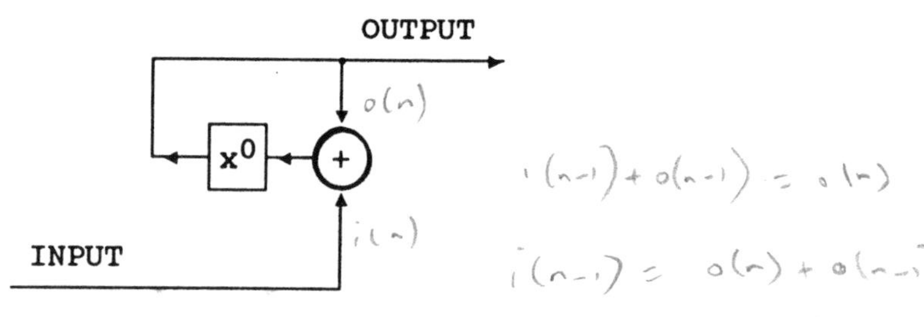

Example #1: Assume the input polynomial to be

$$x^6 + x^5 + x^4 + x^3 + x + 1.$$

Input Bit	Shift Reg State	Output Bit
1 (x^6)	1 (1)	0
1 (x^5)	0	1 (x^5)
1 (x^4)	1 (1)	0
1 (x^3)	0	1 (x^3)
0	0	0
1 (x)	1 (1)	0
1 (1)	0	1 (1)

Example #2: Assume the input polynomial to be $(x^4 + x^3 + 1)$.

Input Bit	Shift Reg State	Output Bit
1 (x^4)	1 (1)	0
1 (x^3)	0	1 (x^3)
0	0	0
0	0	0
1 (1)	1 (1)	0

NOTE: The shift register state is shown after the indicated input bit is clocked.

The circuit below will divide any input bit stream by $(x^3 + x + 1)$.

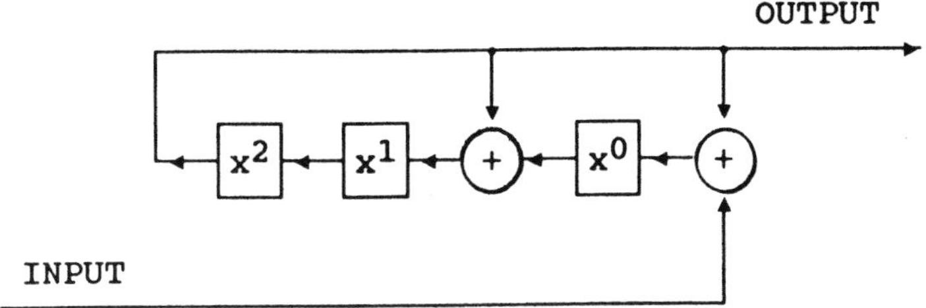

Example #1: Assume the input polynomial to be $(x^5 + 1)$.

Input Bit	Shift Reg State	Output Bit
1 (x^5)	001 (1)	0
0	010 (x)	0
0	100 (x^2)	0
0	011 (x+1)	1 (x^2)
0	110 (x^2+x)	0
1 (1)	110 (x^2+x)	1 (1)

Example #2: Assume the input polynomial to be x^6.

Input Bit	Shift Reg State	Output Bit
1 (x^6)	001 (1)	0
0	010 (x)	0
0	100 (x^2)	0
0	011 (x+1)	1 (x^3)
0	110 (x^2+x)	0
0	111 (x^2+x+1)	1 (x)
0	101 (x^2+1)	1 (1)

NOTE: The shift register state is shown after the indicated input bit is clocked.

A GENERAL DIVISION CIRCUIT

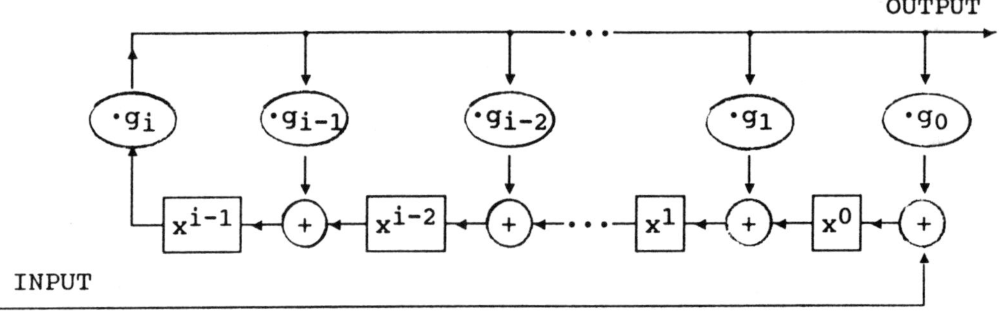

The circuit above divides any input polynomial D(x) by a fixed polynomial P(x). The quotient appears on the output line. The remainder is the final shift register state.

$$P(x) = g_i x^i + g_{i-1} x^{i-1} + \dots + g_1 x + g_0$$

The number of shifts required is equal to the degree of the input polynomial plus one.

DIVIDE CIRCUIT EXAMPLES

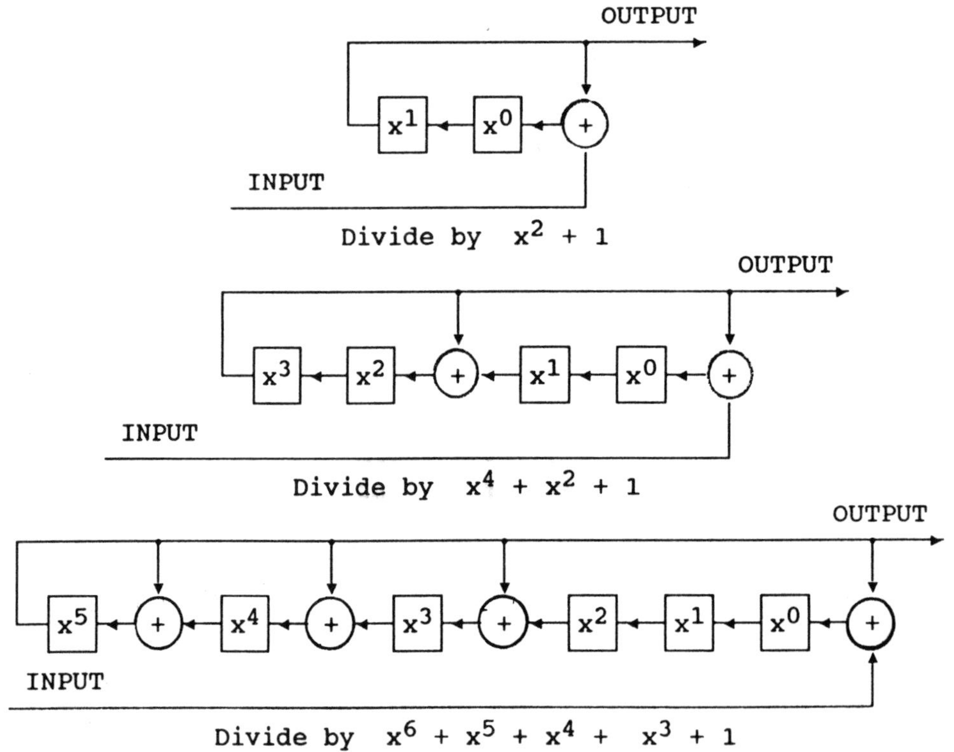

SHIFT REGISTER IMPLEMENTATION OF SIMULTANEOUS MULTIPLICATION AND DIVISION

It is possible to use a shift register to accomplish simultaneous multiplication and division. The circuit below will multiply any input bit stream (input polynomial) by x^3 and simultaneously divide by $(x^3 + x + 1)$. The number of shifts required is equal to the degree of the input polynomial plus one. The quotient appears on output line. The remainder is the final state of shift register.

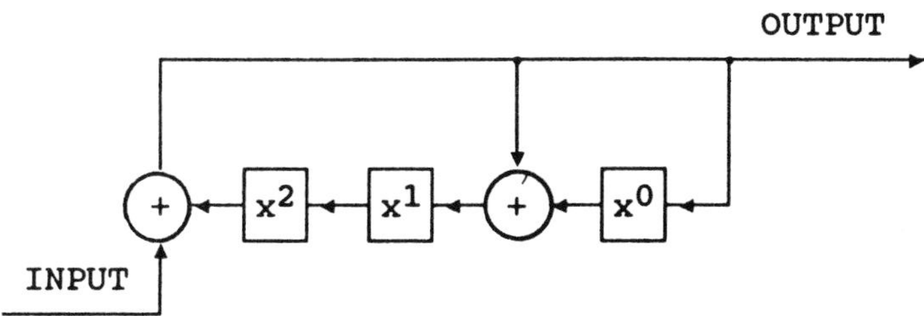

Example #1: Assume the input polynomial to be $(x^5 + 1)$.

Input Bit	Shift Reg State	Output Bit
1 (x^5)	011 $(x+1)$	1 (x^5)
0	110 (x^2+x)	0
0	111 (x^2+x+1)	1 (x^3)
0	101 (x^2+1)	1 (x^2)
0	001 (1)	1 (x)
1 (1)	001 (1)	1 (1)

Example #2: Assume the input polynomial to be x^6.

Input Bit	Shift Reg State	Output Bit
1 (x^6)	011 $(x+1)$	1 (x^6)
0	110 (x^2+1)	0
0	111 (x^2+x+1)	1 (x^4)
0	101 (x^2+1)	1 (x^3)
0	001 (1)	1 (x^2)
0	010 (x)	0
0	100 (x^2)	0

NOTE: The shift register state is shown after the indicated input bit is clocked.

A CIRCUIT TO MULTIPLY AND DIVIDE SIMULTANEOUSLY

A general circuit to accomplish simultaneous multiplication by a polynomial h(x) of degree three and division by a polynomial g(x) of degree two is shown below. The multipliers are all '•1' (connection) or '•0' (no connection).

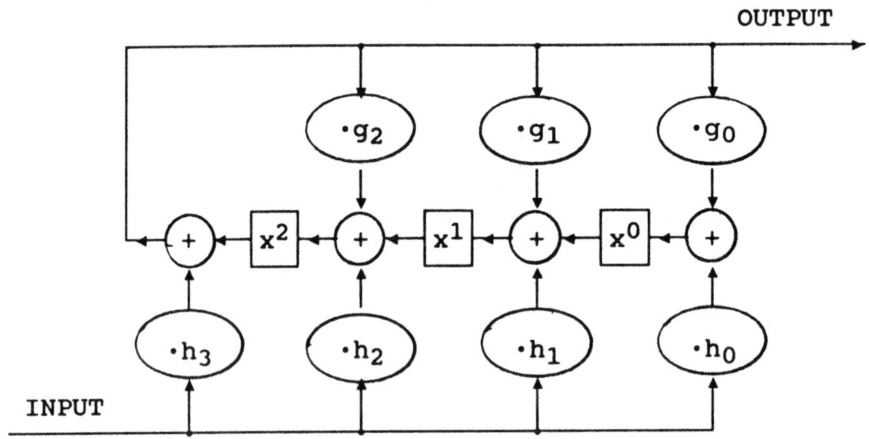

To multiply by x^3, set $h_3=1$ and set all other multipliers to 0.

To multiply by 1 and divide by $(x^3 + x + 1)$, set $h_0=1$, $g_0=1$ and $g_1=1$ and set all other multipliers to 0.

To multiply by x^3 and divide by $(x^3 + x + 1)$, set $h_3=1$ $g_0=1$, and $g_1=1$ and set all other multipliers to 0. This is a form of simultaneous multiplication and division that is encountered frequently in error-correction circuits.

To multiply by $(x + 1)$ and divide by x^3, set $h_0=1$ and $h_1=1$ and set all other multipliers to 0.

A GENERAL CIRCUIT FOR SIMULTANEOUS MULTIPLICATION AND DIVISION

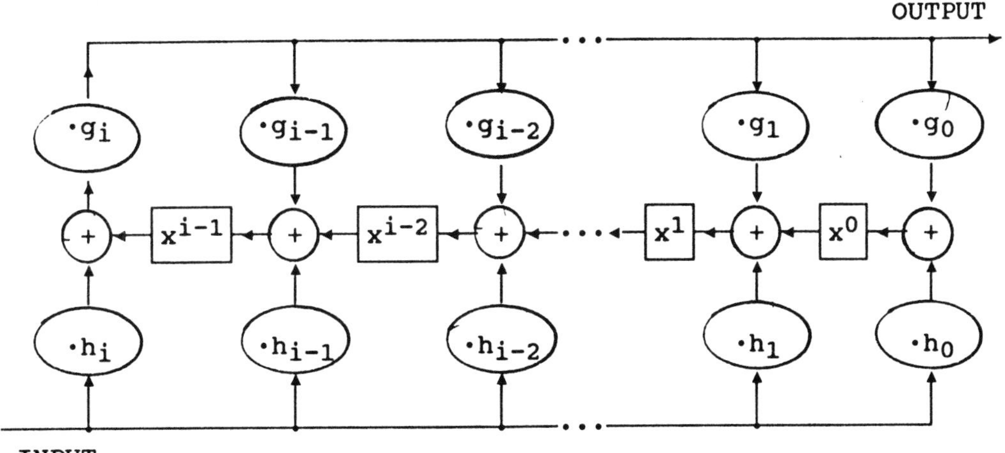

The circuit above multiplies any input polynomial by P1(x) and simultaneously divides by P2(x).

$$P1(x) = h_i x^i + h_{i-1} x^{i-1} + h_{i-2} x^{i-2} + \ldots + h_1 x + h_0$$

$$P2(x) = g_i x^i + g_{i-1} x^{i-1} + g_{i-2} x^{i-2} + \ldots + g_1 x + g_0$$

The number of shifts required is equal to the degree of the input polynomial plus one.

EXAMPLES OF CIRCUITS TO MULTIPLY AND DIVIDE SIMULTANEOUSLY

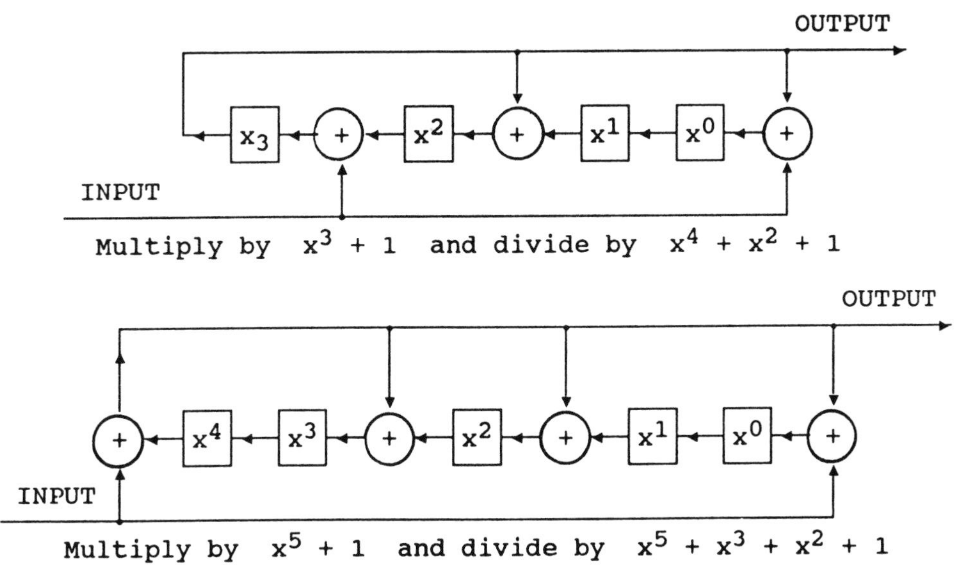

Multiply by $x^3 + 1$ and divide by $x^4 + x^2 + 1$

Multiply by $x^5 + 1$ and divide by $x^5 + x^3 + x^2 + 1$

SIMULTANEOUS MULTIPLICATION AND DIVISION WHEN THE MULTIPLIER POLYNOMIAL HAS A HIGHER DEGREE

The circuit below shows how to construct a shift register to multiply and divide simultaneously when the multiplier polynomial has a higher degree. The number of shifts required is equal to the degree of the input polynomial, plus the degree of the multiplier polynomial, minus the degree of the divider polynomial, plus one. Register states are labeled below for the multiply polynomial and above for the divide polynomial.

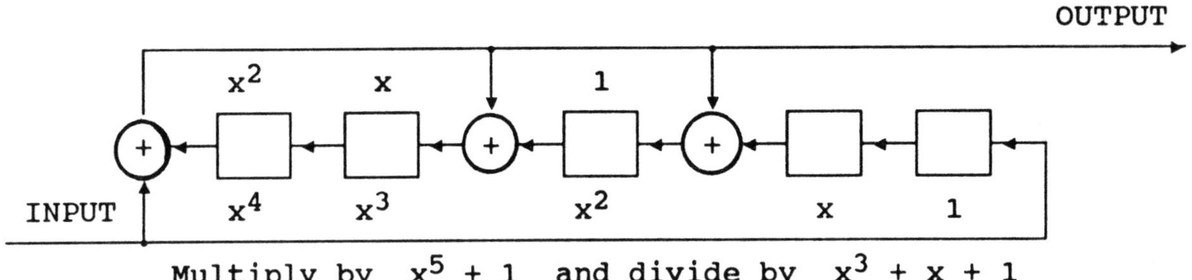

Multiply by $x^5 + 1$ and divide by $x^3 + x + 1$

SHIFT REGISTER IMPLEMENTATION TO COMPUTE A SUM OF PRODUCTS

A single shift register can be used to compute the sum of the products of different variable polynomials with different fixed polynomials *e.g.* $a(x) \bullet h_1(x) + b(x) \bullet h_2(x)$.

The circuit below will multiply an input polynomial a(x) by a fixed polynomial $x^3 + x + 1$ and simultaneously multiply an input polynomial b(x) by the fixed polynomial $x^2 + 1$ and sum the products. The sum of the products appears on the output line. The number of shifts required is equal to the sum of the degrees of the input polynomial of the highest degree and the fixed polynomial of the highest degree plus one.

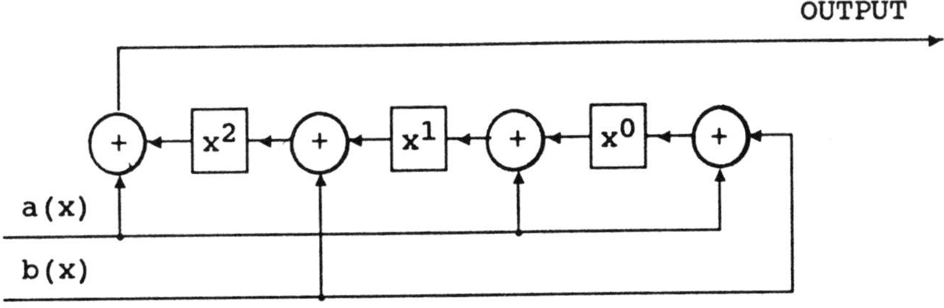

Example #1: Assume a(x) to be x^3 and b(x) to be $(x^5 + x^3 + 1)$.

a(x) Input	b(x) Input	Shift Req State	Output Bit
0	1 (x^5)	101	0
0	0	010	1 (x^7)
1 (x^3)	1 (x^3)	010	1 (x^6)
0	0	100	0
0	0	000	1 (x^4)
0	1 (1)	101	0
0	0	010	1 (x^2)
0	0	100	0
0	0	000	1 (1)

NOTE: The shift register state is shown after the indicated input bit is clocked.

SHIFT REGISTER IMPLEMENTATION TO COMPUTE A SUM OF PRODUCTS MODULO A DIVISOR

A single shift register can be used to compute the remainder of the sum of products of different variable polynomials with different fixed polynomials when divided by another polynomial *e.g.* $[a(x) \cdot h_1(x) + b(x) \cdot h_2(x)]$ MOD $g(x)$.

The circuit below will multiply an input polynomial $a(x)$ by a fixed polynomial $x^2 + x + 1$ and simultaneously multiply an input polynomial $b(x)$ by the fixed polynomial $x^2 + 1$ and sum the products. The sum of the products is reduced modulo $x^3 + x + 1$.

The shift register contents at the end of the operation is the result. The number of shifts required is equal to the degree of the input polynomial of the highest degree plus one.

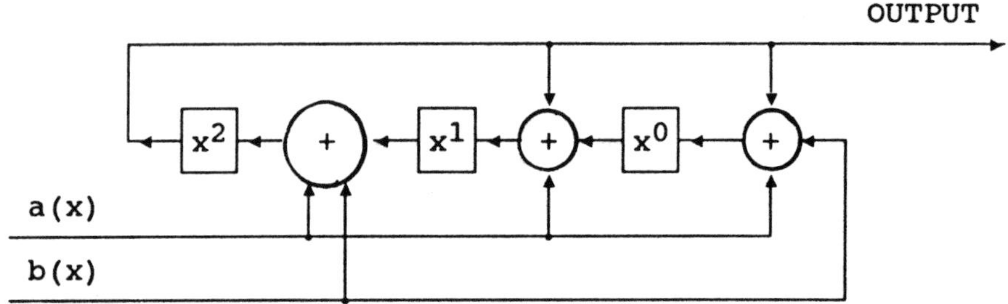

Example #1: Assume $a(x)$ to be x^3 and $b(x)$ to be $(x^5 + x^3 + 1)$.

a(x) Input	b(x) Input	Shift Req State
0	1 (x^5)	101
0	0	001
1 (x^3)	1 (x^3)	000
0	0	000
0	0	000
0	1 (1)	101 (x^2+1)

NOTE: The shift register state is shown after the indicated input bit is clocked.

OTHER FORMS OF THE DIVISION CIRCUIT

The circuit examples below are implemented using the internal-XOR form of shift register.

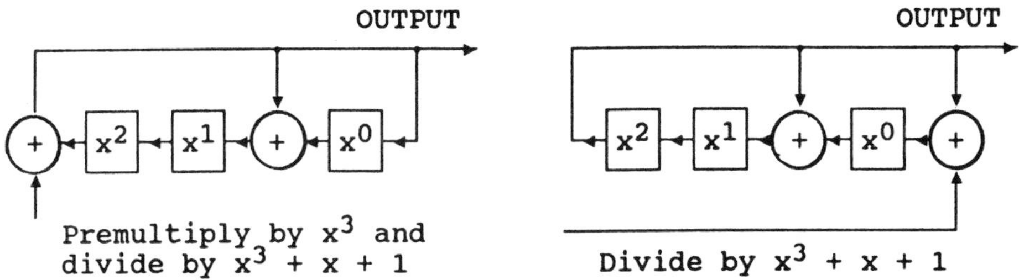

The circuit shown below can accomplish the circuit function of either of the circuits shown above. If the gate is enabled for the entire input polynomial, the circuit function is to premultiply by x^3 and divide by $(x^3 + x + 1)$. However, if the gate is disabled for the last m (m is 3 in this case) bits of the input polynomial, the circuit function is to divide by $(x^3 + x + 1)$ without premultiplying. In the following general discussion, g(x) is the division polynomial and m is the degree of the division polynomial.

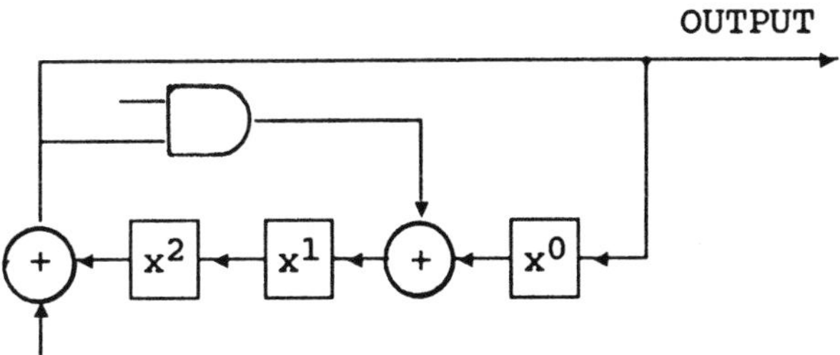

GATE ENABLED DURING LAST m BITS OF INPUT POLYNOMIAL

The circuit function is premultiply by x^m and divide by g(x). The quotient appears on the output line. The remainder is taken from the shift register.

GATE DISABLED DURING LAST m BITS OF INPUT POLYNOMIAL

The circuit function is to divide by g(x) without premultiplying by x^m. The quotient appears on the output line up to the last m bits of the input polynomial. The remainder appears on the output line during the last m bits of the input polynomial. The remainder can also be taken from the shift register.

EXTERNAL-XOR FORM OF SHIFT REGISTER DIVIDE CIRCUIT

There is another form of the shift register divide circuit called the external-XOR form that in many cases can be implemented with less logic than the internal-XOR form. An example is shown below.

External XOR form of shift register divide circuit

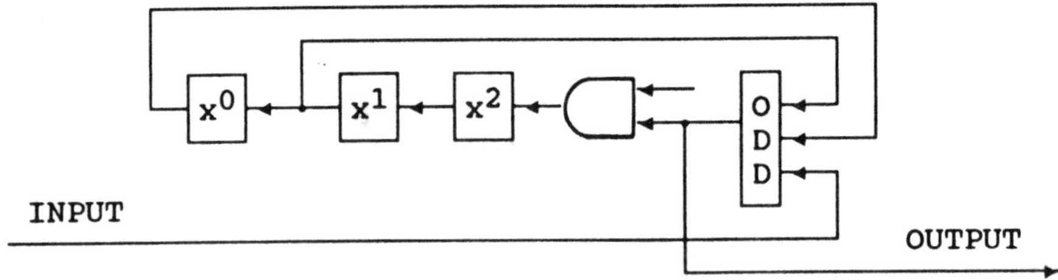

NOTE: The odd circuit is a parity tree.

This circuit is sometimes drawn as shown below.

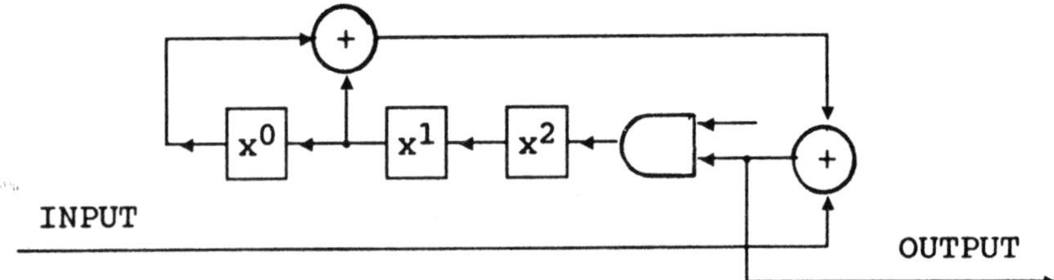

The external-XOR form of the shift register can be implemented two ways.

1. The shift register input is enabled during the entire read of the input polynomial. In this case, the circuit function is premultiply by x^m and divide by g(x).

2. The shift register input is disabled during the last m bits of the input polynomial. In this case, the circuit function is divide by g(x).

Example #1. Input to shift register enabled during entire read of input polynomial.

Circuit function = $a(x)^a x^m / g(x)$
where $a(x) = x^5$ and $g(x) = x^3 + x + 1$

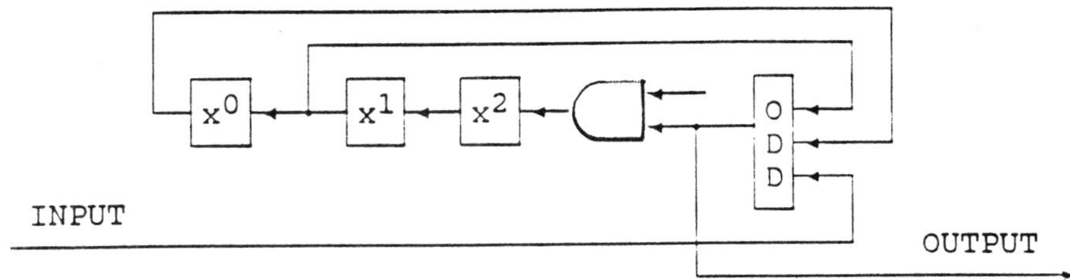

Clocks with gate enabled during read, to get quotient.

DATA	S/R	OUTPUT
1	001	1
0	010	0
0	101	1
0	011	1
0	111	1
0	110	0 ← LSB

Quotient = $x^6 + x^3 + x^2 + x$

Clocks with gate disabled after read, to get remainder.

S/R	OUTPUT
100	0
000	1
000	0 ← LSB

Remainder = x

Output

1. During read, the output is the quotient.

2. After read is complete, disable the gate and clock m more times to place the remainder on the output line.

$$
\begin{array}{r|l}
 & x^5 + x^3 + x^2 + x \\
\hline
x^3 + x + 1 & x^8 \qquad \text{(}x^8\text{ because of premultiply)} \\
 & \underline{x^8 + x^6 + x^5} \\
 & x^6 + x^5 \\
 & \underline{x^6 + x^4 + x^3} \\
 & x^5 + x^4 + x^3 \\
 & \underline{x^5 + x^3 + x^2} \\
 & x^4 + x^2 \\
 & \underline{x^4 + x^2 + x} \\
 & x
\end{array}
$$

Example #2. Input to shift register disabled during last m bits of input polynomial.

Circuit function = $a(x)/g(x)$
where $a(x) = x^5$ and $g(x) = x^3 + x + 1$

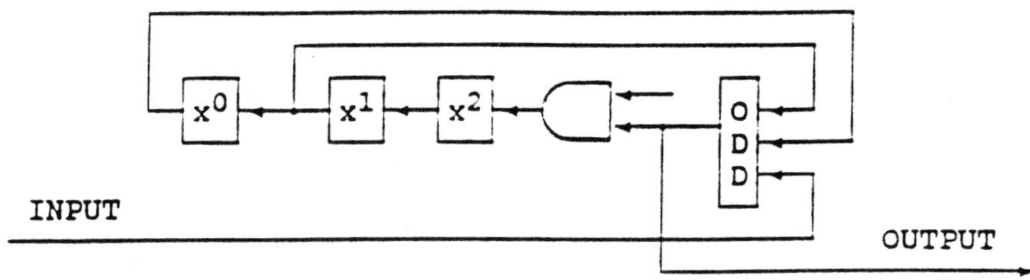

DATA	S/R	OUTPUT	
1	001	1	
0	010	0	Quotient = $x^2 + 1$
0	101	1	← LSB
			Gate disabled at this point.
0	010	1	
0	100	1	Remainder = $x^2 + x + 1$
0	000	1	← LSB

Output

1. Up to the last m bits, the output is the quotient.

2. During the last m bits, the output is the remainder.

$$\begin{array}{r|l} & x^2 + 1 \\ \hline x^3 + x + 1 & x^5 \\ & x^5 + x^3 + x^2 \\ \hline & x^3 + x^2 \\ & x^3 + x + 1 \\ \hline & x^2 + x + 1 \end{array}$$

PERFORMING POLYNOMIAL MULTIPLICATION AND DIVISION WITH COMBINATORIAL LOGIC

Computing parity across groups of data bits using the circuit below was previously studied.

$$a(x) = d6 \cdot x^6 + d5 \cdot x^5 + d4 \cdot x^4 + d3 \cdot x^3 + d2 \cdot x^2 + d1 \cdot x + d0$$

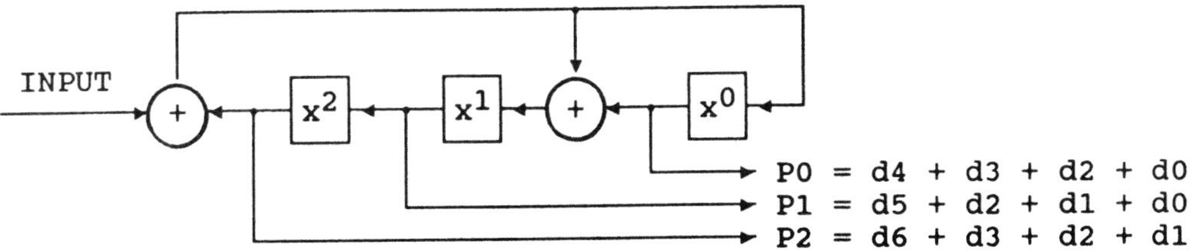

Now that polynomials have been introduced, the function of this circuit can be restated. It premultiplies the input polynomial by x^3 and divides by $(x^3 + x + 1)$. Obviously, the parity check equations can be implemented with combinatorial logic. Therefore, the circuit function can be implemented with combinatorial logic.

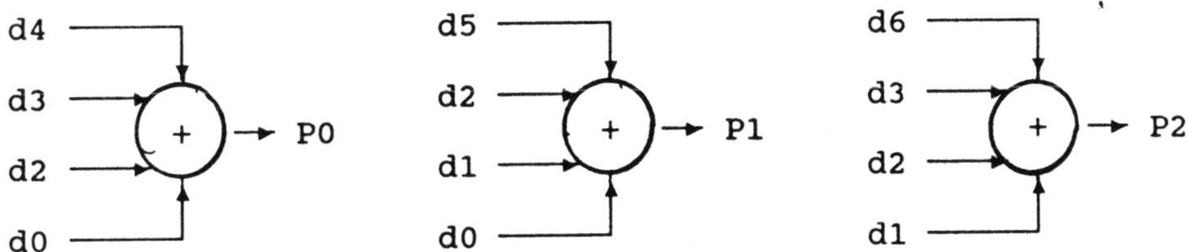

The combinatorial logic circuit above computes the remainder from premultiplying a 7-bit input polynomial by x^3 and dividing by $(x^3 + x + 1)$.

THE SHIFT REGISTER AS A SEQUENCE GENERATOR

Consider the circuit below:

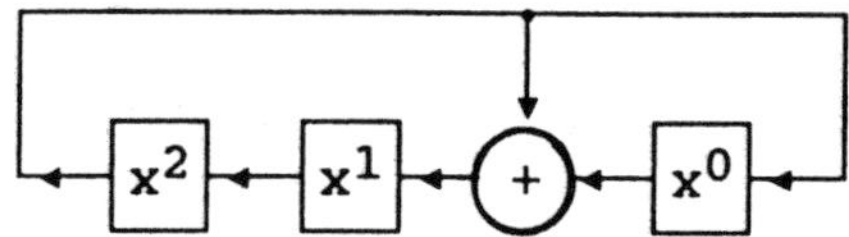

If this circuit is initialized to '001' and clocked, the sequence below will be generated.

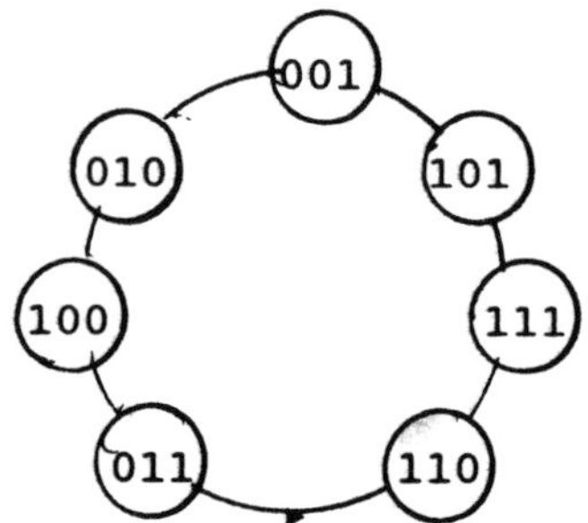

The sequence repeats every seven shifts. The length of the sequence is seven. The maximum length that a shift register can generate is 2^m-1, where m is the shift register length. Shift registers do not always generate the maximum length sequence. The sequence length depends on the implemented polynomial. It will be a maximum length sequence only if the polynomial is primitive.

1.3.3 *MORE ON POLYNOMIALS*

Reciprocal Polynomial. The reciprocal of a polynomial P(x) of degree m with binary coefficients,

$$P(x) = p_m \cdot x^m + p_{m-1} \cdot x^{m-1} + \cdot \cdot \cdot + p_1 \cdot x + p_0$$

is defined as:

$$x^m \cdot P(1/x) = p_0 \cdot x^m + p_1 \cdot x^{m-1} + \cdot \cdot \cdot + p_{m-1} \cdot x + p_m$$

i.e., the coefficients are flipped end-for-end. "Reverse" is a synonym for "reciprocal."

Self-Reciprocal Polynomial. A polynomial is said to be self-reciprocal if it has the same coefficients as its reciprocal polynomial.

Forward Polynomial. A polynomial is called the forward polynomial when it is necessary to distinguish it from its reciprocal (reverse) polynomial. This applies only to polynomials which are not self-reciprocal.

Polynomial Period. The period of a polynomial P(x) is the least positive integer e such that $(x^e + 1)$ is divisible by P(x).

Reducible. A polynomial of degree m is reducible if it is divisible by some polynomial of a degree greater than 0 but less than m.

Irreducible. A polynomial of degree m is said to be irreducible if it is not divisible by any polynomial of degree greater than 0 but less than m. "Prime" is a synonym for "irreducible."

The reciprocal polynomial of an irreducible polynomial is also irreducible.

Primitive Polynomial. A polynomial of degree m is said to be primitive if its period is 2^m-1.

A primitive polynomial is also irreducible.

The reciprocal polynomial of a primitive polynomial is also primitive.

A PROPERTY OF RECIPROCAL POLYNOMIALS

The reciprocal polynomial can be used to generate a sequence in reverse of that generated by the forward polynomial.

Example: Shift register "A" below implements $(x^3 + x + 1)$ and shifts left. Shift register "B" implements $(x^3 + x^2 + 1)$, the reciprocal of $(x^3 + x + 1)$, and shifts right.

Initialize shift register "A" to '001' and clock four times.

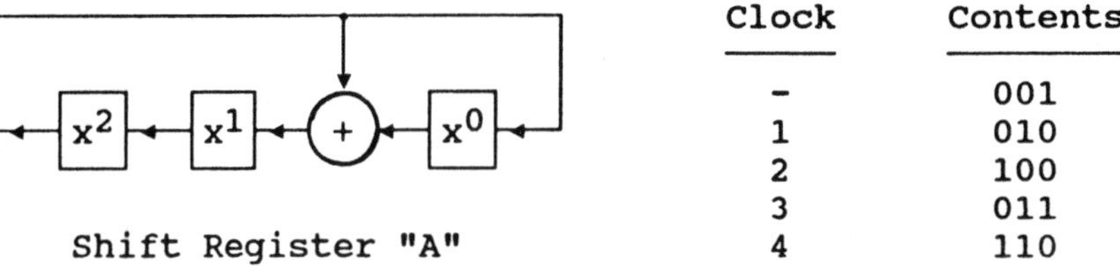

Shift Register "A"

Clock	Contents
-	001
1	010
2	100
3	011
4	110

Transfer the contents of shift register "A" to shift register "B" and clock four times.

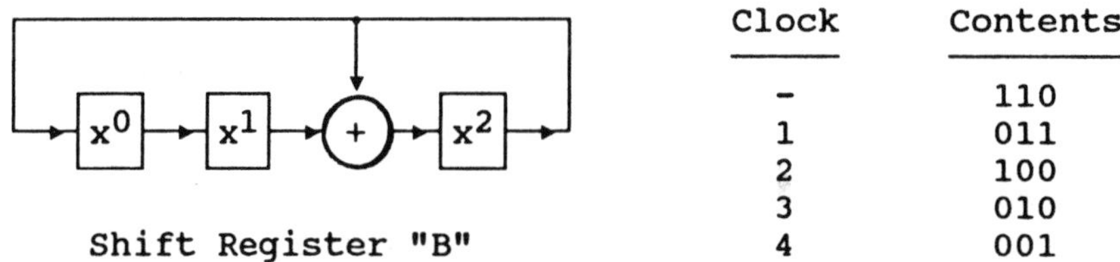

Shift Register "B"

Clock	Contents
-	110
1	011
2	100
3	010
4	001

Shift register "B" retraces in the reverse direction the states of shift register "A".

The property of reciprocal polynomials described above will be used later for decoding some types of error-correcting codes.

DETERMINING THE PERIOD OF AN IRREDUCIBLE POLYNOMIAL WITH BINARY COEFFICIENTS

The algorithm described below for determining the period of an irreducible polynomial g(x) with binary coefficients requires a table. The table is used in determining the residues of powers of x up to (2^m-1).

The table is a list of residues of $x, x^2, x^4, \cdots, x^{2^{m-1}}$ modulo g(x), where m is the degree of the g(x). Each entry in the table can be computed by squaring the prior entry and reducing modulo g(x). The justification is as follows.

$$
\begin{aligned}
x^{2*a} \text{ MOD } g(x) &= (x^a \cdot x^a) \text{ MOD } g(x) \\
&= \{[x^a \text{ MOD } g(x)] \cdot [x^a \text{ MOD } g(x)]\} \text{ MOD } g(x) \\
&= [x^a \text{ MOD } g(x)]^2 \text{ MOD } g(x)
\end{aligned}
$$

The example below illustrates the use of the table for determining the residue of x^{50} modulo g(x).

$$x^{50} \text{ MOD } g(x) = [x^{32+16+2}] \text{ MOD } g(x) = [x^{32} \cdot x^{16} \cdot x^2] \text{ MOD } g(x)$$

$$= \{[x^{32} \text{ MOD } g(x)] \cdot [x^{16} \text{ MOD } g(x)] \cdot [x^2 \text{ MOD } g(x)]\} \text{ MOD } g(x)$$

Select these residues from the table.

The period of an irreducible polynomial of degree m must be a divisor of (2^m-1).

For each e that is a divisor of 2^m-1, compute the residue of x^e modulo g(x) by multiplying together and reducing modulo g(x) an appropriate set of residues from the table.

The period of the polynomial is the least e such that the residue of x^e modulo g(x) is one. If the period is 2^m-1, the polynomial is primitive.

DETERMINING THE PERIOD OF A COMPOSITE POLYNOMIAL WITH BINARY COEFFICIENTS

Let $f_i(x)$ represent the irreducible factors of f(x). If,

$$f(x)=(f_1(x)\cdot f_2(x)\cdot f_3(x)\cdot ....)$$

and there are no repeating factors, the period e of f(x) is given by:

$$e = LCM(e_1,e_2,e_3,....),$$

where the e_i are periods of the irreducible factors.

Example: The period of $(x^3 + 1) = (x + 1)\cdot(x^2 + x + 1)$ is 3.

If f(x) is of the form:

$$f(x) = [f_1(x)^{m_1}]\cdot[f_2(x)^{m_2}]\cdot[f_3(x)^{m_3})]\cdots$$

where the m_i are powers of repeating irreducible factors, then the period e of f(x) is given by:

$$e = k\cdot LCM(e_1,e_2,e_3,\cdots)$$

where k is the least power of two which is not less than any of the m_i.

Example: The period of $(x^3 + x^2 + x + 1) = (x + 1)^3$ is 4.

A SIMPLE METHOD OF COMPUTING PERIOD

A simple method for computing the period of a polynomial is as follows: Initialize a hardware or software shift register implementing the polynomial to '00•••01'. Clock the shift register until it returns to the '00•••01' state. The number of clocks required is the period of the polynomial.

This method can be used to compute the period of composite as well as irreducible polynomials. However, it can be very time consuming when the period is large.

NUMBER OF PRIMITIVE POLYNOMIALS OF GIVEN DEGREE

The divisors (factors) of $(x^{2^m-1} + 1)$ are the polynomials with period 2^m-1 or whose period divides 2^m-1. This may include polynomials of degree less than or greater than m.

The divisors (factors) of $(x^{2^m-1} + 1)$ that are of degree m are the primitive polynomials of degree m.

The number n of primitive polynomials of degree m with binary coefficients is given by:

$$n = \frac{\phi(2^m-1)}{m}$$

where Ú(x) is Euler's phi function and is the number of positive integers equal to or less than x that are relatively prime to x:

$$\phi(x) = \prod_i (p_i)^{e_i-1} \cdot (p_i-1)$$

where

p_i = The prime factors of x
e_i = The powers of prime factors p_i

Example: There are 30 positive integers that are equal to or less than 31 and relatively prime to 31. Therefore, there are 6 primitive polynomials of degree 5.

SHIFT REGISTER SEQUENCES USING A NONPRIMITIVE POLYNOMIAL

Previously, a maximum length sequence generated by a primitive polynomial was studied. Nonprimitive polynomials generate multiple sequences.

The state sequence diagram shown below is for the irreducible nonprimitive polynomial

$$x^4 + x^3 + x^2 + x + 1$$

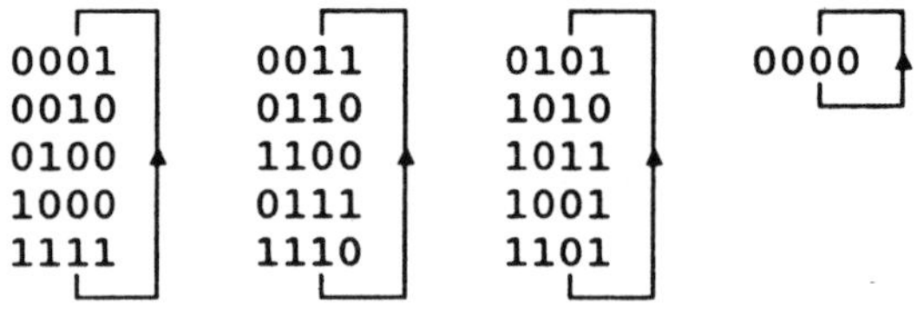

```
0001    0011    0101    0000
0010    0110    1010
0100    1100    1011
1000    0111    1001
1111    1110    1101
```

The state sequence diagram shown below is for the reducible polynomial

$$x^4 + x^3 + x^2 + 1 = (x + 1) \cdot (x^3 + x + 1)$$

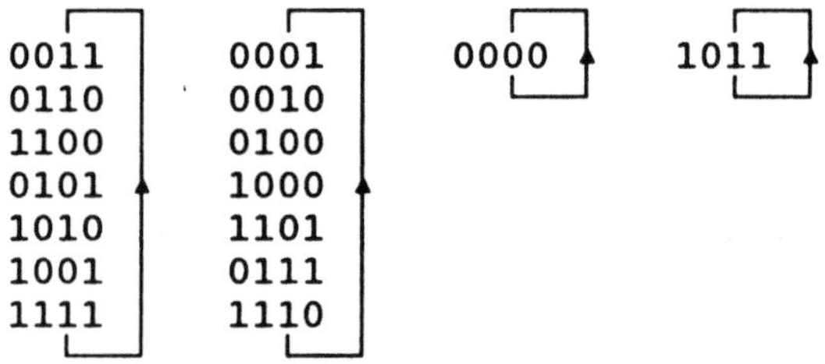

```
0011    0001    0000    1011
0110    0010
1100    0100
0101    1000
1010    1101
1001    0111
1111    1110
```

Each of the four sequences directly above contain states with either an odd number of bits or an even number of bits, but not both. This is caused by the (x + 1) factor.

REDUCTION MODULO A FIXED POLYNOMIAL

It is frequently necessary to reduce an arbitrary polynomial modulo a fixed polynomial, or it may be necessary to reduce the result of an operation modulo a fixed polynomial.

The arbitrary polynomial could be divided by the fixed polynomial and the remainder retained as the modulo result.

Another method is illustrated below. Assume the fixed polynomial to be $(x^3 + x + 1)$. Reduce all terms of the arbitrary polynomial by repeated application of the following relationship.

$$x^{i+3} = x^{i+1} + x^i$$

Suppose the arbitrary polynomial is x^4. Then, using the relationship above with $i=1$ gives:

$$x^4 = x^2 + x.$$

Other examples of arbitrary polynomials reduced modulo $(x^3 + x + 1)$ are shown below.

$$\begin{aligned} x^4 + x^2 &= (x^2 + x) + x^2 \\ &= x \end{aligned}$$

$$\begin{aligned} x^9 &= x^7 + x^6 \\ &= (x^5 + x^4) + (x^4 + x^3) \\ &= x^5 + x^3 \\ &= (x^3 + x^2) + x^3 \\ &= x^2 \end{aligned}$$

DIVIDING BY A COMPOSITE POLYNOMIAL

Sometimes it is necessary to divide a received polynomial C'(x) by a composite polynomial p(x) = p1(x)•p2(x)•p3(x)••••, where p1(x),p2(x),p3(x),••• are relatively prime, in pairs. Assume the remainder is to be checked for zero.

The remainder could be checked for zero after dividing the received polynomial by the composite polynomial. However, dividing the received polynomial by the individual factors of the composite polynomial and checking all individual remainders for zero would be equivalent.

Example #1: p(x) = p1(x)•p2(x)

$$x^4 + x^3 + x^2 + 1 \quad = \quad (x + 1)\cdot(x^3 + x + 1)$$

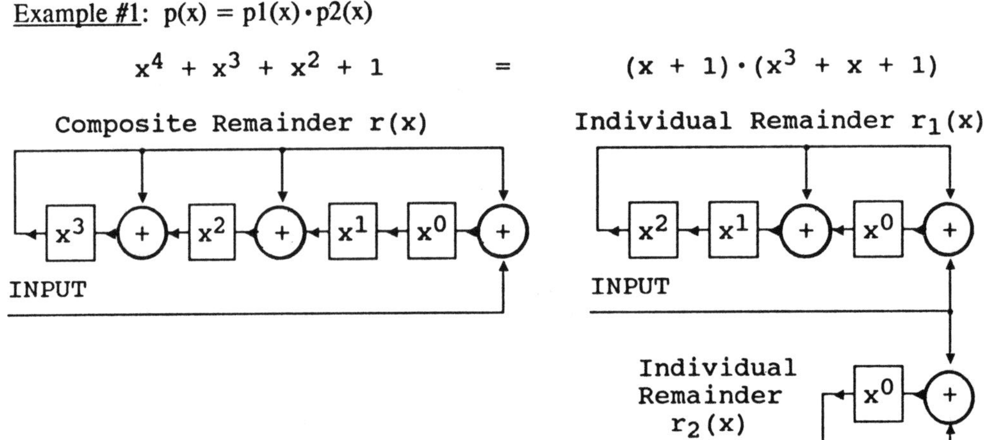

At other times, when the generator polynomial is composite, individual remainders are required for computation.

The received polynomial could be divided directly by each factor of the composite polynomial to get individual remainders. However, the following two-step procedure would be equivalent.

1. Divide the received polynomial by the composite polynomial to get a composite remainder.

2. Divide the composite remainder by factors of the composite polynomial to get individual remainders.

Step 2 could be accomplished by software, sequential logic or combinatorial logic.

In many cases, a slower process can be used in step 2 than in step 1 because fewer cycles are required in dividing the composite remainder.

The diagram below shows an example of computing individual remainders from a composite remainder using combinatorial logic.

Example #2

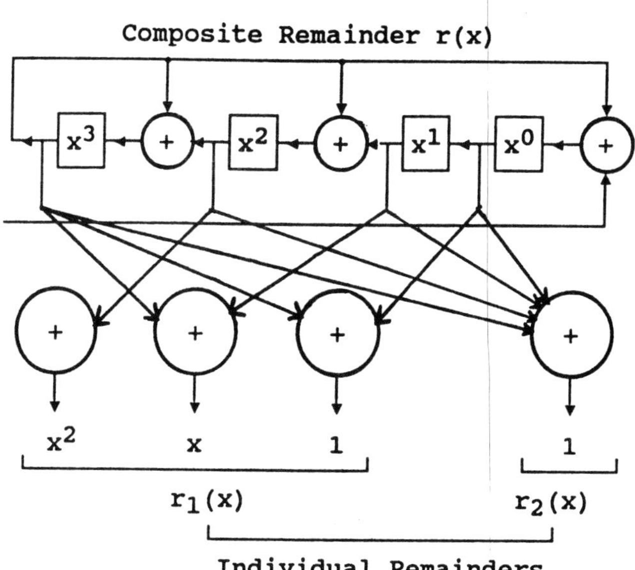

It is also possible to compute a composite remainder from individual remainders, as shown below.

Example #3

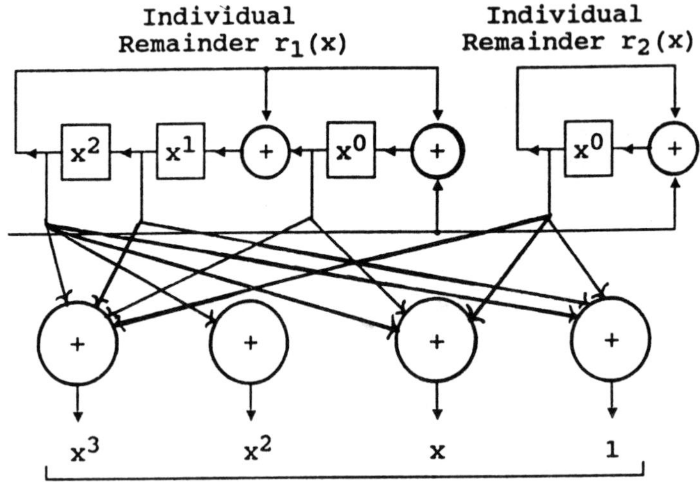

In the examples above, the factors of the composite polynomial are assumed to be relatively prime. If this is the case, the Chinese Remainder Theorem for polynomials guarantees a one-to-one mapping between composite remainders and sets of individual remainders.

To understand how the connections in circuit Examples #2 and #3 were determined, study the mappings below. To generate the first mapping, the individual remainders corresponding to each composite remainder are determined by dividing each possible composite remainder by the factors of the composite polynomial. For the second mapping, the composite remainder corresponding to each set of individual remainders is determined by rearranging the first mapping.

The boxed areas of the first mapping establish the circuit connections for Example #2. The boxed areas of the second mapping establish the circuit connections for Example #3. There are other ways to establish these mappings. The method shown here has been selected for simplicity. However, in a practical sense it is limited to polynomials of a low degree.

FIRST MAPPING

Composite Remainder	Corresponding Individual Remainders
0000	000 0
0001	001 1
0010	010 1
0011	011 0
0100	100 1
0101	101 0
0110	110 0
0111	111 1
1000	011 1
1001	010 0
1010	001 0
1011	000 1
1100	111 0
1101	110 1
1110	101 1
1111	100 0

SECOND MAPPING

Individual Remainders	Corresponding Composite Remainder
000 0	0000
000 1	1011
001 0	1010
001 1	0001
010 0	1001
010 1	0010
011 0	0011
011 1	1000
100 0	1111
100 1	0100
101 0	0101
101 1	1110
110 0	0110
110 1	1101
111 0	1100
111 1	0111

PROBLEMS

1. Write the sequence for the circuit below.

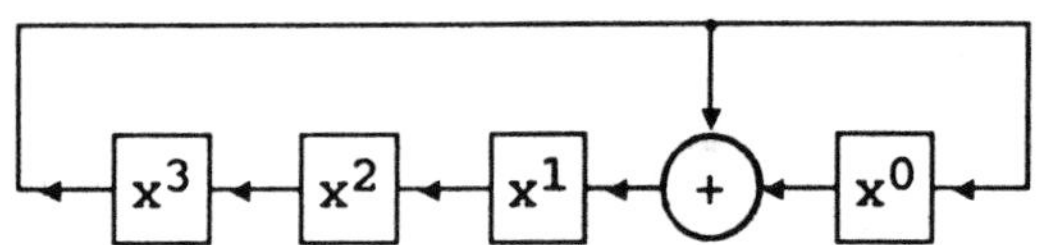

2. Write the polynomial for the circuit above.

3. Perform the multiplication operations below.

$$\begin{array}{r} x^3 + x + 1 \\ \underline{x \quad + 1} \end{array} \qquad \begin{array}{r} x^3 + x^2 + x + 1 \\ \underline{x^3 + 1 \qquad\qquad} \end{array} \qquad \begin{array}{r} x^5 + 1 \\ \underline{x^3 + x + 1} \end{array}$$

4. Perform the division operations below. Show the quotient and the remainder.

$$x^3 + x + 1 \overline{\big)\, x^6 + x + 1} \qquad x^3 + x + 1 \overline{\big)\, x^3 + x}$$

5. Determine the period of the following polynomials:

$$x^3 + 1, \qquad x^3 + x^2 + x + 1, \qquad x^3 + x^2 + 1$$

6. Show a circuit to multiply by $(x^3 + 1)$.

7. Show a circuit to divide by $(x^3 + 1)$.

8. Show a circuit to compute a remainder modulo $(x^3 + x^2 + 1)$ using combinatorial logic. The input polynomial is 7 bits in length.

9. Is $(x^2 + x + 1)$ reducible?

10. Compute the reciprocal polynomial of $(x^4 + x + 1)$.

11. How many primitive polynomials are of degree 4?

CHAPTER 2 - ERROR DETECTION AND CORRECTION FUNDAMENTALS

2.1 DETECTION FUNDAMENTALS

MORE ON POLYNOMIAL SHIFT REGISTERS

The shift register form below is used frequently for error detection and correction. This circuit multiplies by x^m and divides by g(x), where m is the degree of g(x) and also the shift register length. g(x) is the generator polynomial of the error detection/correction code being implemented. For this example, $g(x) = x^3 + x + 1$ and m=3.

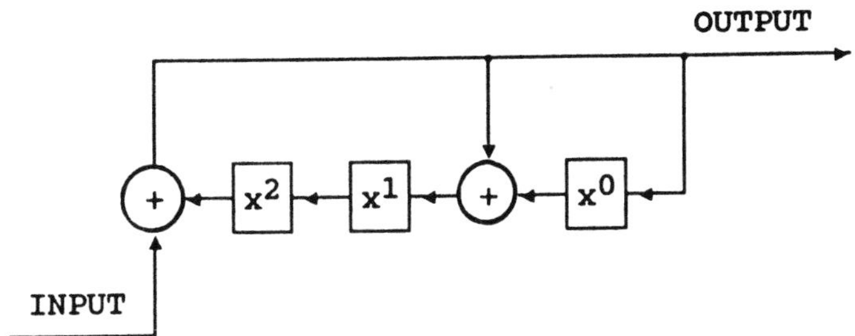

Two properties of this form of shift register are discussed below.

Property #1

If the shift register above is receiving a stream of bits, the last m bits (in this case three) must match the shift register contents in order for the final shift register state to be zero. This is because a difference between the input data bit and the high order shift register stage causes at least the low order stage to be loaded with '1'.

Assume an all-zeros data record. Any burst of length m or fewer bits will leave the shift register in a nonzero state. If an error burst of length greater than m bits is to leave the shift register in its zero state, the last m bits of the burst must match the shift register contents created by the error bits which preceded the last m bits of the burst.

Property #2

Assume the shift register is zero. Receiving an error burst of length m or fewer bits has the same effect as placing the shift register at the state represented by the sequence of error bits.

When reading an all-zeros data record, an error burst of length m or fewer bits sets the shift register to a state on its sequence that is b shifts away from the state representing the error burst, where b is the length of the burst.

SELECTING CHECK BITS

Property #1 implies that for all-zero data, any burst of length m or fewer bits is guaranteed to be detected. Property #2 indicates that for all-zero data, it may be possible to correct some bursts of length less than m bits by clocking the shift register along its sequence until the error burst is contained within the shift register.

Clearly, we must find a way to extend these results to cases of nonzero data if they are to be of any use. The following discussion describes intuitively how check bits must be selected so that on read, the received polynomial leaves the shift register at zero in the absence of error.

Assume a shift register configuration that premultiplies by x^m and divides by g(x). On write, after clocking for all data bits has been completed, the shift register will likely be in a nonzero state if nonzero data bits have been processed. If we transmit as check bits following the data bits, the contents of the shift register created by processing the data bits, then on read in the absence of error, the received data bits will create the same pattern in the shift register, and the received check bits will match this pattern, leaving the shift register in its zero state.

The concatenation of the data bits and their associated check bits is called a codeword polynomial or simply a codeword. A codeword C(x) generated in the manner outlined above by a shift register implementing a generator polynomial g(x) has the property:

$$C(x) \text{ MOD } g(x) = 0$$

This is a mathematical restatement of the condition that processing a codeword must leave the shift register in its zero state.

Theorem 2.1.1. *The Euclidean Division Algorithm*. If D(x) and g(x) are polynomials with coefficients in a field F, and g(x) is not zero, there exists polynomials q(x) (the quotient) and r(x) (the remainder) with coefficients in F such that:

$$D(x) = q(x) \cdot g(x) + r(x)$$

where the degree of r(x) is less than the degree of g(x); r(x) may in fact be zero.

The Euclidean Division Algorithm provides a formal justification for the method of producing check bits outlined above. By the Euclidean Division Algorithm,

$$D(x) = q(x) \cdot g(x) + r(x)$$

where

$$\begin{aligned} D(x) &= \text{Data polynomial} \\ g(x) &= \text{Generator polynomial} \\ q(x) &= \text{Quotient polynomial} \\ r(x) &= \text{Remainder polynomial} \end{aligned}$$

Rearranging gives

$$\frac{D(x) + r(x)}{g(x)} = q(x)$$

This shows that in order to make the data polynomial itself divisible by g(x), r(x) would have to be EXCLUSIVE-OR-ed against D(x). However, this would modify the last m bits of the data polynomial, which is not desirable.

Appending the remainder bits to the input data bits has the effect of premultiplying D(x) by x^m and then dividing by g(x). Then by Euclidean Division Algorithm we have,

$$x^m \cdot D(x) = q(x) \cdot g(x) + r(x)$$

or equivalently,

$$\frac{x^m \cdot D(x) + r(x)}{g(x)} = q(x)$$

This shows that if r(x) is EXCLUSIVE-OR-ed against the data polynomial premultiplied by x^m, the resulting polynomial will be divisible by g(x). This is equivalent to appending r(x) to the end of the original input data polynomial, since coefficients of all x^i terms of $x^m \cdot D(x)$ are zero for $i<m$. The original data polynomial is not modified when check bits are added with this method.

NOTATION

The following symbology will be used in our discussion of error detection and correction codes:

$$\begin{aligned} D(x) &= \text{Data polynomial} \\ k &= \text{Number of information symbols} = \text{degree of } D(x)+1 \\ g(x) &= \text{Code generator polynomial} \\ m &= \text{Number of check symbols} = \text{degree of } g(x) \\ W(x) &= \text{Write redundancy (check) polynomial} \\ &= x^m \cdot D(x) \text{ MOD } g(x) \\ C(x) &= \text{Transmitted codeword polynomial} \\ &= x^m \cdot D(x) + W(x) = x^m \cdot D(x) + [x^m \cdot D(x) \text{ MOD } g(x)] \\ n &= \text{Record length} = \text{degree of } C(x) = k+m \\ E(x) &= \text{Error polynomial} \\ C'(x) &= \text{Received codeword polynomial} \\ &= C(x) + E(x) \end{aligned}$$

An implementation of the encoding process using the internal-XOR form of shift register circuit is shown below. This particular example premultiplies by x^3 and divides by $(x^3 + x + 1)$.

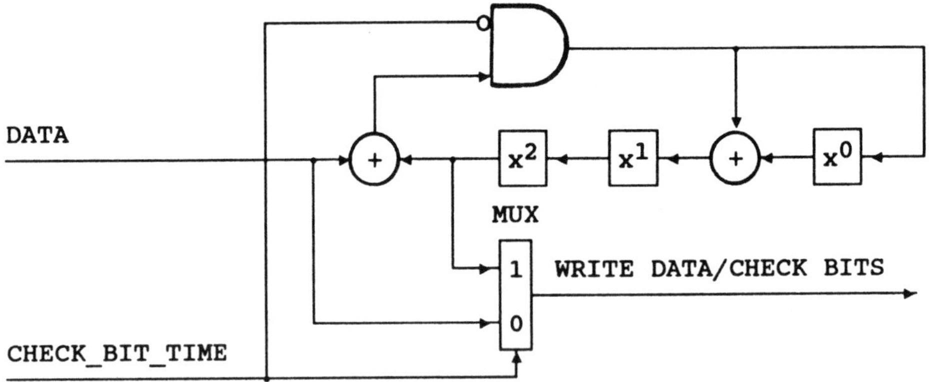

After all DATA bits have been clocked into the shift register, the CHECK-_BIT_TIME signal is asserted. The AND gate then disables feedback, allowing the check bits to be shifted out of the shift register, and the MUX passes the check bits to the device.

An implementation using the external-XOR form of shift register circuit shown below performs the same function. It writes the same check bits for a given data record.

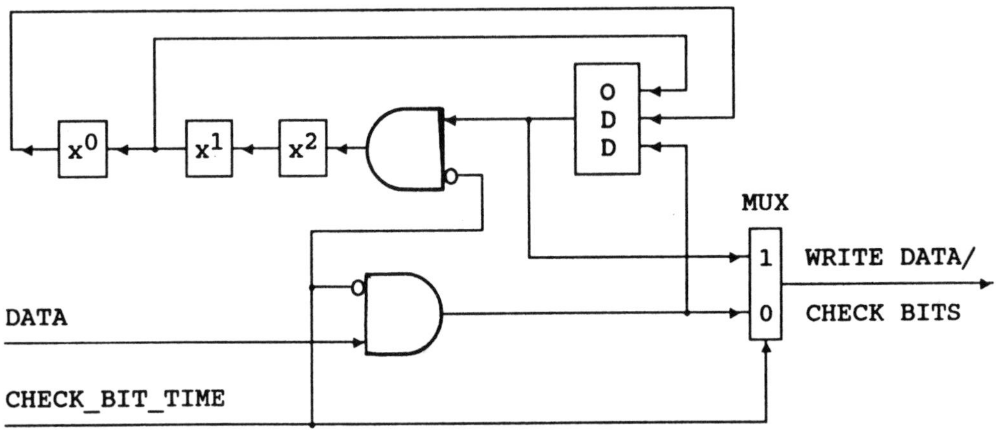

After all DATA bits have been clocked into the shift register, the CHECK-_BIT_TIME signal is asserted. The upper AND gate then disables feedback and the lower AND gate blocks extraneous DATA input to the ODD parity tree, whose output the MUX passes as check bits to the device.

SINGLE-BURST DETECTION SPAN FOR AN ERROR-DETECTION CODE

The single-burst detection span for a detection-only code is equal to the shift register length. This is obvious from Property #1 discussed earlier. Assume a shift register configuration that premultiplies by x^m and divides by g(x). Assume the shift register to be initialized to zero and assume an all zeros data record. The only '1' bits to enter the shift register will be from an error burst. The first bit of the burst sets certain shift register bits to 1, including the low order bit.

In order to set the shift register to zero, the next m error burst bits must match the shift register contents. Therefore, in order for an error burst to set the shift register to zero, it must be longer than the length of the shift register.

This can be also be demonstrated mathematically. It must be shown that the length of an error burst required to leave the shift register at zero is greater than m bits. For an error burst to leave the shift register at zero, it must be divisible by the generator polynomial. It must be shown that to be divisible by the polynomial, a burst must be greater than m bits in length.

Let E(x) contain a single error burst of length m or fewer bits. Let the lowest-order nonzero coefficient of E(x) be the coefficient of the x^j term of C'(x). Then:

$$E(x) = x^j \cdot b(x)$$

where the lowest-order nonzero coefficient of b(x) is that of x^0 and the length of the burst is equal to the degree of b(x) plus one. It is clear that x^j and g(x) are relatively prime, so if g(x) is to divide E(x) it must divide b(x). This is impossible, since if the burst is of length m or fewer bits, b(x) is a polynomial of degree at most (m-1) and is clearly not divisible by g(x), which is of degree m.

THEOREMS FOR ERROR-DETECTION CODES

Theorem 2.1.2. All single-bit errors will be detected by any code whose generator polynomial has more than one term. The simplest example is the code generated by the polynomial (x+1).

Theorem 2.1.3. All cases of an odd number of bits in error will be detected by a code whose generator polynomial has (x^c+1) where c is greater than zero, as a factor.

The check bit generated by (x+1) is simply an overall parity check. All polynomials of the form (x^c+1) are divisible by (x+1). Therefore, any code whose generator polynomial has a factor of the form (x^c + 1) automatically includes an overall parity check.

Theorem 2.1.4. A code will detect all single- and double-bit errors if the record length (including check bits) is no greater than the period of the generator polynomial.

Theorem 2.1.5. A code will detect all single-, double-, and triple-bit errors if its generator polynomial is of the form (x^c + 1)•P(x) and the record length (including check bits) is no greater than the period of the generator polynomial.

Theorem 2.1.6. A code generated by a polynomial of degree m detects all single burst errors of length no greater than m. Note that a burst of length b is defined as any error pattern for which the number of bits between and including the first and last bits in error is b.

Theorem 2.1.7. A code with a generator polynomial of the form (x^c + 1)•P(x) has a guaranteed double-burst detection capability provided the record length (including check bits) is no greater than the period of the generator polynomial. It will detect any combination of double bursts when the length of the shorter burst is no greater than the degree of P(x) and the sum of the burst lengths is no greater than (c+1).

This theorem allows selection of a code by structure for accomplishing double-burst detection. Codes which do double-burst detection can also be selected by a computer evaluation of random polynomials.

Theorem 2.1.8. The misdetection probability P_{md}, defined as the fraction of error bursts of length b>m where m is the degree of the generator polynomial, that go undetected is:

$$P_{md} = \frac{1}{2^m} \quad \text{if } b > (m+1)$$

$$= \frac{1}{2^{m-1}} \quad \text{if } b = (m+1)$$

When all errors are assumed to be possible and equally probable, P_{md} is given by:

$$P_{md} \approx \frac{1}{2^m}$$

If some particular error bursts are more likely to occur than others (which is generally the case), then the misdetection probability depends on the particular polynomial and the nature of the errors.

MULTIPLE-SYMBOL ERROR DETECTION

An error-detection code can be constructed from the binary BCH or Reed-Solomon codes to achieve multiple-bit or multiple-symbol error detection. See Sections 3.3 and 3.4.

CAPABILITY OF A PARTICULAR ERROR-DETECTION CODE: CRC-CCITT CODE

The generator polynomial for the CRC-CCITT code is:

$$x^{16} + x^{12} + x^5 + 1 = (x + 1)\cdot(x^{15} + x^{14} + x^{13} + x^{12} + x^4 + x^3 + x^2 + x + 1)$$

The code's guaranteed capability as determined by its structure is defined below:

a) Detects all occurrences of an odd number of bits in error. (Theorem 2.1.3)

b) Detects all single-, double- and triple-bit errors if the record length (including check bits) is no greater than 32,767 bits. (Theorem 2.1.5)

c) Detects all single-burst errors of sixteen bits or less. (Theorem 2.1.6)

d) Detects 99.99695% of all possible bursts of length 17, and 99.99847% of all possible longer bursts. (Theorem 2.1.8). This property assumes that all errors are possible and equally probable.

The CRC-CCITT polynomial has some double-burst detection capability when used with short records. This capability cannot be determined by its structure. Computer evaluation is required.

When the code is used with a 2088-bit record, it has a guaranteed detection capability for the following double bursts:

Length of First Burst	Length of Second Burst
1	1 to 6
2	1 to 5
3	1 to 4
4	1 to 4
5	1 to 2
6	1

2.2 CORRECTION FUNDAMENTALS

This section introduces single-bit and single-burst error correction from the viewpoint of shift register sequences.

The examples given use very short records and small numbers of check bits. However, the same techniques apply to longer records and greater numbers of check bits as well.

SINGLE-BIT ERROR CORRECTION

The circuit shown below can be used to correct a single-bit error in a seven-bit record (four-data bits and three-check bits). Data bits are numbered d3 through d0. Check bits are numbered p2 through p0. Data and check bits are transmitted and received in the following order:

```
d3 d2 d1 d0 p2 p1 p0
```

Both the encode and decode shift registers premultiply by x^m and divide by g(x). Again m is three and $g(x) = x^3 + x + 1$.

ENCODE CIRCUIT

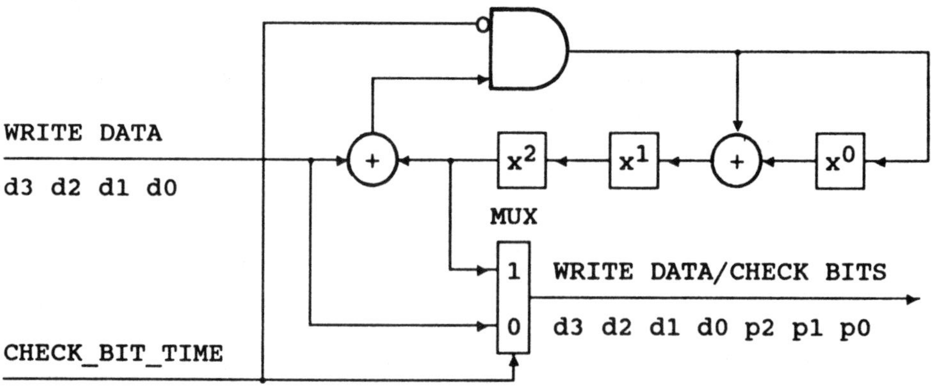

For encoding, the shift register is first cleared. Data bits d3, d2, d1, and d0 are processed and simultaneously passed through the MUX to be sent to the storage device or channel.

After data bits are processed, the gate is disabled and the MUX is switched from data bits to the high order shift register stage. The shift register contents are then sent to the storage device or channel as check bits.

DECODE CIRCUIT

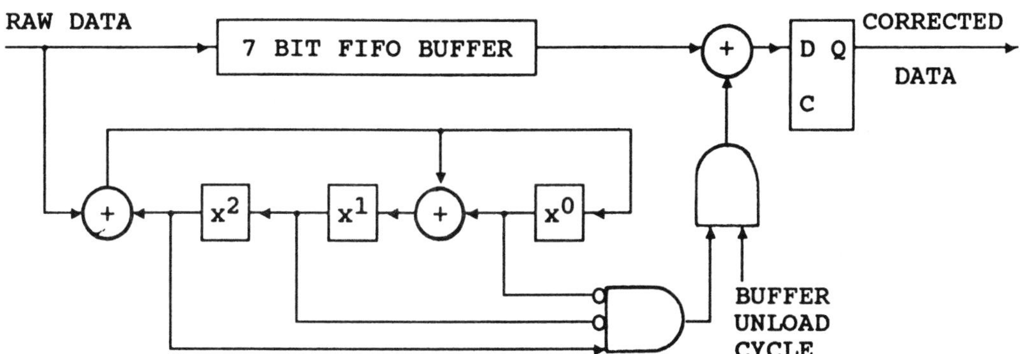

Decoding takes place in two cycles; the buffer load cycle and the buffer unload cycle. A syndrome is generated by the shift register circuit as the buffer is loaded. Correction takes place as the buffer is unloaded. The shift register is cleared just prior to the buffer load cycle.

HOW CORRECTION WORKS

Since g(x) is primitive, it has two sequences: a sequence of length seven and the zero sequence of length one.

```
001          000
010
100
011
110
111
101
```

Assume an all-zeros data record. Assume data bit d1 is in error. The contents of the decode shift register during buffer load would be as shown below.

Clock Number	Error Bits	Shift Register Contents
Initialize		000
d3		000
d2		000
d1	1	011
d0		110
p2		111
p1		101
p0		001

Notice that after the error is processed, the shift register clocks through its sequence until the end of the record is reached. The final shift register state for this example is '001'. This is the syndrome.

The syndrome remains in the shift register as the buffer unload cycle begins. The shift register is clocked as data bits are unloaded from the buffer. As each clock occurs, the shift register clocks through its sequence. Simultaneously, the gate monitors the shift register contents for the '100' state. Correction takes place on the next clock after the '100' state is detected.

The shift register contents during the buffer unload cycle is shown below.

Clock Number	Shift Register Contents
After Read	001
d3	010
d2	100 *
d1	011 **
d0	110
p2	111
p1	101
p0	001

* The three-input gate enables after this clock because the '100' state is detected.

** Correction takes place on this clock.

Consider what happens on the shift register sequence during the buffer load cycle.

```
010
100
011  d1 clock Forces S/R to this point on the sequence.
110  d0 clock Advances S/R to this point on the sequence.
111  p2 clock      "
101  p1 clock      "
001  p0 clock: The final state of the S/R = the syndrome.
```

Since the data record is all zeros, the shift register remains all zeros until the error bit d1 is clocked. The shift register is then set to the '011' state. As each new clock occurs, the shift register advances along its sequence. There is an advance for d0, p2, p1, and p0. After the p0 clock, the shift register is at state '001'. This is the syndrome for the assumed error.

When the error bit occurs, it has the same effect on the shift register as loading the shift register with '100' and clocking once. Regardless of where the error occurs, the first nonzero state of the shift register is '011'.

Error displacement from the end of the record is the number of states between the '100' state and the syndrome. It is determined by the number of times the shift register is clocked between the error occurrence and the end of record.

Consider what happens on the shift register sequence during the buffer unload cycle. The number of states between the syndrome and '100' state represents the error displacement from the front of the record. To determine when to correct, it is sufficient to monitor the shift register for state '100'. Correction occurs on the next clock after this state is detected.

```
001  The syndrome: initial state of the S/R for unload.
010  d3 clock Advances S/R to this point on the sequence.
100  d2 clock The gate is enabled by this S/R state.
011  d1 clock Correction takes place.
110
111
101
```

Consider the case when the data is not all zero. The check bits would have been selected on write such that when the record (data plus check bits) is read without error, a syndrome of zero results. When an error occurs, the operation differs from the all-zeros data case, only while the syndrome is being generated. A given error results in the same syndrome, regardless of data content because the code is linear. Once a syndrome is computed, the operation is the same as previously described for the all-zeros data case.

The code discussed above is a single-error correcting (SEC) Hamming code. It can be implemented with combinatorial logic as well as sequential logic.

SINGLE-BIT ERROR CORRECTION AND DOUBLE-BIT ERROR DETECTION

If an (x + 1) factor is combined with the polynomial of the previous example, the resulting polynomial

$$g(x) = (x + 1)\cdot(x^3 + x + 1) = x^4 + x^3 + x^2 + 1$$

can be used to correct single-bit errors and detect double-bit errors on seven-bit records (three data bits and four check bits). Double-bit errors are detected regardless of the separation between the two error bits.

g(x) has four sequences; the two sequences of length one and two sequences of length seven.

SEQ A	SEQ B		
0001	0011	0000	1011
0010	0110		
0100	1100		
1000	0101		
1101	1010		
0111	1001		
1110	1111		

If a single-bit error occurs, the syndrome will be on sequence A. If a double-bit error occurs, the syndrome will be on sequence B. This gives the code the ability to detect double-bit errors.

The circuit below could be used for decoding. Encoding would be performed with a shift register circuit premultiplying by x^m and dividing by g(x).

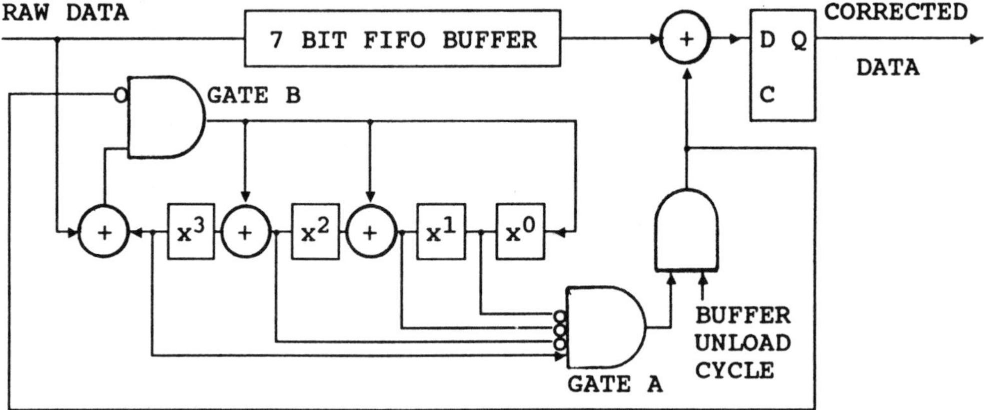

Gate A detects the '1000' state on the clock prior to the clock that corrects the error. Gate B blocks the shift register feedback on the clock following detection of the '1000' state. This causes the shift register to be cleared.

If a double-bit error occurs, the syndrome is on sequence B. The shift register travels around sequence B as it is clocked during the buffer unload cycle. Since the '1000' state is not on this sequence, gate A will not enable and correction will not take place. Since correction does not occur, the shift register remains nonzero. Since the shift register is nonzero at the end of the buffer unload cycle a double error is assumed.

If three bit-errors occur, the syndrome will be on sequence A. During the buffer unload cycle, the shift register state '1000' is detected and a data bit is falsely cleared or set. This is miscorrection because the bit affected is not one of the bits in error.

This code corrects a single-bit error. It detects all occurrences of an even number of bits in error. When more than one bit is in error and the total number of bits in error is odd, miscorrection results.

This code is a single-error correcting (SEC), double-error detecting (DED) Hamming code. It can be implemented with combinatorial logic or with sequential logic.

BURST LENGTH-TWO CORRECTION

The polynomial of the previous example can also be used for burst length-two correction. The circuit is identical except that AND gate A detects '1x00'.

If a burst of length one occurs, the syndrome will be on sequence A. Gate A enables on state '1000'. If a burst of length two occurs, the syndrome will be on sequence B. Gate A enables on state '1100'. When the shift register is clocked from the '1100' state it goes to '1000', due to the action of gate B. Gate A remains enabled. On the next clock, the shift register is cleared due to the action of gate B. Gate A is enabled for two consecutive clock times and therefore two adjacent bits are corrected.

CORRECTION OF LONGER BURSTS

The concepts discussed above can be extended to correction of longer bursts as well.

To construct such a code, select a reducible or irreducible polynomial meeting the following requirements.

1. Each correctable burst must be on a separate sequence.

2. The sequence length must be equal to or greater than the record length (in bits, including check bits) for sequences containing a correctable burst.

3. Any burst that is to be guaranteed detectable must not be on a sequence containing a correctable burst.

Assume a polynomial with multiple sequences and that the bursts '1', '11', '101', and '111' are all on separate sequences of equal length. There may be other sequences as well:

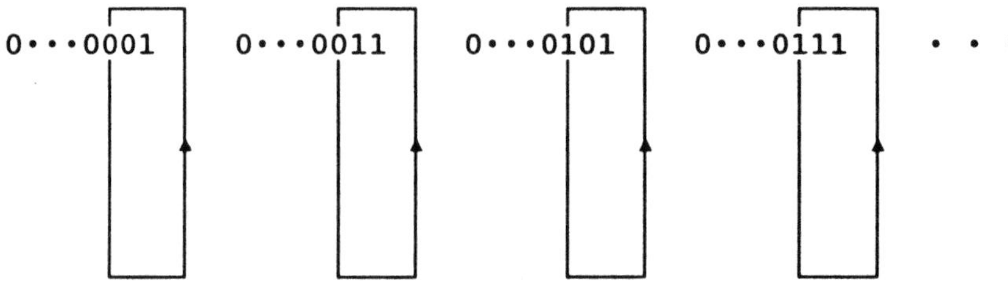

Such a code has at least the following capability: Its correction span can be selected to be one, two, or three bits. In either case, its detection span is guaranteed to be at least three.

Primitive polynomials can also be used for single-burst correction. In this case, the polynomial requirements are:

1. The polynomial period must be equal to or greater than the record length (in bits, including check bits).

2. Correctable bursts must be separated from each other on the sequence by a number of states equal to or greater than the record length (in bits, including check bits).

3. Any burst that is to be guaranteed detectable must be separated from correctable bursts by a number of states equal to or greater than the record length (in bits, including check bits).

It is also possible to state more general requirements for a single-burst correcting code. Any polynomial satisfying either of the two previous sets of requirements would satisfy the more general requirements. Many other polynomials would meet the general requirements as well.

The more general requirements for a single-burst correcting code are:

1. If more than one correctable burst is on a given sequence, these bursts must be separated by a number of states equal to or greater than the record length (in bits, including check bits).
2. If one or more bursts that are to be guaranteed detectable are on a sequence with one or more correctable bursts, they must be separated from each correctable burst by a number of states equal to or greater than the record length (in bits, including check bits).
3. The sequence length must be equal to or greater than the record length (in bits, including check bits) for sequences containing a correctable burst.

ACHIEVING DOUBLE-BURST DETECTION

In order for a computer-generated code to have double-burst detection capability, the following inequality must hold for all i,j, and k such that $0 \leq i,j,k < n$ and $i \neq j$:

$$[x^{i} \cdot b_1(x) + x^{j} \cdot b_2(x)] \text{ MOD } g(x) \neq [x^{k} \cdot b_3(x)] \text{ MOD } g(x)$$

Where

n is the record length (in bits) including check bits

d is the double-burst detection span

s is the single-burst correction span

$b_1(x)$ is any burst of length L_1 such that $0 < L_1 \leq d$

$b_2(x)$ is any burst of length L_2 such that $0 < L_2 \leq d$

$b_3(x)$ is any burst of length L_3 such that $0 < L_3 \leq s$

g(x) is the code generator polynomial

Additionally, if $i>j$ then we require $i>(j+s-L_1)$ and $i\geq(j+L_2)$, while if $i<j$ then we require $i\geq(j-L_1)$ and $i<(j-s+L_2)$.

DST uses special hardware and software to find codes that satisfy these requirements.

SINGLE-BURST CORRECTION VIA STRUCTURED CODES

Fire codes achieve single-burst correction capability by their structure. These codes are generated by the general polynomial form:

$$g(x) = c(x) \cdot p(x) = (x^c + 1) \cdot p(x)$$

where p(x) is any irreducible polynomial of degree z and period e, and e does not divide c. These codes are capable of correcting single bursts of length b and detecting bursts of length $d \geq b$ provided $z \geq b$ and $c \geq (d+b-1)$. The maximum record length in bits, including check bits, is the least common multiple (LCM) of e and c. This is also the period of the generator polynomial g(x).

The structure of Fire code polynomials causes them to have multiple sequences. Each correctable burst is on a separate sequence. Burst error correction with polynomials of this type was discussed earlier in this section. See Section 3.1 for more information on Fire codes.

SINGLE-BURST CORRECTION VIA COMPUTER-GENERATED CODES

The single-burst correction capability of computer-generated codes is achieved by testing.

These codes are based on the fact that if a large number of polynomials of a particular degree are picked at random, some will meet previously defined specifications, provided these specifications are within certain bounds.

There are equations that can be used to predict the probability of success when searching polynomials of particular degree against a particular criteria.

The advantage these codes have over Fire codes is less pattern sensitivity. If miscorrection is to be avoided on certain short double bursts, this can be included as an additional criterion for the computer search. See Section 3.2 for more information on computer-generated codes.

SINGLE-BURST DETECTION SPAN FOR A BURST-CORRECTING CODE

Let n represent the record length in bits (including check bits). Let m represent the shift register length in bits. Assume an all-zeros data record. Assume a shift register configuration that premultiplies by x^m and divides by g(x).

An error burst, m bits or less in length, has the same effect as loading the shift register with the burst. Therefore, a particular error burst will place the shift register at a particular point in the sequence.

If the point in the sequence is far away from any correctable pattern, the shift register will not sequence to a correctable pattern in n shifts and there is no possibility of miscorrection. However, if the particular error burst places the shift register at a point in the sequence that is near a correctable pattern, the correctable pattern may be detected in n shifts and miscorrection will result. It follows that the error bursts of length m or less that have the exposure of miscorrection, are those bursts that force the shift register to points in the sequence near correctable patterns.

The result of having a particular pattern (or state) in the shift register is the same as if the same pattern were an input-error burst. It follows that the list of shift register states near the correctable patterns also represents a list of error bursts, of length m or less, that may result in miscorrection.

The search software shifts a simulated shift register more than n times forward and reverse from each correctable pattern. After each shift, the burst length in the shift register is determined. One less than the minimum burst length found over the entire process represents the single-burst detection span.

PROBABILITY OF MISCORRECTION

Let

b = correction span
n = record length including check bits
m = number of check bits

The total number of possible syndromes is then 2^m. The total number of valid syndromes must be equal to the total number of correctable bursts, which is $n \cdot 2^{b-1}$.

Assume that all error bursts are possible and equally probable and that when random bursts are received, one syndrome is just as likely as another. If all syndromes have equal probability and there are $n \cdot 2^{b-1}$ valid syndromes out of 2^m total possible syndromes, then the probability of miscorrection for bursts exceeding the code's guaranteed detection capability is:

$$Pmc \approx \frac{n \cdot 2^{b-1}}{2^m}$$

This equation provides a measure for comparing the effect that record length, correction span, and number of check bits have on miscorrection probability.

One must be careful using this equation. A very simple assumption is made, which is that all error bursts are possible and equally probable. This is unlikely to be the case except for particular types of errors such as synchronization errors. To accurately calculate the probability of miscorrection requires a detailed knowledge of the types of errors that occur and detailed information on the capability and characteristics of the polynomial.

PATTERN SENSITIVITY OF A BURST-CORRECTING CODE

Some burst-correcting codes have pattern sensitivity. The Fire code, for example, has a higher miscorrection probability on short double bursts than on all possible error bursts.

Pattern sensitivity is discussed in greater detail in Sections 4.4 and 4.6.

2.3 DECODING FUNDAMENTALS

The following pages show various examples of decoding single-burst-error-correcting codes. These points will help in understanding the examples.

1. Forward displacements are counted from the first data bit to the first bit in error. The first data bit is counted as zero.

2. Reverse displacements are counted from the last check bit to the first bit in error. The last check bit is counted as zero.

3. If a negative displacement is computed, add the record length (seven in all examples) to the displacement. If a displacement greater than the record length minus one is computed, subtract the record length from the displacement.

4. Shift register states are shown after the indicated clock.

5. For all examples, the final error pattern is in the register from left to right. The left-most bit of pattern represents the first bit in error from the front of the record.

6. In these simple examples, check bits are corrected as well as data bits.

7. In these examples, only the read decode circuit is shown. The write circuit always premultiplies by x^m and divides by g(x).

8. Each suffix A example is the same as the prior example, except that a different error has been assumed.

9. In examples 1 through 4A, it is not necessary to have additional hardware that detects shift register nonzero at the end of a read. In examples 5 through 8A, this additional hardware is required.

10. In these simple examples, if an error occurs that exceeds the correction capability of the code, miscorrection results. In a real world implementation, excess redundancy would be added to keep miscorrection probability low.

11. The following abbreviations are used in the decoding examples.

    CLK - Clock
    CNT - Count
    ERR - Error
    FIFO - First in, first out
    S/R - Shift register

Example #1:

- Correction in hardware, forward clocking.
- Single-bit-correcting code, single-bit error, data all zeros.
- Spaced data blocks, on-the-fly correction (data delay = 1 block).
- Internal-XOR form of shift register.
- $g(x) = x^3 + x + 1$.
- Detect zeros in right-most bits of shift register.
- Premultiply by x^3.

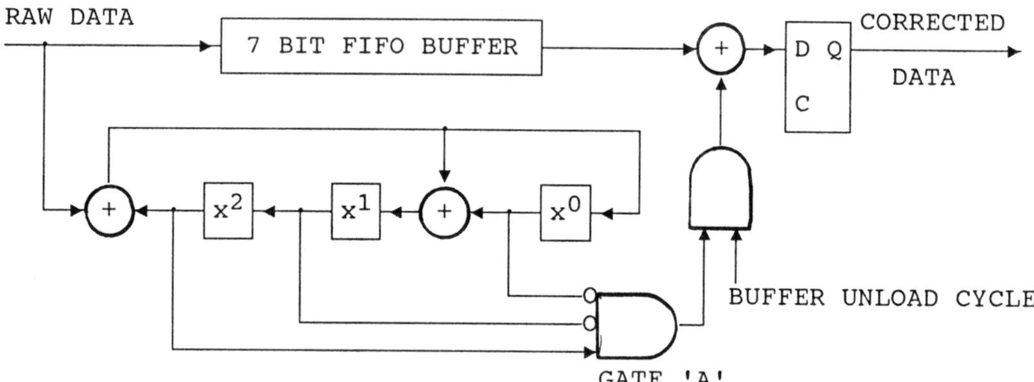

READ CYCLE (BUFFER LOAD)	ERR	S/R	CORRECT CYCLE (BUFFER UNLOAD)	S/R	
				010	*
d3	0	000	d3	100	**
d2	1	011	d2	011	***
d1	0	110	d1	110	
d0	0	111	d0	111	
p2	0	101	p2	101	
p1	0	001	p1	001	
p0	0	010	p0	010	

* Shift register contents at start of correction cycle.
** Gate A enables after the d3 clock.
*** Correction takes place on d2 clock.

Example #2:

- Correction in hardware, forward clocking.
- Single-bit-correcting code, single-bit error, data all zeros.
- Spaced data blocks, on-the-fly correction (data delay = 1 block).
- Internal-XOR form of shift register.
- $g(x) = x^3 + x + 1$.
- Detect zeros in left-most bits of shift register.
- No premultiplication.

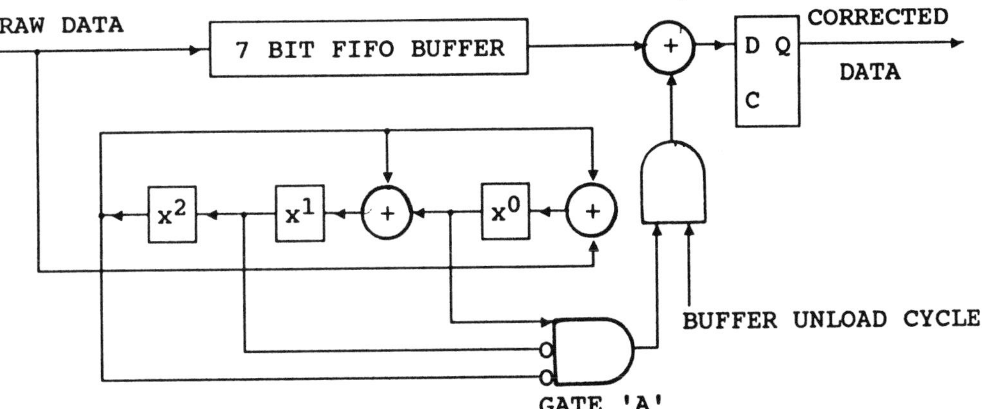

READ CYCLE (BUFFER LOAD)	ERR	S/R	CORRECT CYCLE (BUFFER UNLOAD)	S/R	
				111	*
d3	0	000	d3	101	**
d2	1	001	d2	001	***
d1	0	010	d1	010	
d0	0	100	d0	100	
p2	0	011	p2	011	
p1	0	110	p1	110	
p0	0	111	p0	111	

* Shift register contents at start of correction cycle.
** Gate A enables after the d2 clock.
*** Correction takes place on d2 delayed clock.

Example #3:

- Correction in hardware, forward clocking.
- Burst length-two correcting code, two-adjacent error, data all zeros.
- Spaced data blocks, on-the-fly correction (data delay = 1 block).
- Internal-XOR form of shift register.
- $g(x) = (x + 1)\cdot(x^3 + x + 1) = x^4 + x^3 + x^2 + 1$.
- Detect zeros in right-most bits of shift register.
- Premultiply by x^4.

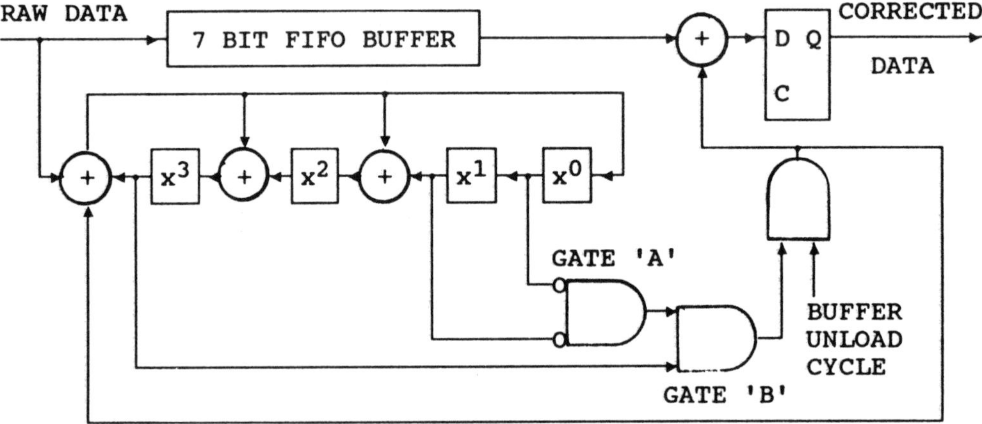

READ CYCLE (BUFFER LOAD)			CORRECT CYCLE (BUFFER UNLOAD)		
	ERR	S/R		S/R	
				0110	*
d2	0	0000	d2	1100	**
d1	1	1101	d1	1000	***
d0	1	1010	d0	0000	****
p3	0	1001	p3	0000	
p2	0	1111	p2	0000	
p1	0	0011	p1	0000	
p0	0	0110	p0	0000	

* Shift register contents at start of correction cycle.
** Gates A and B enable after the d2 clock.
*** Bit d1 is corrected on the d1 clock.
**** Bit d0 is corrected on the d0 clock.

Example #3A:

- Correction in hardware, forward clocking.
- Burst length-two correcting code, single-bit error, data all zeros.
- Spaced data blocks, on-the-fly correction (data delay = 1 block).
- Internal-XOR form of shift register.
- $g(x) = (x + 1) \cdot (x^3 + x + 1) = x^4 + x^3 + x^2 + 1$.
- Detect zeros in right-most bits of shift register.
- Premultiply by x^4.

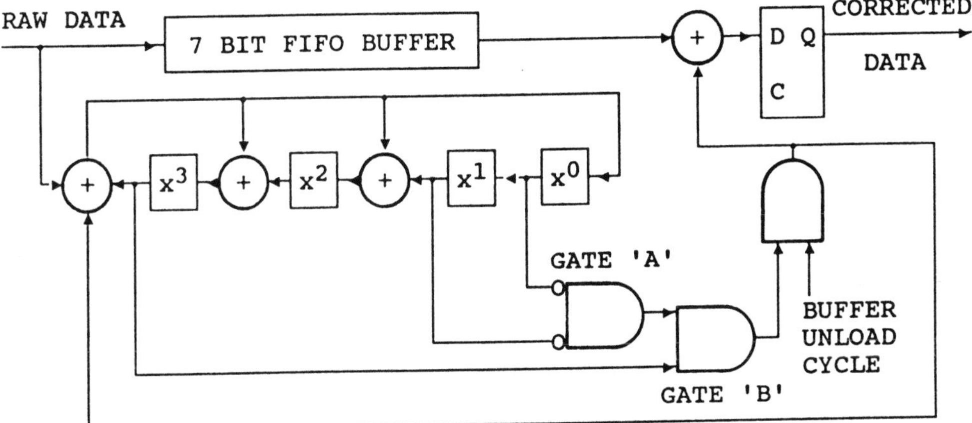

READ CYCLE (BUFFER LOAD)	ERR	S/R	CORRECT CYCLE (BUFFER UNLOAD)	S/R	
				0010	*
d2	0	0000	d2	0100	**
d1	0	0000	d1	1000	***
d0	1	1101	d0	0000	****
p3	0	0111	p3	0000	
p2	0	1110	p2	0000	
p1	0	0001	p1	0000	
p0	0	0010	p0	0000	

* Shift register contents at start of correction cycle.

** Gate A enables after the d2 clock.

*** Gate B enables after the d1 clock. No correction takes place on the d1 clock because gate B is disabled at the time of the clock.

**** Bit d0 is corrected on the d0 clock.

Example #4:

- Correction in hardware, forward clocking.
- Burst length-two correcting code, two adjacent error, data all zeros.
- Consecutive data blocks, on-the-fly correction (delay = 1 block).
- Internal-XOR form of shift register.
- $g(x) = (x + 1) \cdot (x^3 + x + 1) = x^4 + x^3 + x^2 + 1$.
- Detect zeros in right-most bits of shift register.
- Premultiply by x^4.

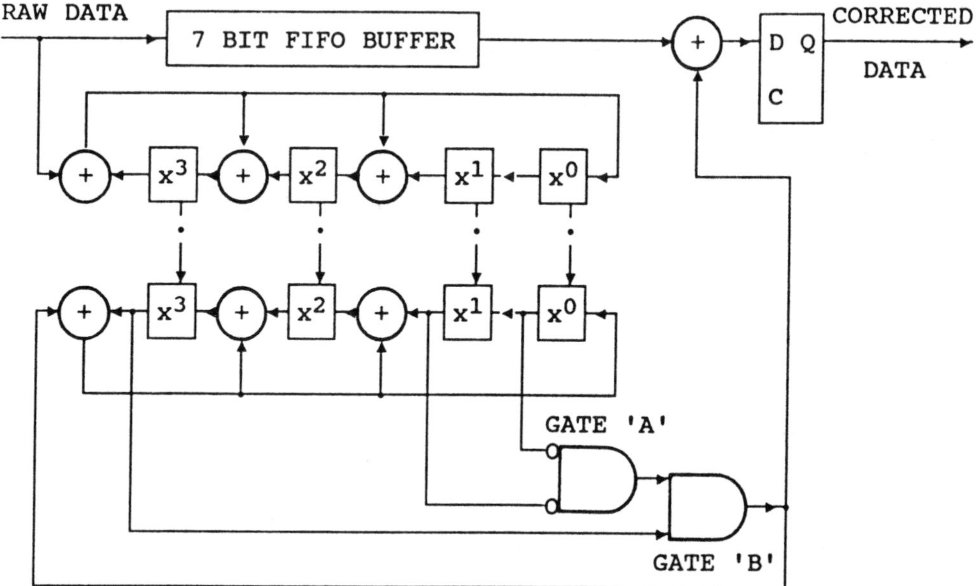

READ CYCLE (BUFFER LOAD)	ERR	S/R	CORRECT CYCLE (BUFFER UNLOAD)	S/R	
				0110	*
d2	0	0000	d2	1100	**
d1	1	1101	d1	1000	***
d0	1	1010	d0	0000	****
p3	0	1001	p3	0000	
p2	0	1111	p2	0000	
p1	0	0011	p1	0000	
p0	0	0110	p0	0000	

* Shift register contents at start of correction cycle.
** Gates A and B enable after the d2 clock.
*** Bit d1 is corrected on the d1 clock.
**** Bit d0 is corrected on the d0 clock.

Example #4A:

- Correction in hardware, forward clocking.
- Burst length-two correcting code, single-bit error, data all zeros.
- Consecutive data blocks, on-the-fly correction (delay = 1 block).
- Internal-XOR form of shift register.
- $g(x) = (x + 1) \cdot (x^3 + x + 1) = x^4 + x^3 + x^2 + 1$.
- Detect zeros in right-most bits of shift register.
- Premultiply by x^4.

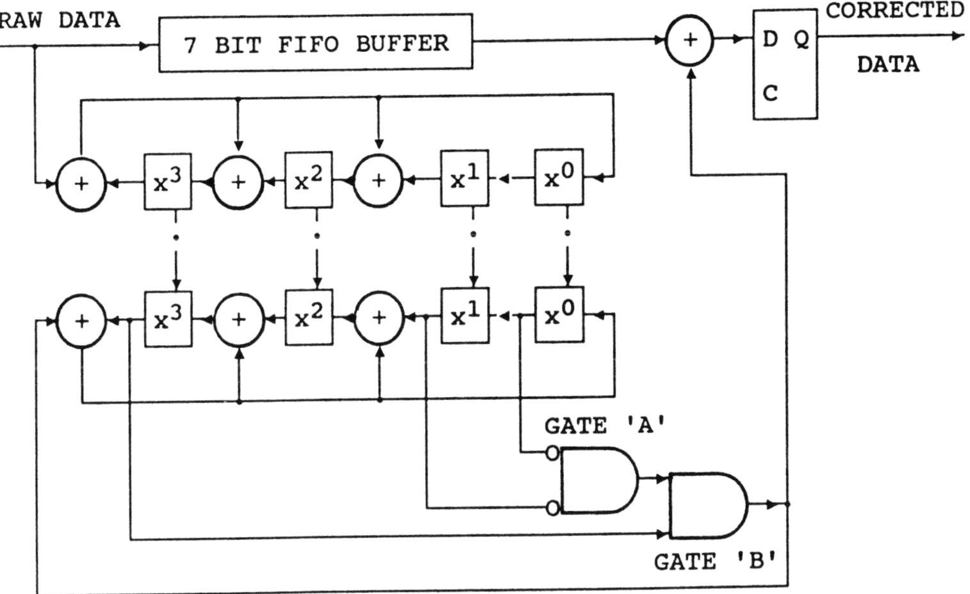

READ CYCLE (BUFFER LOAD)	ERR	S/R	CORRECT CYCLE (BUFFER UNLOAD)	S/R	
				0100	*
d2	0	0000	d2	1000	**
d1	1	1101	d1	0000	***
d0	0	0111	d0	0000	
p3	0	1110	p3	0000	
p2	0	0001	p2	0000	
p1	0	0010	p1	0000	
p0	0	0100	p0	0000	

* Shift register contents at start of correction cycle.

** Gate B enables after the d2 clock.

*** Bit d1 is corrected on the d1 clock.

Example #5:

- Correction in hardware, forward clocking, software assist.
- Burst length-two correcting code, two adjacent error, data all zeros.
- Time delay required when an error occurs.
- Internal-XOR form of shift register.
- $g(x) = (x + 1) \cdot (x^3 + x + 1) = x^4 + x^3 + x^2 + 1$.
- Detect zeros in right-most bits of shift register.
- Premultiply by x^4.

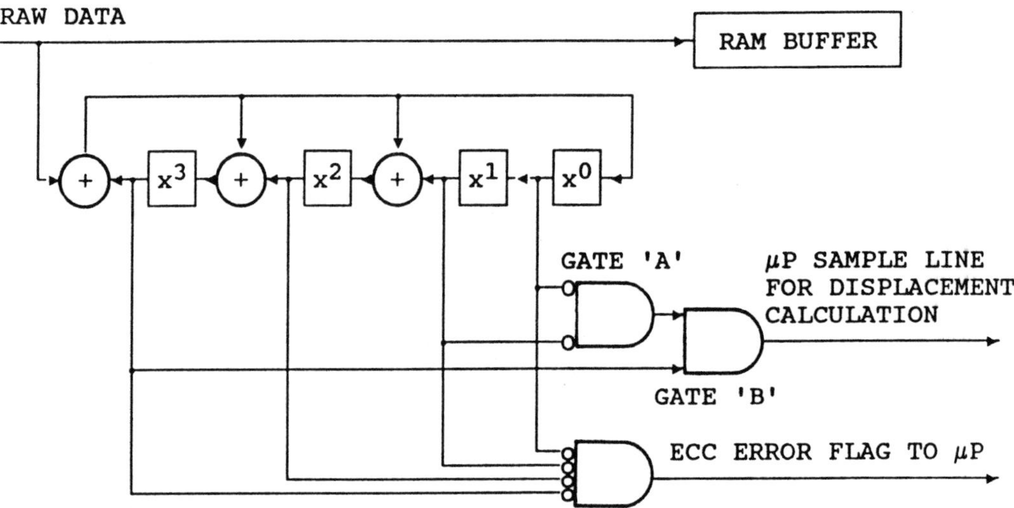

SOFTWARE CORRECTION ALGORITHM

1. Clock the shift register in a software loop until high output on gate B.
2. Forward displacement to first bit in error is clock count plus one.
3. Pattern is in left-most two bits of shift register.
4. Use pattern and displacement to correct RAM buffer.

READ CYCLE (BUFFER LOAD)			CORRECT CYCLE (CORRECT BUFFER)		
	ERR	S/R	SOFTWARE CLK CNT	S/R	
d2	0	0000	-	1001	*
d1	0	0000	0	1111	
d0	0	0000	1	0011	
p3	0	0000	2	0110	
p2	1	1101	3	1100	**
p1	1	1010			
p0	0	1001			

* Shift register contents at start of software algorithm.
** Gate B enables, software stops clocking.

Example #5A:

- Correction in hardware, forward clocking, software assist.
- Burst length-two correcting code, single-bit error, data all zeros.
- Time delay required when an error occurs.
- Internal-XOR form of shift register.
- $g(x) = (x + 1) \cdot (x^3 + x + 1) = x^4 + x^3 + x^2 + 1$.
- Detect zeros in right-most bits of shift register.
- Premultiply by x^4.

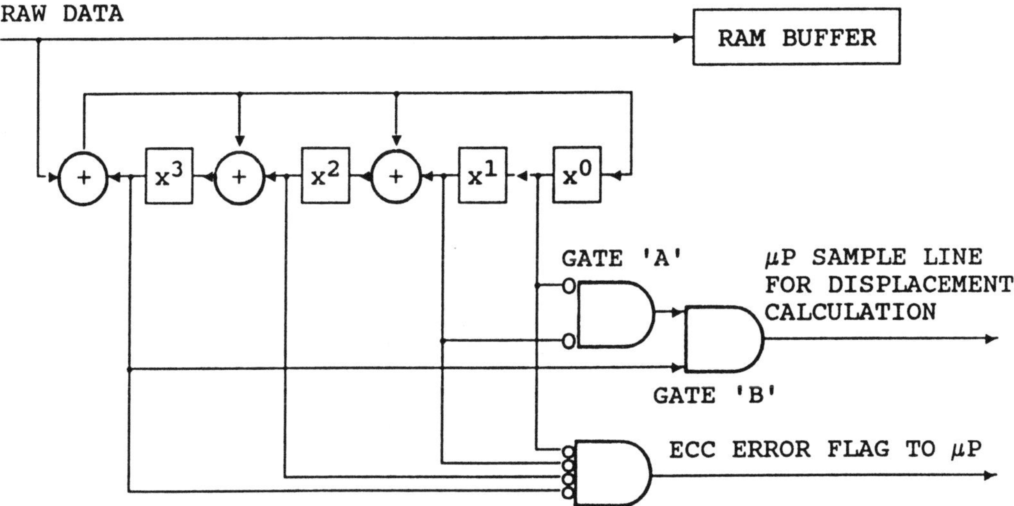

SOFTWARE CORRECTION ALGORITHM

1. Clock the shift register in a software loop until high output on gate B.
2. Forward displacement to first bit in error is clock count plus one.
3. Pattern is in left-most two bits of shift register.
4. Use pattern and displacement to correct RAM buffer.

READ CYCLE (BUFFER LOAD)			CORRECT CYCLE (CORRECT BUFFER)		
	ERR	S/R	SOFTWARE CLK CNT	S/R	
d2	0	0000	-	1110	*
d1	0	0000	0	0001	
d0	0	0000	1	0010	
p3	0	0000	2	0100	
p2	1	1101	3	1000	**
p1	0	0111			
p0	0	1110			

* Shift register contents at start of software algorithm.
** Gate B enables, software stops clocking.

Example #6:

- Correction in hardware, forward clocking, software assist.
- Burst length-two correcting code, two adjacent error, data all zeros.
- Time delay required when an error occurs.
- Internal-XOR form of shift register.
- $g(x) = (x + 1) \cdot (x^3 + x + 1) = x^4 + x^3 + x^2 + 1$.
- Detect zeros in left-most bits of shift register.
- No premultiplication.

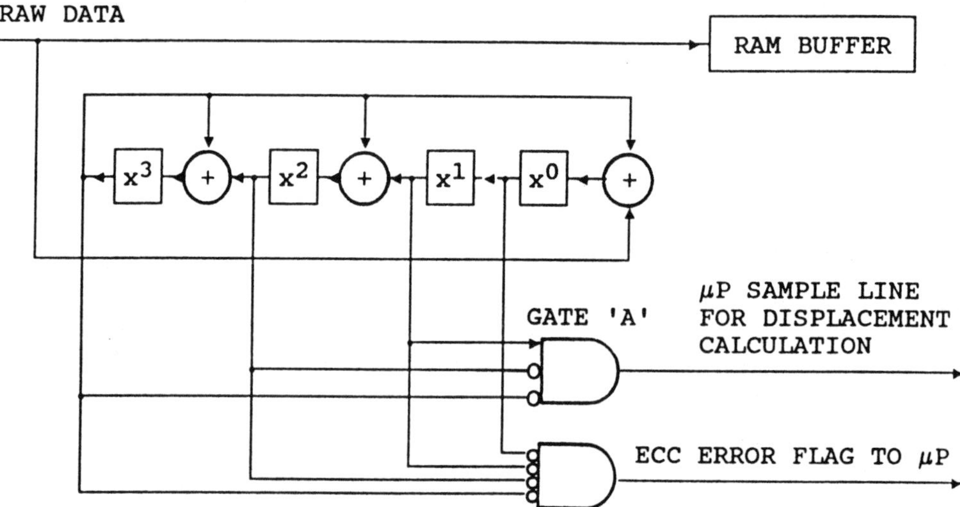

SOFTWARE CORRECTION ALGORITHM

1. Clock the shift register in a software loop until high output on gate A.
2. Forward displacement to first bit in error is clock count minus one.
3. Pattern is in right-most two bits of shift register.
4. Use pattern and displacement to correct RAM buffer.

READ CYCLE (BUFFER LOAD)			CORRECT CYCLE (CORRECT BUFFER)		
	ERR	S/R	SOFTWARE CLK CNT	S/R	
d2	0	0000	-	0101	*
d1	0	0000	0	1010	
d0	1	0001	1	1001	
p3	1	0011	2	1111	
p2	0	0110	3	0011	**
p1	0	1100			
p0	0	0101			

* Shift register contents at start of software algorithm.
** Gate A enables, software stops clocking.

Example #6A:

- Correction in hardware, forward clocking, software assist.
- Burst length-two correcting code, single-bit error, data all zeros.
- Time delay required when an error occurs.
- Internal-XOR form of shift register.
- $g(x) = (x + 1)\cdot(x^3 + x + 1) = x^4 + x^3 + x^2 + 1$.
- Detect zeros in left-most bits of shift register.
- No premultiplication.

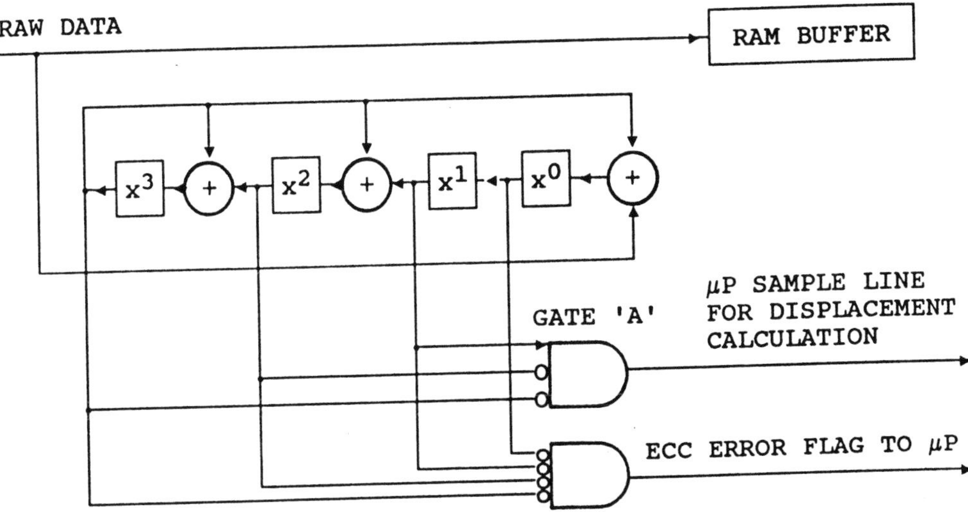

SOFTWARE CORRECTION ALGORITHM

1. Clock the shift register in a software loop until high output on gate A.
2. Forward displacement to first bit in error is clock count minus one.
3. Pattern is in right-most two bits of shift register.
4. Use pattern and displacement to correct RAM buffer.

READ CYCLE (BUFFER LOAD)

	ERR	S/R
d2	0	0000
d1	1	0001
d0	0	0010
p3	0	0100
p2	0	1000
p1	0	1101
p0	0	0111

CORRECT CYCLE (CORRECT BUFFER)

SOFTWARE CLK CNT	S/R	
-	0111	*
0	1110	
1	0001	
2	0010	**

* Shift register contents at start of software algorithm.
** Gate A enables, software stops clocking.

Example #7:

- Correction in hardware, reverse clocking, software assist.
- Burst length-two correcting code, two adjacent error, data all zeros.
- Time delay required when an error occurs.
- Internal-XOR form of shift register.
- $g(x) = (x + 1)\cdot(x^3 + x + 1) = x^4 + x^3 + x^2 + 1$.
- Detect zeros in right-most bits of shift register.
- Premultiply by x^4.

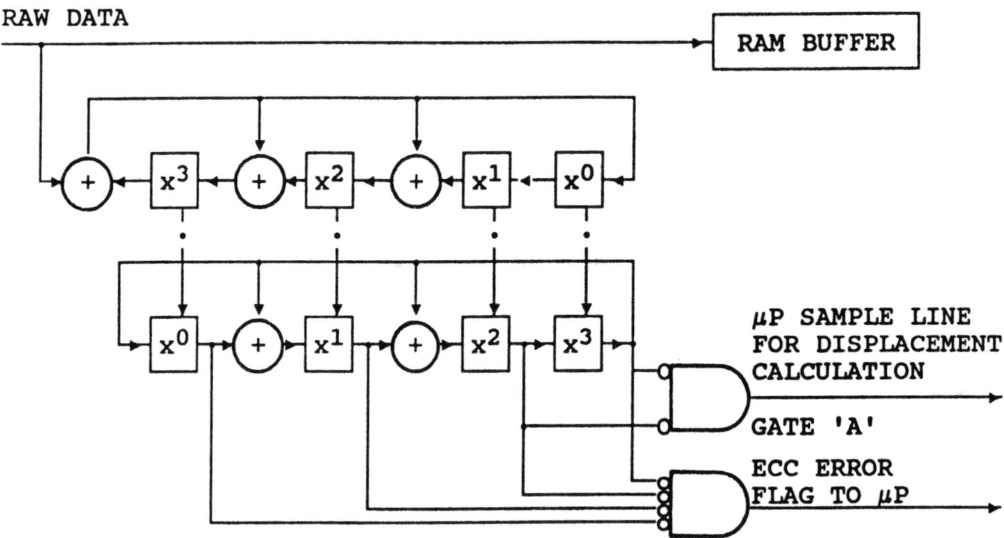

SOFTWARE CORRECTION ALGORITHM

1. Clock the shift register in a software loop until high output on gate A.
2. Reverse displacement to first bit in error is clock count.
3. Pattern is in left-most two bits of shift register.
4. Use pattern and displacement to correct RAM buffer.

READ CYCLE (BUFFER LOAD)			CORRECT CYCLE (CORRECT BUFFER)		
	ERR	S/R	SOFTWARE CLK CNT	S/R	
d2	0	0000	-	1001	*
d1	0	0000	0	1010	
d0	0	0000	1	0101	
p3	0	0000	2	1100	**
p2	1	1101			
p1	1	1010			
p0	0	1001			

* Shift register contents at start of software algorithm.

** Gate A enables, software stops clocking.

Example #7A:

- Correction in hardware, reverse clocking, software assist.
- Burst length-two correcting code, single-bit error, data all zeros.
- Time delay required when an error occurs.
- Internal-XOR form of shift register.
- $g(x) = (x + 1)\cdot(x^3 + x + 1) = x^4 + x^3 + x^2 + 1$.
- Detect zeros in right-most bits of shift register.
- Premultiply by x^4.

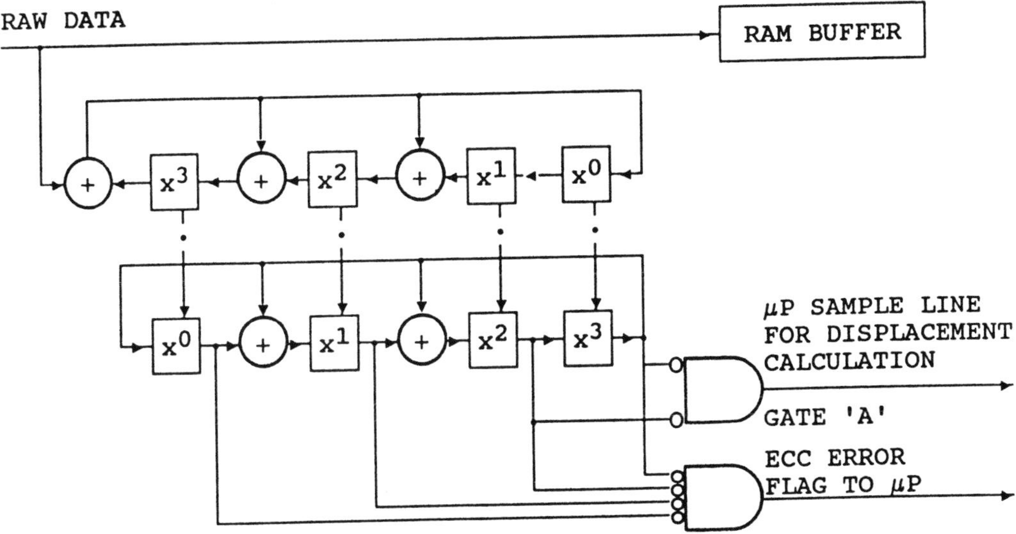

SOFTWARE CORRECTION ALGORITHM

1. Clock the shift register in a software loop until high output on gate A.
2. Reverse displacement to first bit in error is clock count.
3. Pattern is in left-most two bits of shift register.
4. Use pattern and displacement to correct RAM buffer.

READ CYCLE (BUFFER LOAD)

	ERR	S/R
d2	0	0000
d1	0	0000
d0	0	0000
p3	0	0000
p2	1	1101
p1	0	0111
p0	0	1110

CORRECT CYCLE (CORRECT BUFFER)

SOFTWARE CLK CNT	S/R	
-	1110	*
0	0111	
1	1101	
2	1000	**

* Shift register contents at start of software algorithm.

** Gate A enables, software stops clocking.

Example #8:

- Correction in hardware, reverse clocking, software assist.
- Burst length-two correcting code, two adjacent error, data all zeros.
- Time delay required when an error occurs.
- Internal-XOR form of shift register.
- $g(x) = (x + 1) \cdot (x^3 + x + 1) = x^4 + x^3 + x^2 + 1.$
- Detect zeros in left-most bits of shift register.
- No premultiplication.

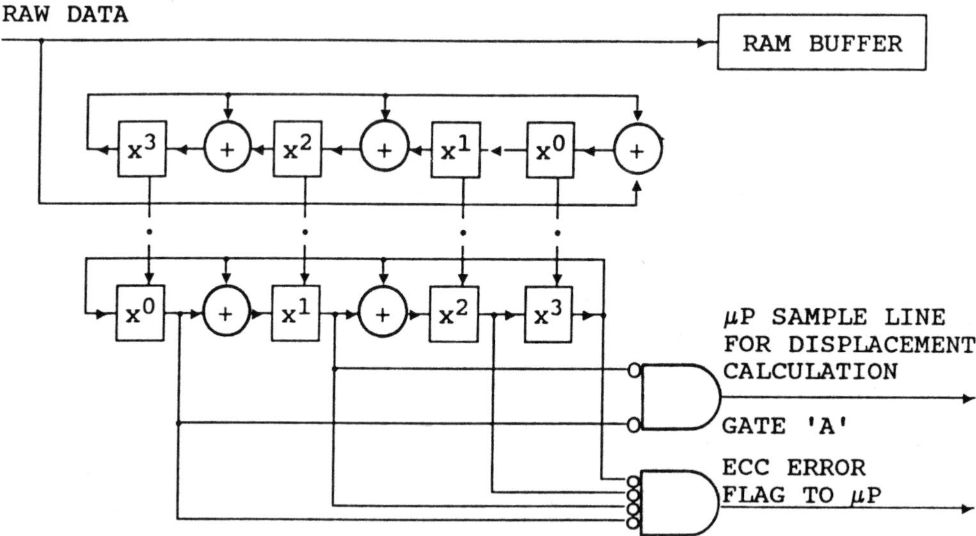

SOFTWARE CORRECTION ALGORITHM

1. Clock the shift register in a software loop until high output on gate A.
2. Reverse displacement to first bit in error is clock count plus two.
3. Pattern is in right-most two bits of shift register.
4. Use pattern and displacement to correct RAM buffer.

READ CYCLE (BUFFER LOAD)			CORRECT CYCLE (CORRECT BUFFER)		
	ERR	S/R	SOFTWARE CLK CNT	S/R	
d2	0	0000	-	1100	*
d1	0	0000	0	0110	
d0	0	0000	1	0011	**
p3	1	0001			
p2	1	0011			
p1	0	0110			
p0	0	1100			

* Shift register contents at start of software algorithm.

** Gate A enables, software stops clocking.

Example #8A:

- Correction in hardware, reverse clocking, software assist.
- Burst length-two correcting code, single-bit error, data all zeros.
- Time delay required when an error occurs.
- Internal-XOR form of shift register.
- $g(x) = (x + 1) \cdot (x^3 + x + 1) = x^4 + x^3 + x^2 + 1$.
- Detect zeros in left-most bits of shift register.
- No premultiplication.

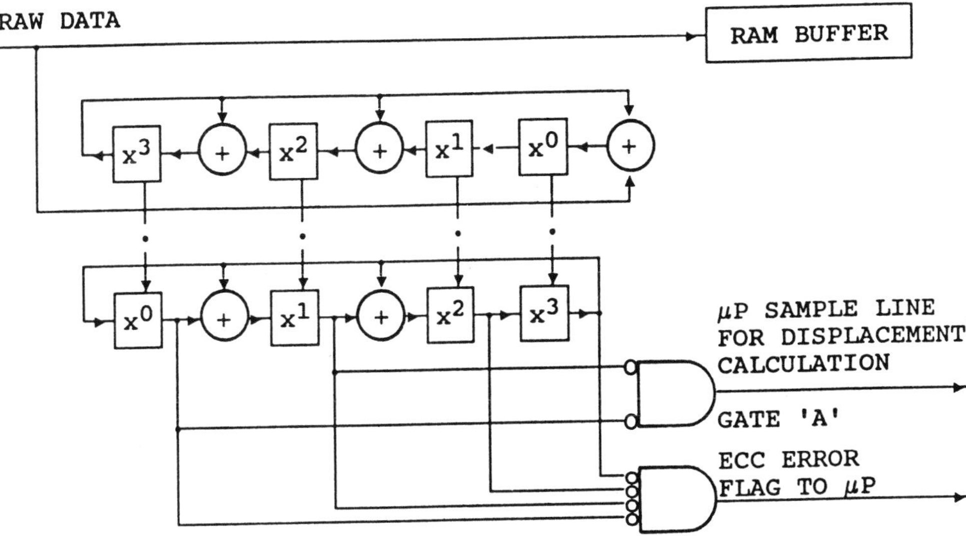

SOFTWARE CORRECTION ALGORITHM

1. Clock the shift register in a software loop until high output on gate A.
2. Reverse displacement to first bit in error is clock count plus two.
3. Pattern is in right-most two bits of shift register.
4. Use pattern and displacement to correct RAM buffer.

READ CYCLE (BUFFER LOAD)

	ERR	S/R
d2	0	0000
d1	0	0000
d0	0	0000
p3	1	0001
p2	0	0010
p1	0	0100
p0	0	1000

CORRECT CYCLE (CORRECT BUFFER)

SOFTWARE CLK CNT	S/R	
-	1000	*
0	0100	
1	0010	**

* Shift register contents at start of software algorithm.

** Gate A enables, software stops clocking.

2.4 DECODING SHORTENED CYCLIC CODES

In the decoding examples of the previous section, the record length was equal to the polynomial period. The method discussed in this section allows forward clocking to be used in searching for the correctable pattern when the record length is shorter than the polynomial period. Shortening does not change code properties.

The method assumes that the error pattern is detected when it is justified to the high order end of the shift register. If this is not the case, the method must be modified.

Let,

$g(x)$	= the code generator polynomial
$g'(x)$	= reciprocal polynomial of $g(x)$
$P_{mult}(x)$	= Premultiply polynomial for decoding
n	= number of information plus check bits
m	= number of check bits [the degree of $g(x)$]
e	= the period of $g(x)$

Use a shift register to multiply and divide simultaneously. On write, premultiply by x^m and divide by $g(x)$. On read, premultiply by $P_{mult}(x)$ and divide by $g(x)$. $P_{mult}(x)$ is computed using either of the following equations:

$$P_{mult}(x) = x^{e-n+m} \text{ MOD } g(x)$$

or

$$P_{mult}(x) = x^{m-1} \cdot F(1/x) \quad \text{where} \quad F(x) = x^{n-1} \text{ MOD } g'(x)$$

i.e. $P_{mult}(x)$ is the reciprocal polynomial of [(the highest power of x in a codeword) modulo (the reciprocal polynomial of the code generator polynomial)].

EXAMPLES OF COMPUTING THE MULTIPLIER POLYNOMIAL FOR SHORTENED CYCLIC CODES

$g(x) = x^4 + x + 1, \quad g'(x) = x^4 + x^3 + 1$

Tables of x^r MOD $g(x)$ and x^r MOD $g'(x)$

r	x^r MOD g(x)
0	0001
1	0010
2	0100
3	1000
4	0011
5	0110
6	1100
7	1011
8	0101
9	1010
10	0111
11	1110
12	1111
13	1101
14	1001

r	x^r MOD g'(x)
0	0001
1	0010
2	0100
3	1000
4	1001
5	1011
6	1111
7	0111
8	1110
9	0101
10	1010
11	1101
12	0011
13	0110
14	1100

Example #1: n=10, m=4, e=15

$$\begin{aligned} P_{mult} &= x^{e-n+m} \text{ MOD } g(x) \\ &= x^9 \text{ MOD } (x^4 + x + 1) \\ &= x^3 + x \end{aligned}$$

or

$$\begin{aligned} P_{mult} &= x^{m-1} \cdot F(1/x) && \text{where } F(x) = x^{n-1} \text{ MOD } g'(x) \\ &= x^3 \cdot F(1/x) && \text{where } F(x) = x^9 \text{ MOD } (x^4 + x^3 + 1) \\ &= x^3 \cdot (x^{-2} + 1) \\ &= x^3 + x \end{aligned}$$

Example #2: n=8, m=4, e=15

$$\begin{aligned} P_{mult} &= x^{e-n+m} \text{ MOD } g(x) \\ &= x^{11} \text{ MOD } (x^4 + x + 1) \\ &= x^3 + x^2 + x \end{aligned}$$

or

$$\begin{aligned} P_{mult} &= x^{m-1} \cdot F(1/x) && \text{where } F(x) = x^{n-1} \text{ MOD } g'(x) \\ &= x^3 \cdot F(1/x) && \text{where } F(x) = x^7 \text{ MOD } (x^4 + x^3 + 1) \\ &= x^3 \cdot (x^{-2} + x^{-1} + 1) \\ &= x^3 + x^2 + x \end{aligned}$$

CORRECTION EXAMPLE FOR A SHORTENED CODE

The code is single-bit correcting only.

Interlaced sectors are assumed.

$g(x) = x^4 + x + 1$

$g'(x) = x^4 + x^3 + 1$

$n = 8,\ m = 4,\ e = 15$

$P_{mult} = x^3 + x^2 + x$

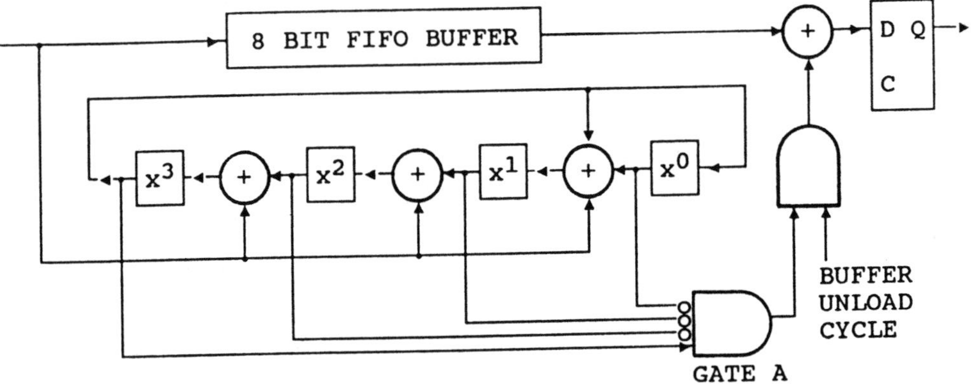

	READ SECTOR (READ CYCLE)			SKIPPED SECTOR (CORRECT CYCLE)	
	ERR	SR		0010	
d3	0	0000	d3	0100	
d2	0	0000	d2	1000	*
d1	1	1110	d1	0011	**
do	0	1111	d0	0110	
p3	0	1101	p3	1100	
p2	0	1001	p2	1011	
p1	0	0001	p1	0101	
p0	0	0010	p0	1010	

* GATE A gate enables.

** Correction takes place on d1 clock.

CORRECTION EXAMPLE FOR A SHORTENED BURST LENGTH-TWO CODE

The code of this example corrects bursts of length one or two.

Interlaced sectors assumed.

$g(x) = (x + 1)\cdot(x^4 + x + 1) = x^5 + x^4 + x^2 + 1$

$g'(x) = x^5 + x^3 + x + 1$

$n = 9, e = 15, m = 5$

$P_{mult}(x) = x^3 + x^2 + x$

Tables of x^r MOD g(x) and x^r MOD g'(x)

r	x^r MOD g(x)
0	00001
1	00010
2	00100
3	01000
4	10000
5	10101
6	11111
7	01011
8	10110
9	11001
10	00111
11	01110
12	11100
13	01101
14	11010

r	x^r MOD g'(x)
0	00001
1	00010
2	00100
3	01000
4	10000
5	01011
6	10110
7	00111
8	01110
9	11100
10	10011
11	01101
12	11010
13	11111
14	10101

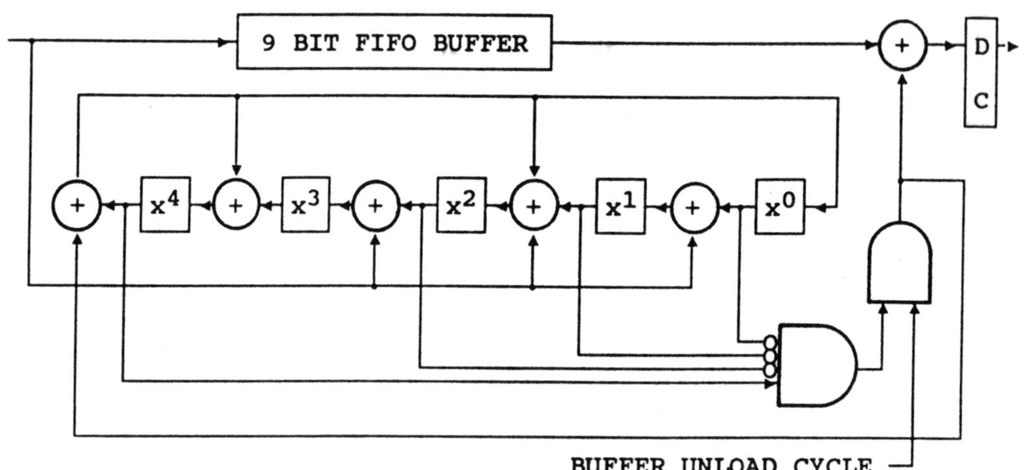

READ SECTOR (READ CYCLE)	ERR	S/R	SKIPPED SECTOR (CORRECTION CYCLE)	01100	
d3	0	00000	d3	11000	*
d2	1	01110	d2	10000	**
d1	1	10010	d1	00000	***
do	0	10001	d0	00000	
P4	0	10111	P4	00000	
P3	0	11011	P3	00000	
P2	0	00011	P2	00000	
P1	0	00110	P1	00000	
P0	0	01100	P0	00000	

* Gates A and B enable on this clock.
** Bit d2 is corrected on the d2 clock.
*** Bit d1 is corrected on the d1 clock.

2.5 INTRODUCTION TO FINITE FIELDS

A knowledge of finite fields is required for the study of many codes, including BCH and Reed-Solomon codes.

Before discussing finite fields, the definition of a field must be stated. This definition is reprinted from NTIS document AD717205.

DEFINITION OF A FIELD. A field is a set F of at least two elements together with a pair of operations, (+) and (•), which have the following properties:

a. *Closure*: For all x and y ϵ F,

$$(x + y) \epsilon F \text{ and } (x \cdot y) \epsilon F$$

b. *Associativity*: For all x, y, and z ϵ F,

$$(x + y) + z = x + (y + z) \text{ and } (x \cdot y) \cdot z = x \cdot (y \cdot z)$$

c. *Commutativity*: For all x and y ϵ F,

$$x + y = y + x \text{ and } x \cdot y = y \cdot x$$

d. *Distributivity*: For all x, y and z ϵ F,

$$x \cdot (y + z) = (x \cdot y) + (x \cdot z)$$

e. *Identities*: There exist an additive identity, zero (0), and a multiplicative identity, one (1), ϵ F such that for all x ϵ F,

$$x + 0 = x \text{ and } x \cdot 1 = x$$

f. *Inverses*: For each x ϵ F, there exists a unique element y ϵ F such that

$$x + y = 0$$

and for each non-zero x ϵ F, there exists a unique element y ϵ F such that

$$x \cdot y = 1$$

The set of positive and negative rational numbers together with ordinary addition and multiplication comprise a field with an infinite number of elements, therefore it is called an infinite field. The set of positive and negative real numbers together with ordinary addition and multiplication and the set of complex numbers together with complex addition and multiplication also comprise infinite fields.

FINITE FIELDS

Fields with a finite number of elements are called finite fields. These fields are also called Galois fields, in honor of the French mathematician Evariste Galois.

The order of a finite field is the number of elements it contains. A finite field of order p^n, denoted $GF(p^n)$, exists for every prime p and every positive integer n. The prime p of a finite field $GF(p^n)$ is called the characteristic of the field. The field GF(p) is referred to as the ground field and $GF(p^n)$ is called an extension field of GF(p). The field $GF(p^n)$ can also be denoted GF(q), where $q=p^n$.

Let β represent an arbitrary field element, that is, an arbitrary power of α. Then the order e of β is the least positive integer for which $\beta^e = 1$. More simply, the order of β is the number of terms in the sequence $(\beta, \beta^2, \beta^3, \cdots)$ before it begins to repeat. Elements of order 2^n-1 in $GF(2^n)$ are called primitive elements. They are also called generators of the field. Do not confuse the order of a field element with the order of a field, which is defined in the previous paragraph.

Two fields are said to be isomorphic if one can be obtained from the other by some appropriate one-to-one mapping of elements and operations. Any two finite fields with the same number of elements (the same order) are isomorphic. Therefore, for practical purposes there is only one finite field of order p^n.

FIELDS OF CHARACTERISTIC TWO

Most error-correcting codes of a practical interest are defined over fields of characteristic two. Such fields have interesting properties. First, every element is its own additive inverse *i.e.* $x + x = 0$. Secondly, the square and square root functions are linear *i.e.*

$$f(x + y + \cdots) = f(x) + f(y) + \cdots$$

Therefore, in a field of characteristic two the following identities hold.

$$(x + y + \cdots)^2 = x^2 + y^2 + \cdots$$

$$(x + y + \cdots)^{\frac{1}{2}} = x^{\frac{1}{2}} + y^{\frac{1}{2}} + \cdots$$

$$(x + y + \cdots)^{2^k} = x^{2^k} + y^{2^k} + \cdots$$

$$(x + y + \cdots)^{1/2^k} = x^{1/2^k} + y^{1/2^k} + \cdots$$

These identities will be helpful in performing finite field computations in fields $GF(2^n)$.

GENERATION OF A FIELD

The finite field GF(2) has only two elements (0,1). Larger fields can be defined by polynomials with coefficients from GF(2).

Let p(x) be a polynomial of degree n with coefficients from GF(2). Let α be a root of p(x). If p(x) is primitive, the powers of α up through 2^n-2 will all be unique. Appropriately selected operations of addition and multiplication together with the field elements:

$$0,1,\alpha,\alpha^2,\cdots,\alpha^{2^n-2}$$

define a field of 2^n elements GF(2^n).

Assume a finite field is defined by $p(x) = x^3 + x + 1$. Since α is a root of p(x), $p(\alpha)=0$. Therefore,

$$\alpha^3 + \alpha + 1 = 0 \quad \text{and} \quad \alpha^3 = \alpha + 1$$

The field elements for this field are:

0	MOD	$(\alpha^3 + \alpha + 1)$	$= 0$
α^0	"	"	$= \alpha^0 = 1$
α^1	"	"	$= \alpha^1$
α^2	"	"	$= \alpha^2$
α^3	"	"	$\alpha + 1$
α^4	"	"	$\alpha\cdot\alpha^3 = \alpha\cdot(\alpha + 1) = \alpha^2 + \alpha^1$
α^5	"	"	$\alpha\cdot\alpha^4 = \alpha\cdot(\alpha^2 + \alpha) = \alpha^3 + \alpha^2 = \alpha^2 + \alpha^1 + 1$
α^6	"	"	$\alpha\cdot\alpha^5 = \alpha\cdot(\alpha^2 + \alpha + 1) = \alpha^3 + \alpha^2 + \alpha = \alpha^2 + 1$
α^7	"	"	$= \alpha^0$
α^8	"	"	$= \alpha^1$
.	.	.	. .
.	.	.	. .

The elements of the field can be represented in binary fashion by using one bit to represent each of the three powers of α whose sum comprises an element. For the field constructed above, we generate the following table:

	α^2	α^1	α^0
0	0	0	0
α^0	0	0	1
α^1	0	1	0
α^2	1	0	0
α^3	0	1	1
α^4	1	1	0
α^5	1	1	1
α^6	1	0	1

Figure 2.5.1

This list can also be viewed as the zero state plus the sequential nonzero states of a shift register implementing the polynomial

$$x^3 + x + 1$$

The number of elements in the field of Figure 2.5.1, including the zero element, is eight. This field is called GF(8) or GF(2^3).

OPERATIONS IN A FIELD GF(2^n) (Examples use GF(2^3))

+ Addition: Form the modulo-2 (EXCLUSIVE-OR) sum of the components of the addends to obtain the components of the sum, *e.g.*:

$$\begin{aligned}\alpha^0 + \alpha^3 &= \text{'001'} \oplus \text{'011'} \\ &= \text{'010'} \\ &= \alpha^1\end{aligned}$$

\- Subtraction: In GF(2^n), subtraction is the same as addition, since each element is its own additive inverse. This is not the case in all finite fields.

• Multiplication: If either multiplicand is zero, the product is zero. Otherwise add exponents modulo seven (the field size minus one) *e.g.*:

$$0 \cdot \alpha^4 = 0$$

$$\begin{aligned}\alpha^3 \cdot \alpha^5 &= \alpha^{(3+5) \bmod 7} \\ &= \alpha^1\end{aligned}$$

/ Division: If the divisor is zero, the quotient is undefined. If the dividend is zero, the quotient is zero. Otherwise subtract exponents modulo seven *e.g.*:

$$\alpha^5/\alpha^3 = \alpha^{(5-3)} = \alpha^2$$

$$\begin{aligned}\alpha^3/\alpha^5 &= \alpha^{(3-5)} = \alpha^{-2} \\ &= \alpha^{(-2+7)} = \alpha^5\end{aligned}$$

By convention, multiplication and division take precedence over addition and subtraction except where parentheses are used.

LOG Logarithm: Take the logarithm to the base α, *e.g.*:

$$\text{LOG}(\alpha^n) = n$$

ANTILOG Antilogarithm: Raise α to the given power, *e.g.*:

$$\text{ANTILOG}(n) = \alpha^n$$

FINITE FIELD COMPUTATION

From the list of field elements above, α^3 represents the vector '011' and α^5 represents the vector '111'. The integer 6 is the exponent of α^6.

The log function in this field produces an exponent from a vector while the antilog function produces a vector from an exponent. The log of α^4 ('110') is 4. The antilog of 3 is α^3 ('011'). The familiar properties of logarithms hold.

Finite field computation is frequently performed by a computer. At times, field elements are stored in the computer in vector form. At other times, the logs of field elements are stored instead of the field elements themselves. For example, consider finite field math implemented on a computer with an eight-bit wide data path. Assume the finite field of Figure 2.5.1. If a memory location storing α^4 is examined, the binary value '0000 0110' is observed. This binary value represents the vector '110' or $\alpha^2 + \alpha$. If a memory location storing the log of α^4 is examined, the binary value '0000 0100' is observed. This value represents the integer 4 which is the exponent and log of α^4. Finite field computers frequently employ log and antilog tables to convert from one representation to the other.

Finite field addition for this field is modulo-2 addition (bit-wise EXCLUSIVE-OR operation). The sum of α^4 ('110') and α^5 ('111') is α^0 ('001'). The sum of α^3 ('011') and α^6 ('101') is α^4 ('110'). Subtraction in this field, as in all finite fields of characteristic two, is the same as addition. The '+' symbol will be used to represent modulo-2 addition (bit-wise EXCLUSIVE-OR operation). The '+' symbol will also continue to be used for ordinary addition, such as adding exponents. In most cases, when '+' represents modulo-2 addition, it will be preceded and followed by a space, and when used to represent ordinary addition, its operands will immediately precede and follow it. Usage should be clear from the context.

There are two basic ways to accomplish finite field multiplication for the field of Figure 2.5.1. The vectors representing the field elements can be multiplied and the result reduced modulo $(x^3 + x + 1)$. Alternatively, the product may be computed by first taking logs of the finite field elements being multiplied; then taking the antilog of the sum of the logs modulo 7 (field size minus one). The '•' symbol will be used to represent finite field multiplication. The '*' symbol will be used to represent ordinary multiplication, such as for multiplying an exponent, which is an ordinary number and not a finite field element, by another ordinary number.

The examples below multiply α^4 ('110') times α^5 ('111') using the methods described above.

Example #1

1. Multiply the vectors '110' (α^4) and '111' (α^5) to get the vector '10010'.
2. Reduce the vector '10010' modulo $\alpha^3 + \alpha + 1$ to get the vector '100' (α^2).

Example #2

1. Take the logs base α of α^4 and α^5 to get exponents 4 and 5.
2. Add exponents 4 and 5 modulo 7 to get the exponent 2.
3. Take the antilog of the exponent 2 to get the vector α^2 ('100').

Division is accomplished by inverting (multiplicative inversion) and multiplying. The inverse of any element in the field of Figure 2.5.1, other than the zero element, is given by:

$$\frac{1}{\alpha^j} = \alpha^{(-j)\text{ MOD } 7}$$

The inverse of the zero element is undefined. α^0 is its own inverse.

Inversion Examples:

$$\frac{1}{\alpha^3} = \alpha^{(-3)\text{ MOD } 7} = \alpha^4$$

$$\frac{1}{\alpha^1} = \alpha^{(-1)\text{ MOD } 7} = \alpha^6$$

Division Examples:

$$\frac{\alpha^2}{\alpha^3} = \alpha^2 \cdot \frac{1}{\alpha^3} = \alpha^2 \cdot \alpha^4 = \alpha^6$$

$$\frac{\alpha^4}{\alpha^2} = \alpha^{4-2} = \alpha^2$$

Examples of finite field computation in the field of Figure 2.5.1 are shown below. To provide greater insight, some examples use different approaches than others with various levels of details being shown. Note that all operations on exponents are performed modulo 7 (field size minus one).

$$\begin{aligned} y &= \alpha^3 + \alpha^4 \\ &= \text{'011'} + \text{'110'} \\ &= \text{'101'} \\ &= \alpha^6 \end{aligned}$$

$$\begin{aligned} y &= \alpha^1 \cdot \alpha^4 \\ &= \text{'010'} \cdot \text{'110'} \\ &= (\alpha^1) \cdot (\alpha^2 + \alpha^1) \\ &= \alpha^3 + \alpha^2 \\ &= (\alpha + 1) + \alpha^2 \\ &= \alpha^2 + \alpha + 1 \\ &= \text{'111'} = \alpha^5 \end{aligned}$$

$$\begin{aligned} y &= \alpha^2 \cdot \alpha^6 \\ &= \alpha^{(2+6) \text{ MOD } 7} \\ &= \alpha^1 \end{aligned}$$

$$\begin{aligned} y &= \frac{1}{\alpha^4} \\ &= \alpha^{(-4) \text{ MOD } 7} \\ &= \alpha^3 \end{aligned}$$

$$\begin{aligned} y &= \frac{\alpha^2}{\alpha^5} \\ &= \alpha^{(2-5) \text{ MOD } 7} \\ &= \alpha^4 \end{aligned}$$

$$\begin{aligned} (x + \alpha^0) \cdot (x + \alpha^1) &= x^2 + (\alpha^0 + \alpha^1) \cdot x + \alpha^0 \cdot \alpha^1 \\ &= x^2 + \alpha^3 x + \alpha^1 \end{aligned}$$

The modulo operations shown above for adding and subtracting exponents are understood for finite field computation and will not be shown for the remainder of the book.

Other examples are:

$$y = \alpha^2 \cdot \alpha^6 = \alpha^{2+6} = \alpha^1$$

$$y = \alpha^1 + \alpha^2 = \alpha^4$$

$$y = \mathrm{LOG}_\alpha\left[\frac{\alpha^2}{\alpha^5}\right] = \mathrm{LOG}_\alpha(\alpha^{2-5}) = \mathrm{LOG}_\alpha(\alpha^4) = 4$$

$$y = \mathrm{LOG}_\alpha\left[\frac{\alpha^2}{\alpha^5}\right] = \mathrm{LOG}_\alpha(\alpha^2) - \mathrm{LOG}_\alpha(\alpha^5) = (2-5) \text{ MOD } 7 = 4$$

$$y = \frac{\alpha^3}{\alpha^2 \cdot (\alpha^4 + \alpha^3)} = \frac{\alpha^3}{\alpha^1} = \alpha^{3-1} = \alpha^2$$

$$y = (\alpha^3)^3 = \alpha^{3*3} = \alpha^2$$

$$y = (x + \alpha^0) \cdot (x + \alpha^1) \cdot (x + \alpha^2)$$

$$= x^3 + (\alpha^0 + \alpha^1 + \alpha^2) \cdot x^2 + (\alpha^0 \cdot \alpha^1 + \alpha^0 \cdot \alpha^2 + \alpha^1 \cdot \alpha^2) \cdot x + \alpha^0 \cdot \alpha^1 \cdot \alpha^2$$

$$= x^3 + \alpha^5 \cdot x^2 + \alpha^6 \cdot x + \alpha^3$$

FIELD PROPERTY EXAMPLES

ASSOCIATIVITY

$$(x + y) + z = x + (y + z)$$
$$(\alpha^2 + \alpha^3) + \alpha^4 = \alpha^2 + (\alpha^3 + \alpha^4)$$
$$(\text{'100'} + \text{'011'}) + \text{'110'} = \text{'100'} + (\text{'011'} + \text{'110'})$$
$$\text{'111'} + \text{'110'} = \text{'100'} + \text{'101'}$$
$$\text{'001'} = \text{'001'}$$

$$(x \cdot y) \cdot z = x \cdot (y \cdot z)$$
$$(\alpha^4 \cdot \alpha^5) \cdot \alpha^6 = \alpha^4 \cdot (\alpha^5 \cdot \alpha^6)$$
$$\alpha^{(4+5 \text{ MOD } 7)} \cdot \alpha^6 = \alpha^4 \cdot \alpha^{(5+6 \text{ MOD } 7)}$$
$$\alpha^2 \cdot \alpha^6 = \alpha^4 \cdot \alpha^4$$
$$\alpha^{(2+6 \text{ MOD } 7)} = \alpha^{(4+4 \text{ MOD } 7)}$$
$$\alpha^1 = \alpha^1$$

COMMUTATIVITY

$$x + y = y + x$$
$$\alpha^3 + \alpha^4 = \alpha^4 + \alpha^3$$
$$\text{'011'} + \text{'110'} = \text{'110'} + \text{'011'}$$
$$\text{'101'} = \text{'101'}$$

$$x \cdot y = y \cdot x$$
$$\alpha^5 \cdot \alpha^6 = \alpha^6 \cdot \alpha^5$$
$$\alpha^{(5+6 \text{ MOD } 7)} = \alpha^{(6+5 \text{ MOD } 7)}$$
$$\alpha^4 = \alpha^4$$

DISTRIBUTIVITY

$$x \cdot (y + z) = (x \cdot y) + (x \cdot z)$$
$$\alpha^4 \cdot (\alpha^5 + \alpha^6) = (\alpha^4 \cdot \alpha^5) + (\alpha^4 \cdot \alpha^6)$$
$$\alpha^4 \cdot (\text{'111'} + \text{'101'}) = \alpha^{(4+5 \text{ MOD } 7)} + \alpha^{(4+6 \text{ MOD } 7)}$$
$$\alpha^4 \cdot \text{'010'} = \alpha^2 + \alpha^3$$
$$\alpha^4 \cdot \alpha^1 = \text{'100'} + \text{'011'}$$
$$\alpha(4+1 \text{ MOD } 7) = \text{'111'}$$
$$\alpha^5 = \alpha^5$$

SIMULTANEOUS LINEAR EQUATIONS IN A FIELD

Simultaneous linear equations in GF(2^n) can be solved by determinants. For example, given:

$$a \cdot x + b \cdot y = c$$
$$d \cdot x + e \cdot y = f$$

where x and y are independent variables and a, b, c, d, e, and f are constants, then:

$$x = \frac{\begin{vmatrix} c & b \\ f & e \end{vmatrix}}{\begin{vmatrix} a & b \\ d & e \end{vmatrix}} = \frac{c \cdot e + b \cdot f}{a \cdot e + b \cdot d}$$

$$y = \frac{\begin{vmatrix} a & c \\ d & f \end{vmatrix}}{\begin{vmatrix} a & b \\ d & e \end{vmatrix}} = \frac{a \cdot f + c \cdot d}{a \cdot e + b \cdot d}$$

POLYNOMIALS IN A FIELD

Polynomials can be written with variables and coefficients from GF(2^n) and manipulated in much the same manner as polynomials involving rational or real numbers.

Polynomial Multiplication Example:

$$\begin{array}{rrrr}
 & \alpha^4 \cdot x^2 & + \alpha^5 \cdot x & + \alpha \\
 & & x & + \alpha \\
\hline
 & \alpha^5 \cdot x^2 & + \alpha^6 \cdot x & + \alpha^2 \\
\alpha^4 \cdot x^3 & + \alpha^5 \cdot x^2 & + \alpha^1 \cdot x & \\
\hline
\alpha^4 \cdot x^3 & & + \alpha^5 \cdot x & + \alpha^2
\end{array}$$

Polynomial Division Example:

$$\begin{array}{r|lllll}
 & \alpha^6 \cdot x & + \alpha^2 & & & \\
\hline
x^3 + \alpha^4 \cdot x + \alpha^2 & \alpha^6 \cdot x^4 & + \alpha^2 \cdot x^3 & + \alpha^4 \cdot x^2 & + \alpha^1 \cdot x & + \alpha^2 \\
 & \alpha^6 \cdot x^4 & & + \alpha^3 \cdot x^2 & + \alpha^1 \cdot x & \\
\hline
 & & \alpha^2 \cdot x^3 & + \alpha^6 \cdot x^2 & & + \alpha^2 \\
 & & \alpha^2 \cdot x^3 & & + \alpha^6 \cdot x & + \alpha^4 \\
\hline
 & & & \alpha^6 \cdot x^2 & + \alpha^6 \cdot x & + \alpha
\end{array}$$

Thus

$$(\alpha^6 \cdot x^4 + \alpha^2 \cdot x^3 + \alpha^4 \cdot x^2 + \alpha^1 \cdot x + \alpha^2) \text{ MOD } (x^3 + \alpha^4 \cdot x + \alpha^2)$$
$$= \alpha^6 \cdot x^2 + \alpha^6 \cdot x + \alpha$$

QUADRATIC SOLUTION DIFFICULTY IN A FIELD OF CHARACTERISTIC 2

The correlation between finite field, of characteristic 2, algebra and algebra involving real numbers does not include the quadratic formula:

$$x = \frac{-b \pm \sqrt{b^2 - 4ac}}{2a}$$

The 2 in the denominator must be interpreted as an integer, but:

$$2a = a + a = 0$$

and division by zero is undefined.

DIFFERENTIATION IN A FIELD OF CHARACTERTISTIC 2

The derivative of x^n in $GF(2^n)$ is:

$$nx^{(n-1)}$$

where n is interpreted as an integer, not as a finite field element. Thus the derivative of any even power is zero and the derivative of any odd power is $x^{(n-1)}$. For example,

$$d(x^2)/dx = 2x = x + x = 0$$

$$d(x^3)/dx = 3x^2 = x^2 + x^2 + x^2 = x^2$$

etc.

FINITE FIELDS AND SHIFT REGISTER SEQUENCES

The shift register below implements the polynomial $x^3 + x + 1$, which defines the field of Figure 2.5.1.

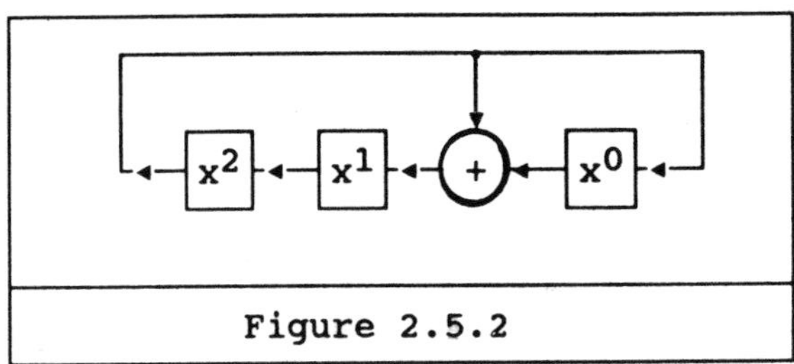

Figure 2.5.2

This shift register has two sequences, a sequence of length seven and the zero sequence of length one.

STATE NUMBER	SHIFT REGISTER CONTENTS
---	000
0	001
1	010
2	100
3	011
4	110
5	111
6	101

Notice the similarity of the sequences above to the field definition of Figure 2.5.1. The consecutive shift register states correspond to the consecutive list of field elements. The state numbers correspond to the exponents of powers of α.

Advancing the shift register once is identical to multiplying its contents by α. Advancing the shift register twice is identical to multiplying its contents by α^2, and so on.

COMPUTING IN A SMALLER FIELD

We have been representing powers of α by components. For example, in the field of Figure 2.5.1, the components of α^3 are α and 1. The components of α^4 are α^2 and α. An arbitrary power of α can also be represented by its components. Let X represent any arbitrary power of α from the field of Figure 2.5.1; then

$$X = X_2 \cdot \alpha^2 + X_1 \cdot \alpha + X_0$$

The coefficients X_2, X_1, and X_0 are from GF(2), the field of two elements, 0 and 1.

In performing finite field operations in a field such as GF(2^3), it is frequently necessary to perform multiple operations in a smaller field such as GF(2). For example, multiplication of an arbitrary field element X by α, might be accomplished as follows:

$$
\begin{aligned}
Y &= \alpha \cdot X \\
&= \alpha \cdot (X_2 \cdot \alpha^2 + X_1 \cdot \alpha + X_0) \\
&= X_2 \cdot \alpha^3 + X_1 \cdot \alpha^2 + X_0 \cdot \alpha
\end{aligned}
$$

But $\alpha^3 = \alpha + 1$, so

$$
\begin{aligned}
Y &= X_2 \cdot (\alpha + 1) + X_1 \cdot \alpha^2 + X_0 \cdot \alpha \\
&= X_1 \cdot \alpha^2 + (X_2 + X_0) \cdot \alpha + X_2
\end{aligned}
$$

The result Y can also be expressed in component form, therefore:

$$Y_2 \cdot \alpha^2 + Y_1 \cdot \alpha + Y_0 = X_1 \cdot \alpha^2 + (X_2 + X_0) \cdot \alpha + X_2$$

Equating coefficients on like powers of α gives

$$
\begin{aligned}
Y_2 &= X_1 \\
Y_1 &= X_2 + X_0 \\
Y_0 &= X_2
\end{aligned}
$$

These results have been used to design the combinatorial logic circuit shown below. This circuit uses a compute element (modulo-2 adder) from GF(2) to construct a circuit to multiply any arbitrary field element from the field of Figure 2.5.1 by α.

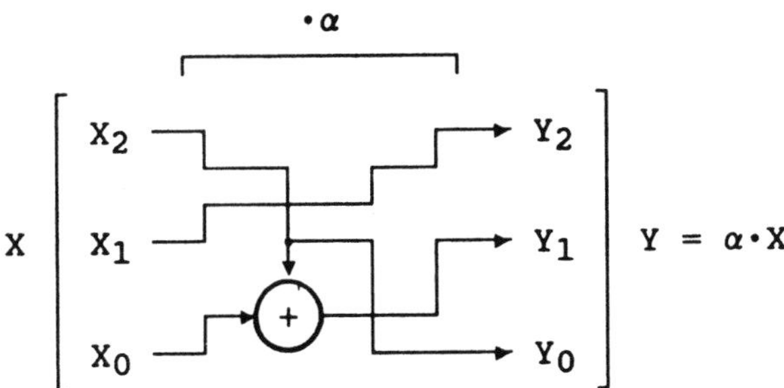

+ = Finite field addition in GF(2)

ANOTHER LOOK AT THE SHIFT REGISTER

The shift register of Figure 2.5.2 has been redrawn below to show that it contains a circuit to multiply by α.

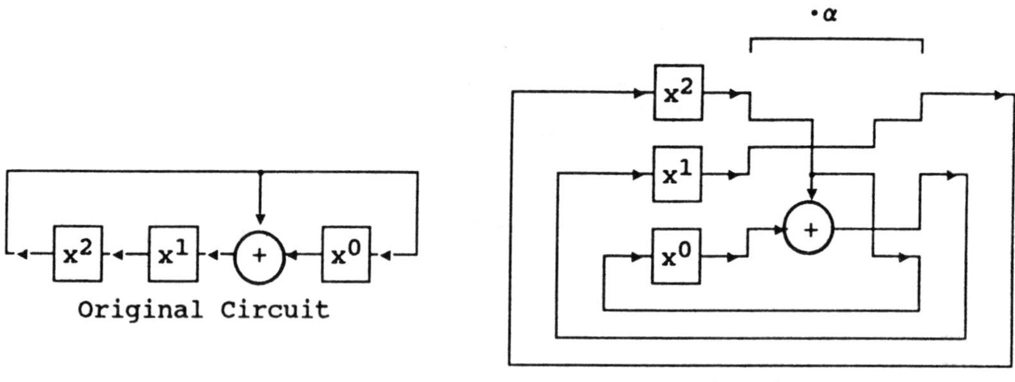

MORE ON FIELD GENERATION

Let β represent the primitive element α^2 from the field of Figure 2.5.1. The field can be redefined as follows:

	α^2	α^1	α^0
0	0	0	0
β^0	0	0	1
β^1	1	0	0
β^2	1	1	0
β^3	1	0	1
β^4	0	1	0
β^5	0	1	1
β^6	1	1	1

All the properties of a field still apply. A multiply example:

$$\beta^2 \cdot \beta^4 = (\text{'110'}) \cdot (\text{'010'})$$

$$= (\alpha^2 + \alpha) \cdot (\alpha)$$

$$= \alpha^3 + \alpha^2$$

But, $\alpha^3 = \alpha + 1$, so

$$\beta^2 \cdot \beta^4 = \alpha^2 + \alpha + 1$$

$$= (\text{'111'})$$

$$= \beta^6$$

This definition of the field could be viewed as having been generated by the circuit below.

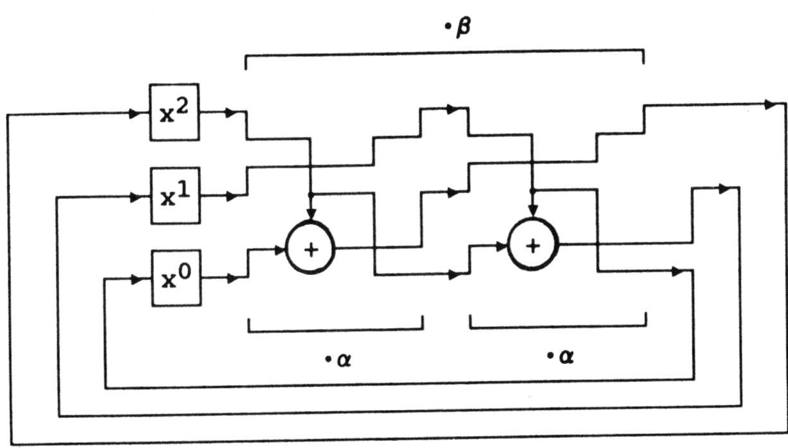

A similar redefinition of the field could be accomplished by letting β represent any primitive element of the field of Figure 2.5.1.

DEFINING FIELDS WITH POLYNOMIALS OVER FIELDS LARGER THAN GF(2)

A polynomial over GF(q) where $q=p^n$ is a polynomial with coefficients from GF(q). So far, we have worked with a field GF(8) that is defined by the polynomial $x^3 + x + 1$ over GF(2). It is also possible to define a field by a polynomial over GF(4) or GF(8), and so on.

A primitive polynomial of degree m over $GF(2^n)$ can define a field $GF(2^{m*n})$.

Fields $GF(2^{2*n})$ are particularly interesting. Operations in these fields can be accomplished by performing several simple operations in $GF(2^n)$. These fields will be studied in Section 2.7.

COMPUTING IN GF(2)

Consider the field of two elements.

	α^0
0	0
α^0	1

An element of this field is either 0 or 1. The result of a multiplication is either 0 or 1. The result of raising any element to a power is either 0 or 1, and so on.

Let β represent an arbitrary element of this field; then,

$$\beta \cdot \beta = \beta \qquad (\beta)^3 = \beta$$

$$(\beta)^2 = \beta \qquad (\beta)^n = \beta$$

Let a and b represent arbitrary elements of this field; then,

$a \cdot b = 0$ if either $a = 0$ or $b = 0$

$a \cdot b = 1$ if both $a = 1$ and $b = 1$

Clearly, multiplication in GF(2) can be accomplished with an AND gate:

a, b → $a \cdot b$

Let $\bar{b}$ represent the logical NOT of b; then, in GF(2),

$$a + a \cdot b = a \cdot (1 + b)$$
$$= a \cdot \bar{b}$$

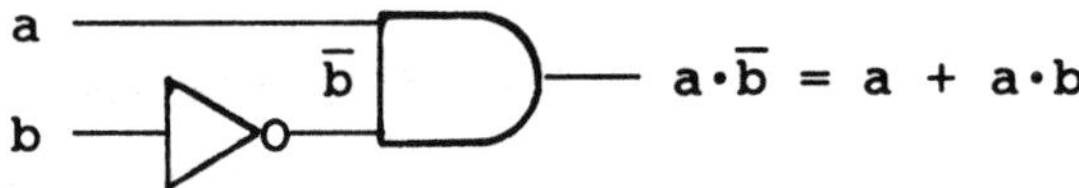

Let V represent the INCLUSIVE-OR operation; then, in GF(2)

$$a + a \cdot b + b = a \vee b$$

a, b → $a \vee b = a + a \cdot b + b$

2.6 FINITE FIELD CIRCUITS FOR FIELDS OF CHARACTERISTIC 2

This section introduces finite field circuits for finite fields of characteristic 2 with examples. The notation for various GF(8) finite field circuits is shown below. The field of Figure 2.5.1 is assumed.

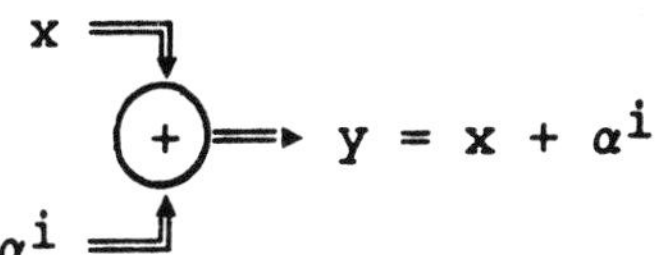

Fixed field element adder

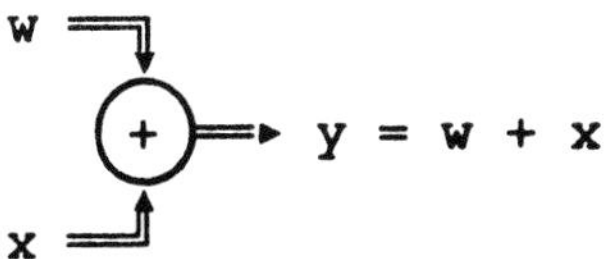

Arbitrary field element adder

Fixed field element multiplier

x ⟹ $(\cdot\alpha^{-i})$ ⟹ $y = \alpha^{-i} \cdot x$

Fixed field element multiplier

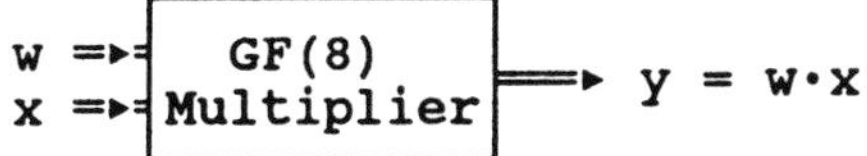

Arbitrary field element multiplier

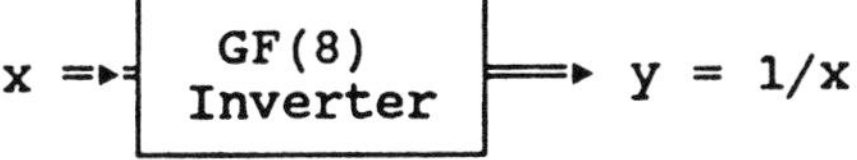

Multiplicative inversion

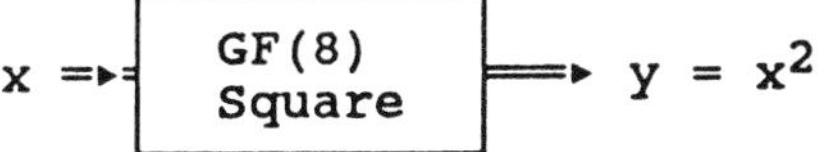

Square an arbitrary field element

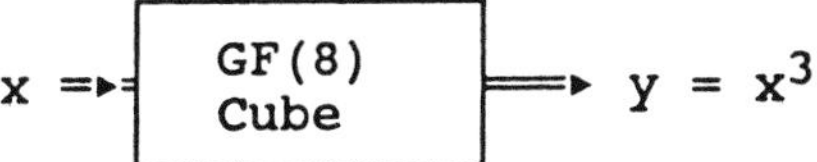

Cube an arbitrary field element

x ⟹ [GF(8) Log] ⟹ $j = \log_\alpha(x)$

Compute $\log_\alpha$ of an arbitrary field element

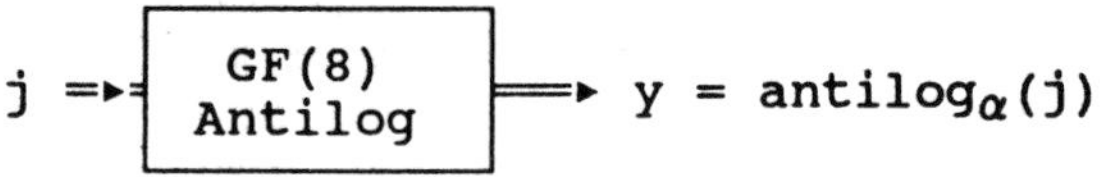

Compute antilog$_\alpha$ of an arbitrary integer

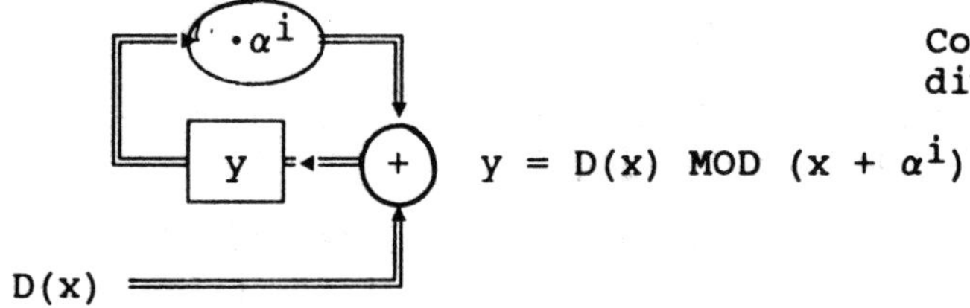

Compute the remainder from dividing D(x) by (x + α^i)

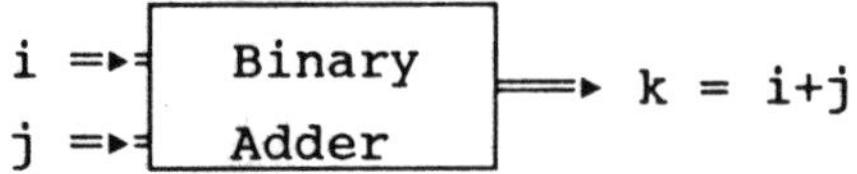

Add logs of finite field elements modulo the field size minus one

COMBINING FINITE FIELD CIRCUITS

Finite field circuits can be combined for computing. For illustration, assume that:

$$Y = \frac{X + W^3}{W^3}$$

must be computed. This can be accomplished with the circuit below:

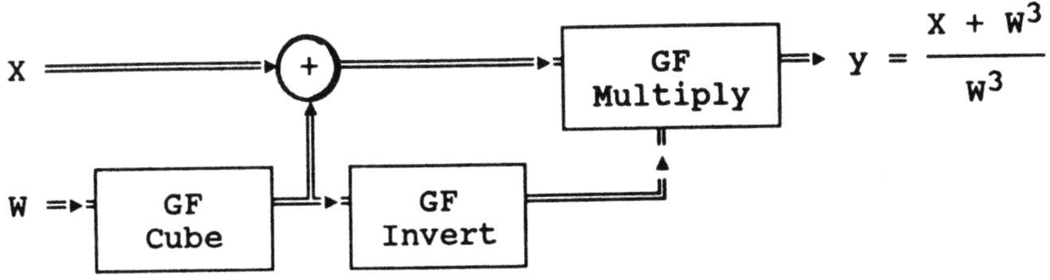

Another circuit solution becomes obvious when the equation is rearranged as follows:

$$Y = \frac{X + W^3}{W^3} = \frac{X}{W^3} + 1 = \frac{X}{W^3} + \alpha^0$$

X
GF Multiply
+
$Y = \frac{X}{W^3} + \alpha^0$
α^0
W
GF Cube
GF Invert

Another example of combining finite field circuits in GF(2^3) is shown below.

X
$\cdot\alpha$
+

$$\begin{aligned} Y &= \alpha \cdot X + X \\ &= (\alpha + 1) \cdot X \\ &= \alpha^3 \cdot X \end{aligned}$$

This example shows how a circuit to multiply by the fixed field element α^3 can be constructed using two other GF(2^3) circuits: a circuit to add two arbitrary field elements and a circuit to multiply an arbitrary field element by α. Later, circuits will be shown that accomplish this type of operation with GF(2) circuits.

Still another example of combining finite field circuits follows:

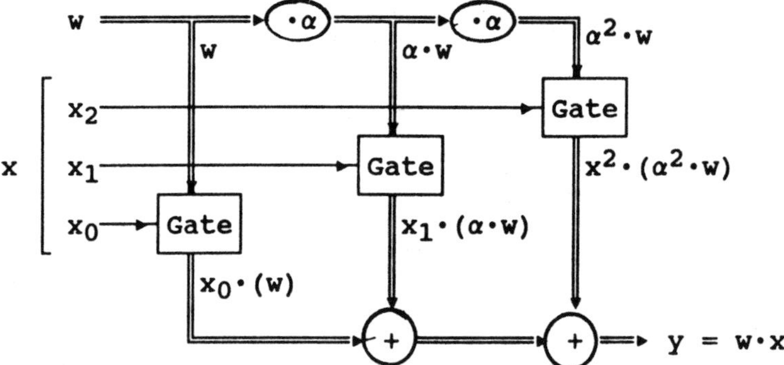

This circuit is called an array multiplier and is based on the following finite field math:

$$
\begin{aligned}
y &= x \cdot w \\
&= (x_2 \cdot \alpha^2 + x_1 \cdot \alpha + x_0) \cdot w \\
&= x_2 \cdot \alpha^2 \cdot w + x_1 \cdot \alpha \cdot w + x_0 \cdot w \\
&= x_2 \cdot (\alpha^2 \cdot w) + x_1 (\alpha \cdot w) + x_0 \cdot (w)
\end{aligned}
$$

Fixed field element adder:

$$\begin{aligned} y &= x + \alpha^3 \\ &= (x_2 \cdot \alpha^2 + x_1 \cdot \alpha + x_0) + (\alpha + 1) \\ &= x_2 \cdot \alpha^2 + (x_1 + 1) \cdot \alpha + (x_0 + 1) \end{aligned}$$

But, y can also be expressed in component form, therefore:

$$y = y_2 \cdot \alpha^2 + y_1 \cdot \alpha + y_0 = x_2 \cdot \alpha^2 + (x_1 + 1) \cdot \alpha + (x_0 + 1)$$

Equating coefficients on like powers of α gives:

$$\begin{aligned} y_2 &= x_2 \\ y_1 &= x_1 + 1 \\ y_0 &= x_0 + 1 \end{aligned}$$

This is realized by the following circuit:

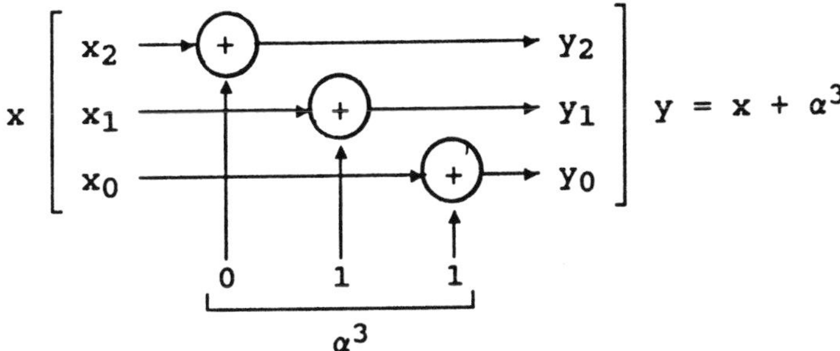

A simpler fixed field element adder:

$$\begin{aligned} y &= x + \alpha^3 \\ &= (x_2 \cdot \alpha^2 + x_1 \cdot \alpha + x_0) + (\alpha + 1) \\ &= x_2 \cdot \alpha^2 + (x_1 + 1) \cdot \alpha + (x_0 + 1) \end{aligned}$$

But $(x_1 + 1) = \overline{x_1}$ and $(x_0 + 1) = \overline{x_0}$, so:

$$y = x_2 \cdot \alpha^2 + \overline{x_1} \cdot \alpha + \overline{x_0}$$

Again expressing y in component form, we have:

$$y_2 \cdot \alpha^2 + y_1 \cdot \alpha + y_0 = x_2 \cdot \alpha^2 + \overline{x_1} \cdot \alpha + \overline{x_0}$$

and equating coefficients of like powers of α gives:

$$\begin{aligned} y_2 &= x_2 \\ y_1 &= \overline{x_1} \\ y_0 &= \overline{x_0} \end{aligned}$$

which is realized by the following circuit:

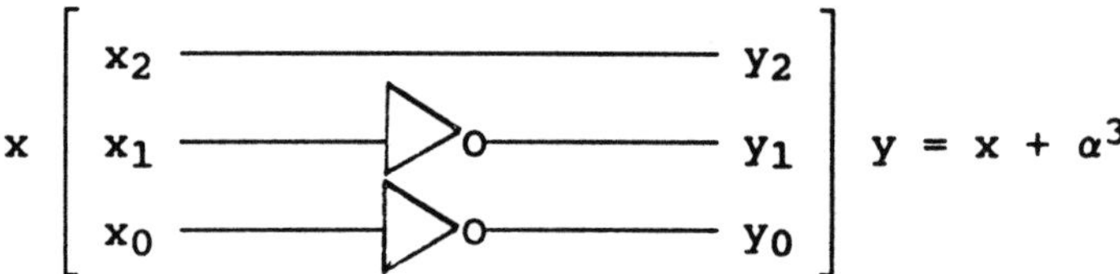

The arbitrary finite field adder:

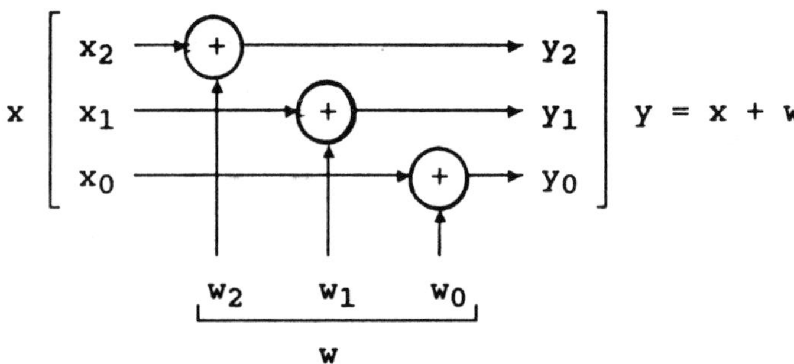

may be implemented using bit-serial techniques:

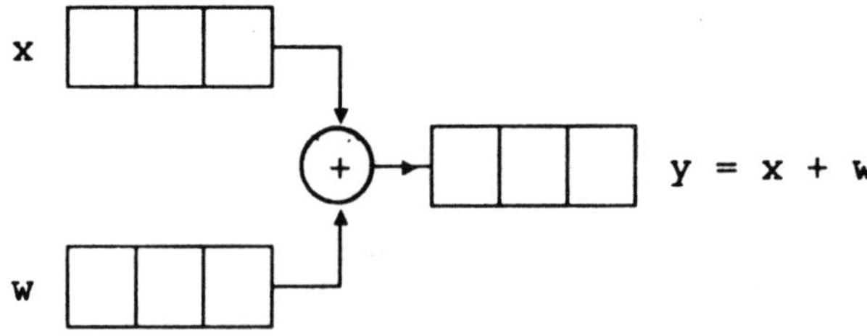

Fixed field element multiplier to multiply by α.

$$y = \alpha \cdot x$$

$$= \alpha \cdot (x_2 \cdot \alpha^2 + x_1 \alpha + x_0)$$

$$= x_2 \cdot \alpha^3 + x_1 \cdot \alpha^2 + x_0 \cdot \alpha$$

But, $\alpha^3 = \alpha + 1$, so:

$$y = x_1 \cdot \alpha^2 + (x_2 + x_0) \cdot \alpha + x_2$$

Expressing y in component form:

$$y_2 \cdot \alpha^2 + y_1 \cdot \alpha + y_0 = x_1 \cdot \alpha^2 + (x_2 + x_0) \cdot \alpha + x_2$$

Equating coefficients of like powers of α:

$$y_2 = x_1$$

$$y_1 = x_2 + x_0$$

$$y_0 = x_2$$

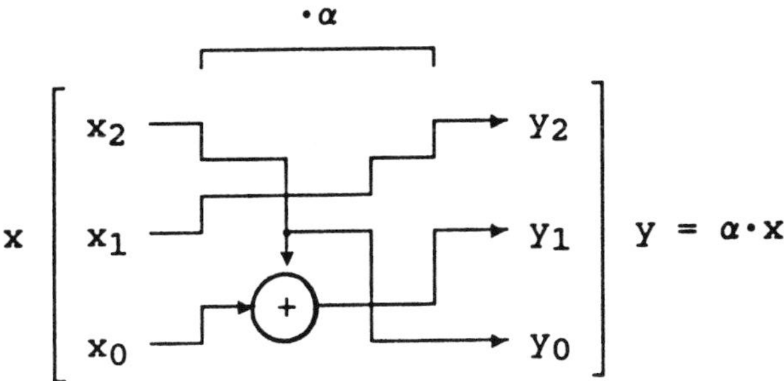

Fixed field element multiplier to multiply by α^{-1}.

$$y = \alpha^{-1} \cdot x$$
$$= \alpha^{6} \cdot x$$
$$= \alpha^{6} \cdot (x_2 \cdot \alpha^2 + x_1 \cdot \alpha + x_0)$$
$$= x_2 \cdot \alpha^8 + x_1 \cdot \alpha^7 + x_0 \cdot \alpha^6$$
$$= x_0 \cdot \alpha^2 + x_2 \cdot \alpha + (x_1 + x_0)$$

Expressing y in component form:

$$y_2 \cdot \alpha^2 + y_1 \cdot \alpha + y_0 = x_0 \cdot \alpha^2 + x_2 \cdot \alpha + (x_1 + x_0)$$

Equating coefficients:

$$y_2 = x_0$$
$$y_1 = x_2$$
$$y_0 = x_1 + x_0$$

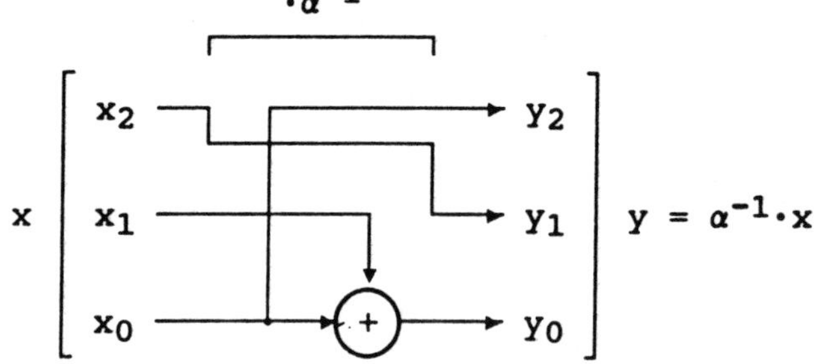

Fixed field element multiplier to multiply by α^2. The finite field math for this circuit is similar to the math for the α and α^{-1} multipliers above.

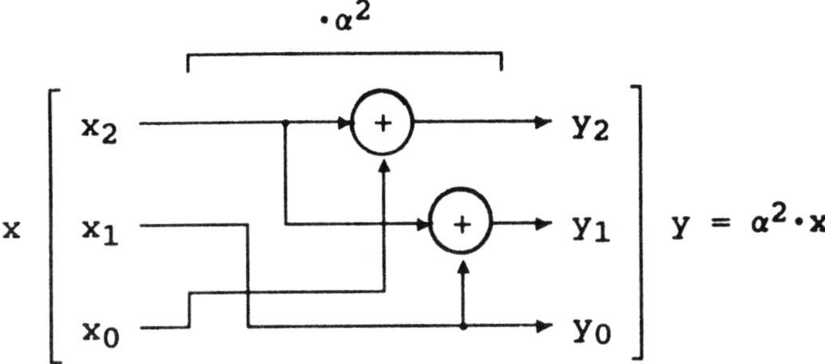

Fixed field element multiplier to multiply by α^2 using two circuits that multiply by α:

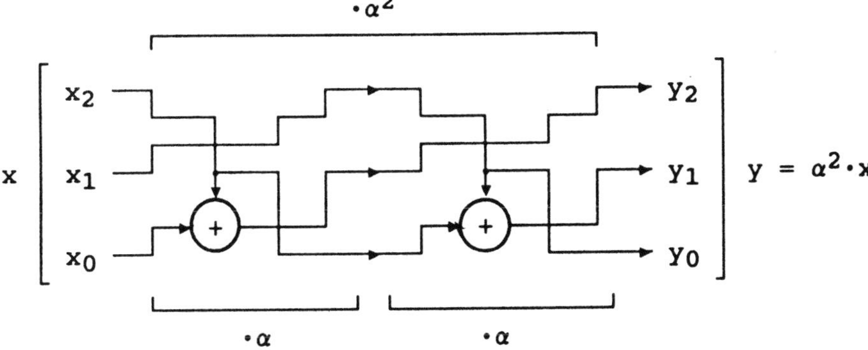

Fixed field element multiplier to multiply by α using bit serial techniques.

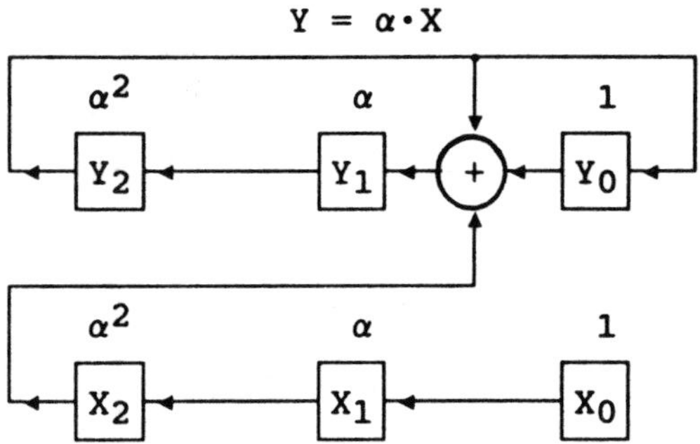

PROCEDURE:

1. Clear the Y register.
2. Load the X register.
3. Apply three clocks. In $GF(2^n)$ apply n clocks.
4. Accept the result from the Y register.

Fixed field element multiplier to multiply by α^4. To understand the input connections, recall that $\alpha^4 = \alpha^2 + \alpha$.

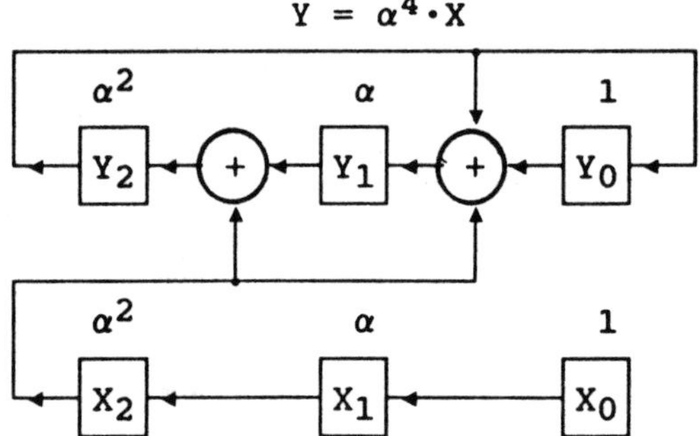

PROCEDURE:

Same as above.

Finite field circuit to compute $Y = \alpha \cdot X + \alpha^4 \cdot W$ using bit serial techniques.

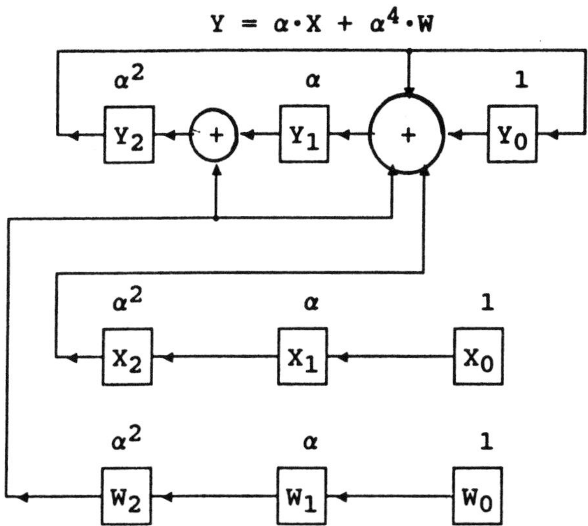

Arbitrary field element multiplier using combinatorial logic.

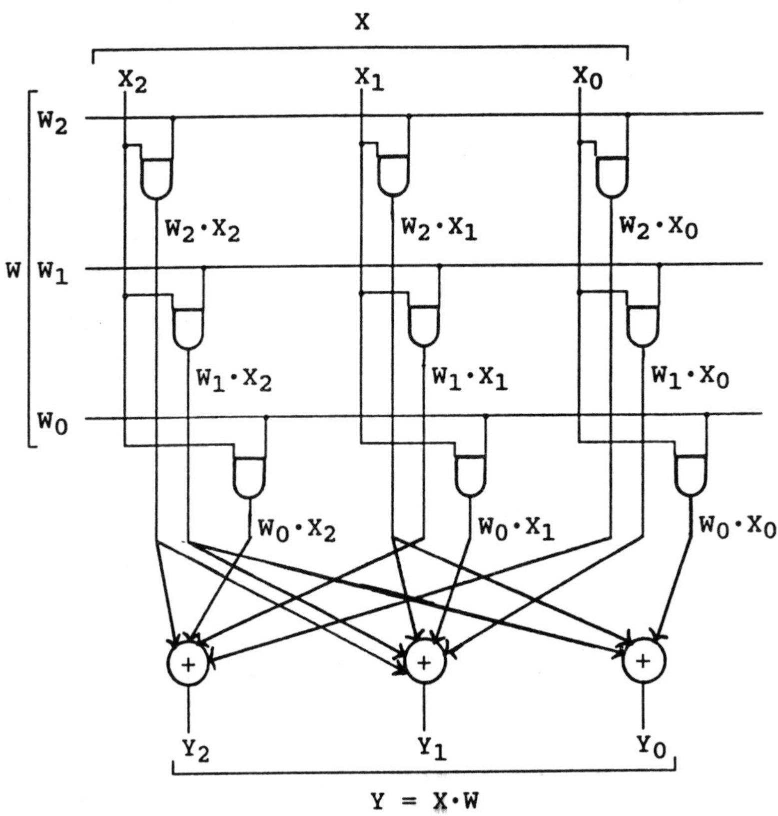

$$
\begin{aligned}
Y &= X \cdot W \\
&= (X_2 \cdot \alpha^2 + X_1 \cdot \alpha + X_0) \cdot (W_2 \cdot \alpha^2 + W_1 \cdot \alpha + W_0) \\
&= \quad (X_2 \cdot W_2) \cdot \alpha^4 + (X_2 \cdot W_1 + X_1 \cdot W_2) \cdot \alpha^3 \\
&\quad + (X_2 \cdot W_0 + X_1 \cdot W_1 + X_0 \cdot W_2) \cdot \alpha^2 + (X_1 \cdot W_0 + X_0 \cdot W_1) \cdot \alpha + X_0 \cdot W_0
\end{aligned}
$$

But $\alpha^4 = \alpha^2 + \alpha$ and $\alpha^3 = \alpha + 1$, so

$$
\begin{aligned}
Y = &\quad (X_2 \cdot W_2 + X_2 \cdot W_0 + X_1 \cdot W_1 + X_0 \cdot W_2) \cdot \alpha^2 \\
&+ (X_2 \cdot W_2 + X_2 \cdot W_1 + X_1 \cdot W_2 + X_1 \cdot W_0 + X_0 \cdot W_1) \cdot \alpha \\
&+ (X_2 \cdot W_1 + X_1 \cdot W_2 + X_0 \cdot W_0)
\end{aligned}
$$

Expressing Y in component form and equating coefficients on like powers of α gives:

$$
\begin{aligned}
Y2 &= X_2 \cdot W_2 + X_2 \cdot W_0 + X_1 \cdot W_1 + X_0 \cdot W_2 \\
Y1 &= X_2 \cdot W_2 + X_2 \cdot W_1 + X_1 \cdot W_2 + X_1 \cdot W_0 + X_0 \cdot W_1 \\
Y0 &= X_2 \cdot W_1 + X_1 \cdot W_2 + X_0 \cdot W_0
\end{aligned}
$$

Array multiplier - another arbitrary field element multiplier using combinatorial logic.

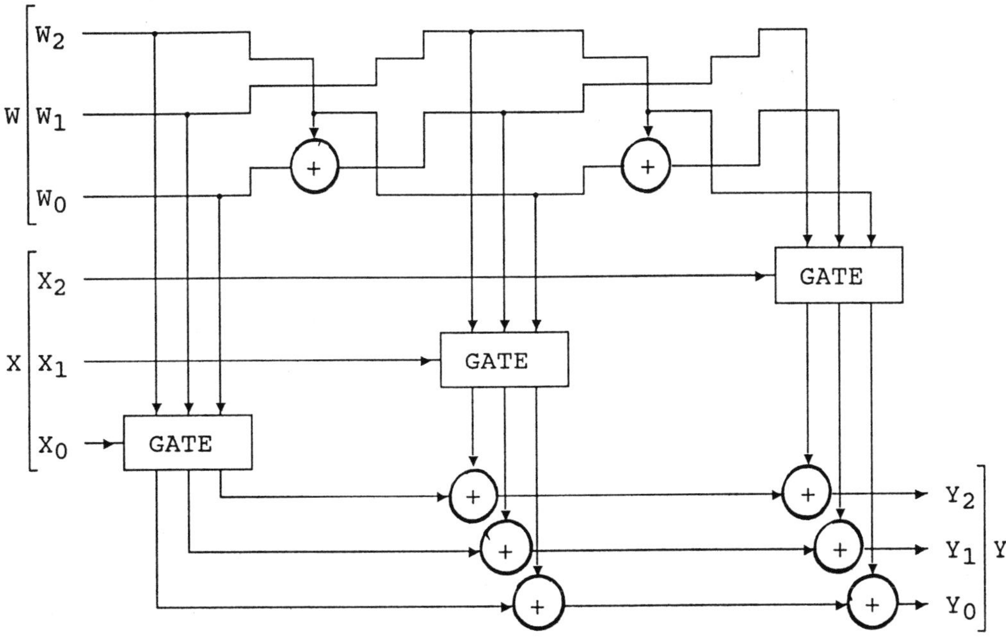

$$Y = X \cdot W$$

$$= (X_2 \cdot \alpha^2 + X_1 \cdot \alpha + X_0) \cdot W$$

$$= X_2 \cdot \alpha^2 \cdot W + X_1 \cdot \alpha \cdot W + X_0 \cdot W$$

$$= X_2 \cdot (\alpha^2 \cdot W) + X_1 \cdot (\alpha \cdot W) + (W)$$

Arbitrary field element multiplier using bit serial techniques.

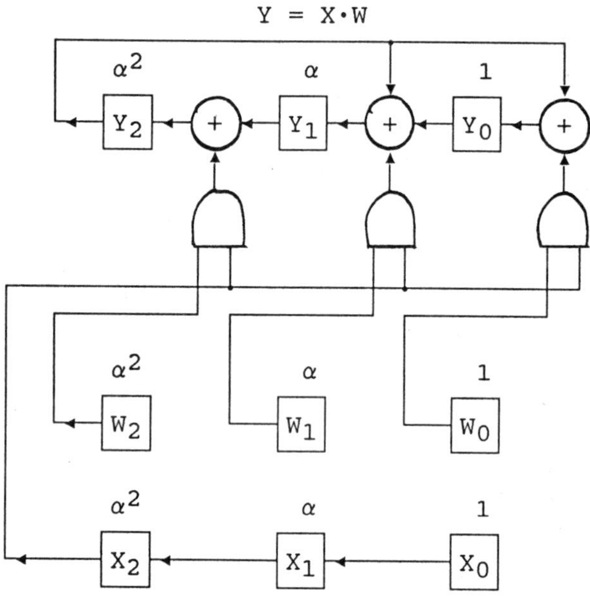

The X register is a shift register. The W register is composed of flip-flops that hold their value until reloaded.

PROCEDURE:

1. Clear the Y register
2. Load the W register with multiplicand.
3. Load the X register with multiplier.
4. Clock the circuit three times. For GF(2^n), clock n times.
5. Accept the result from the Y register.

DEVELOPMENT

$$
\begin{aligned}
Y &= X \cdot W \\
&= (X_2 \cdot \alpha^2 + X_1 \cdot \alpha + X_0) \cdot W \\
&= X_2 \cdot \alpha^2 \cdot W + X_1 \cdot \alpha \cdot W + X_0 \cdot W \\
&= [(X_2 \cdot W) \cdot \alpha + X_1 \cdot W] \cdot \alpha + X_0 \cdot W
\end{aligned}
$$

Another arbitrary field element multiplier using bit serial techniques.

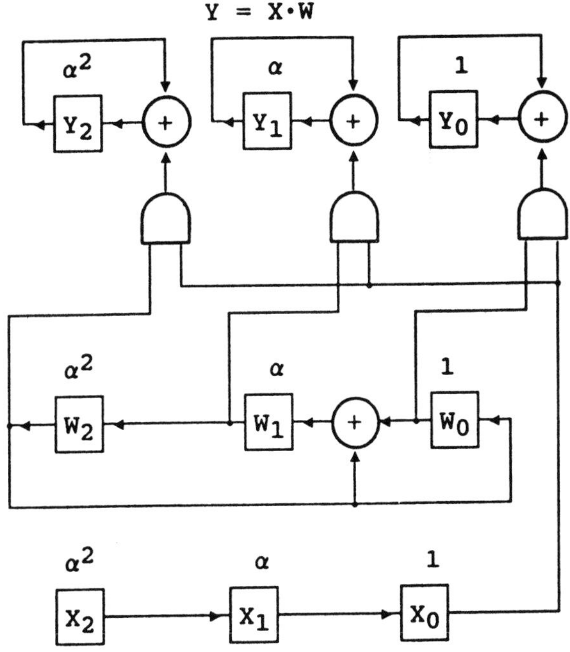

PROCEDURE

1. Clear the Y register.
2. Load the W register with multiplicand.
3. Load the X register with multiplier.
4. Clock the circuit three times. For $GF(2^n)$, clock n times.
5. Accept the result from the Y register.

DEVELOPMENT

$$Y = X \cdot W$$

$$= (X_2 \cdot \alpha^2 + X_1\alpha + X0) \cdot W$$

$$= X_2 \cdot \alpha^2 \cdot W + X_1 \cdot \alpha \cdot W + X_0 \cdot W$$

$$= X_2 \cdot (\alpha^2 \cdot W) + X_1 \cdot (\alpha \cdot W) + X_0 \cdot (W)$$

Arbitrary field element multiplier using log and antilog tables.

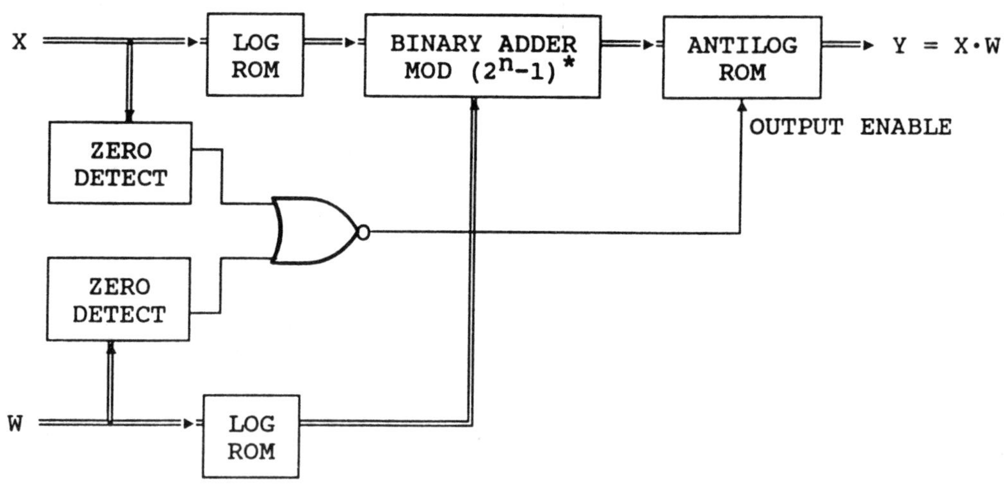

DEVELOPMENT

```
Y = X•W
IF (X=0) OR (W=0) THEN
  Y = 0
ELSE
  Y = ANTILOGα[ LOGα(X•W) ]
    = ANTILOGα[ (LOGα(X)+LOGα(W)) MOD (2^n-1)* ]
END IF
```

* For n-bit symbols, 2^n is the field size of GF(2^n), so (2^n-1) is the field size minus one.

Circuit to cube an arbitrary field element.

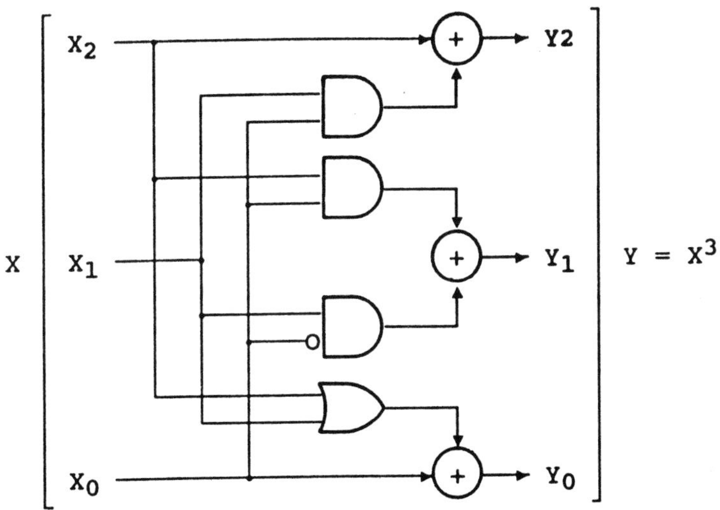

DEVELOPMENT

$$
\begin{aligned}
Y &= X^3 \\
&= (X_2\cdot\alpha^2 + X_1\cdot\alpha + X_0)^3 \\
&= (X_2\cdot\alpha^2 + X_1\cdot\alpha + X_0)^2\cdot(X_2\cdot\alpha^2 + X_1\cdot\alpha + X_0) \\
&= [(X_2\cdot\alpha^2)^2 + (X_1\cdot\alpha)^2 + (X_0)^2]\cdot(X_2\cdot\alpha^2 + X_1\cdot\alpha + X_0) \\
&= (X_2\cdot\alpha^4 + X_1\cdot\alpha^2 + X_0)\cdot(X_2\cdot\alpha^2 + X_1\cdot\alpha + X_0) \\
&= \quad X_2\cdot\alpha^6 + X_1\cdot X_2\cdot\alpha^5 + X_0\cdot X_2\cdot\alpha^4 + X_1\cdot X_2\cdot\alpha^4 \\
&\quad + X_1\cdot\alpha^3 + X_0\cdot X_1\cdot\alpha^2 + X_0\cdot X_2\cdot\alpha^2 + X_0\cdot X_1\cdot\alpha + X_0 \\
&= X_2\cdot(\alpha^2 + 1) + X_1\cdot X_2\cdot(\alpha^2 + \alpha + 1) + X_0\cdot X_2\cdot(\alpha^2 + \alpha) \\
&\quad + X_1\cdot X_2\cdot(\alpha^2 + \alpha) + X_1\cdot(\alpha + 1) + X_0\cdot X_1\cdot(\alpha^2) \\
&\quad + X_0\cdot X_2\cdot(\alpha^2) + X_0\cdot X_1(\alpha) + X_0 \\
&= \quad (X_2 + X_0\cdot X_1)\cdot\alpha^2 \\
&\quad + (X_0\cdot X_2 + X_1 + X_0\cdot X_1)\cdot\alpha \\
&\quad + (X_2 + X_1\cdot X_2 + X_1 + X_0)
\end{aligned}
$$

Expressing Y in component form and equating components of like powers of Á gives:

$$
\begin{aligned}
Y_2 &= X_2 + X_0X_1 \\
Y_1 &= X_0\cdot X2 + X_1 + X_0\cdot X_1 = X_0\cdot X_2 + X_1\cdot(1 + X_0) \\
&= X_0\cdot X_2 + X_1\cdot\overline{X_0} \\
Y_0 &= X_2 + X_1\cdot X_2 + X_1 + X_0 = (X_2 \vee X_1) + X_0
\end{aligned}
$$

where v is the INCLUSIVE-OR operator.

IMPLEMENTING GF(2^n) FINITE FIELD CIRCUITS WITH ROMS

In many cases, finite field circuits can be implemented with ROMs. For example, a GF(256) inverter is an 8-bit-in, 8-bit-out function and can be implemented with a 256:8 ROM.

Other examples:

1. The square function in GF(256) can be implemented with a 256:8 ROM. The same is true for any power or root function in GF(256).

2. A GF(16) arbitrary field element multiplier can be implemented with a 256:4 ROM. A GF(256) arbitrary field element multiplier can be implemented with a 65536:8 ROM. It is also possible to implement a GF(256) multiplier with four 256:4 ROMs and several finite field adders. (See Section 2.7.)

3. A GF(256) fixed field element multiplier can be implemented with a 256:8 ROM.

When back-to-back functions are required, it is sometimes possible to combine them in a single ROM. For example, the equation:

$$Y = [1/X]^3 \cdot \alpha^2$$

in GF (256) can be solved for Y when X is known with a single 256:8 ROM.

SOLVING FINITE FIELD EQUATIONS

Finding a power of a finite field element results in a single solution, but the same solution may be obtained by raising other finite field elements to the same power.

Finding the root(s) of a finite field element may result in a single solution, multiple solutions or no solution.

Finding the root(s) of a finite field equation may result in a single solution, multiple solutions or no solution.

FINDING ROOTS OF FINITE FIELD EQUATIONS

In decoding the Reed-Solomon and binary BCH codes, it is frequently necessary to find the roots of nonlinear equations whose coefficients are from a finite field. These roots provide error-location information. The degree of the equation and the number of roots are equal to the number of errors that occur. Examples of these equations are shown below:

$$x + \sigma_1 = 0$$

$$x^2 + \sigma_1 \cdot x + \sigma_2 = 0$$

$$x^3 + \sigma_1 \cdot x^2 + \sigma_2 \cdot x + \sigma_3 = 0$$

$$x^4 + \sigma_1 \cdot x^3 + \sigma_2 \cdot x^2 + \sigma_3 \cdot x + \sigma_4 = 0$$

One way to find the roots of such an equation is to substitute all possible finite field values for x. The equation evaluates to zero for any finite field elements that are roots.

Two methods which perform the substitution will be discussed. The first method uses "brute force", and is shown only to illustrate the idea of substitution.

The second method is the Chien search. This is a practical method that can be used to find the roots of equations of a low degree or high degree.

After discussing the Chien search, alternatives will be explored for finding roots of nonlinear equations of a low degree.

SUBSTITUTION METHOD - BRUTE FORCE

Assume the roots of $X^3 + \sigma_1 X^2 + \sigma_2 X + \sigma_3 = 0$ must be found. The circuit below could be used:

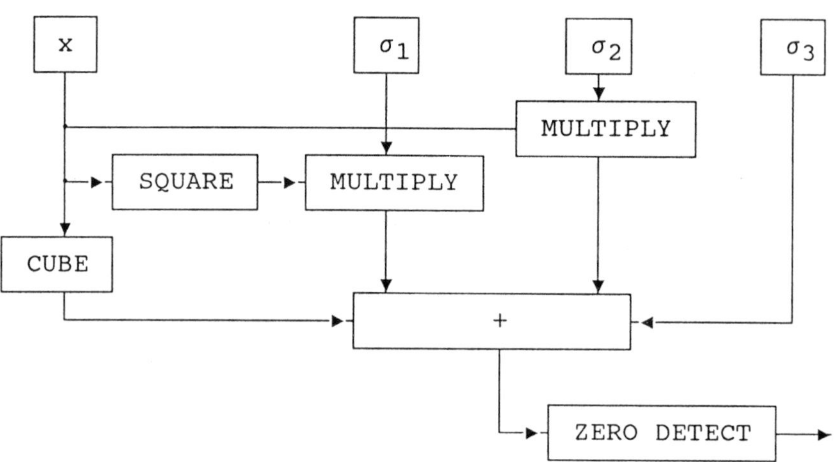

Each possible finite field value must be substituted for x while checking the output of the zero detector.

This circuit is easy to understand, although it is not practical because of circuit complexity.

SUBSTITUTION METHOD - CHIEN SEARCH

Assume the roots of

$$x^3 + \sigma_1 \cdot x^2 + \sigma_2 \cdot x + \sigma_3 = 0$$

must be found. The Chien search circuit below could be used:

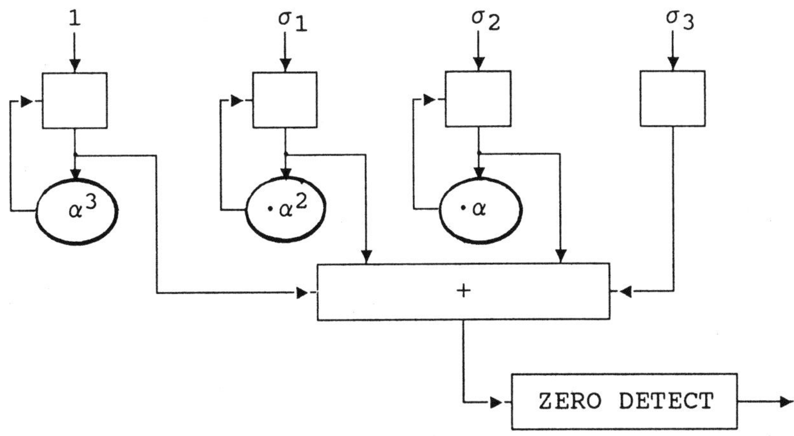

The circuit is initialized as shown. If the zero detect output is immediately asserted, α^0 is a root. The circuit is clocked. If the zero detect output is then asserted, α^1 is a root. The circuit is clocked again. If the zero detect output is then active, α^2 is a root. Operation continues as described until all finite field values have been substituted and all roots recorded.

This method uses less complex circuits than the "brute force" method.

The example circuit above finds roots of finite field equations of degree three. The circuit can be extended in a logical fashion to find the roots of equations of a higher degree.

RECIPROCAL ROOTS

There are times when the reciprocals of roots of finite field equations are required. If

$$x^3 + \sigma_1 \cdot x^2 + \sigma_2 \cdot x + \sigma_3 = 0$$

is an equation for which reciprocal roots are required, then

$$\sigma_3 \cdot x^3 + \sigma_2 \cdot x^2 + \sigma_1 \cdot x + 1 = 0$$

is an equation whose roots are the reciprocals of the roots of the first equation. The Chien search circuit below can be used to find reciprocal roots.

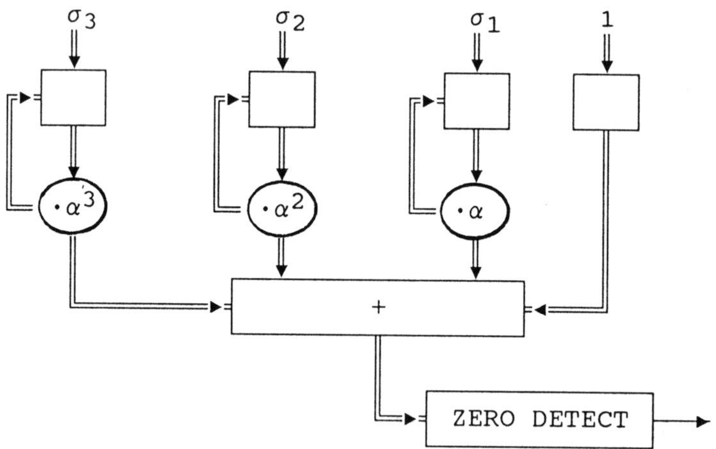

In this circuit, the inputs to the XOR circuit are from the multipliers instead of the registers because the equation is evaluated at α^1 first.

FINDING ROOTS OF EQUATIONS OF DEGREE 2

AN EXAMPLE

We illustrate the method by generating a quadratic table for solving $y^2 + y = C$ in the field $GF(2^3)$ generated by the polynomial $x^3 + x + 1$ over GF(2).

First generate the antilog table for the field. Next construct a table giving C when y is known. Then construct a table giving y when C is known (Table A below).

Antilog Table

Exponent	Vector
---	000
0	001
1	010
2	100
3	011
4	110
5	111
6	101

'y' is known

y	C
000	000
001	000
010	110
011	110
100	010
101	010
110	100
111	100

Table A. 'C' is known

C	y
000	000,001
001	No solution
010	100,101
011	No Solution
100	110,111
101	No Solution
110	010,011
111	No Solution

We may verify the validity of Table A by using it to solve the following equations:

$y^2 + y = \alpha^2 \Rightarrow y = 110,111$

$y^2 + y = \alpha^5 \Rightarrow y =$ No Solution

FINITE FIELD PROCESSORS

Finite field processors are programmable or microprogrammable processors, which are designed especially for finite field computation. An example for computing in GF(256) is shown below. Except where noted, all paths are eight bits wide.

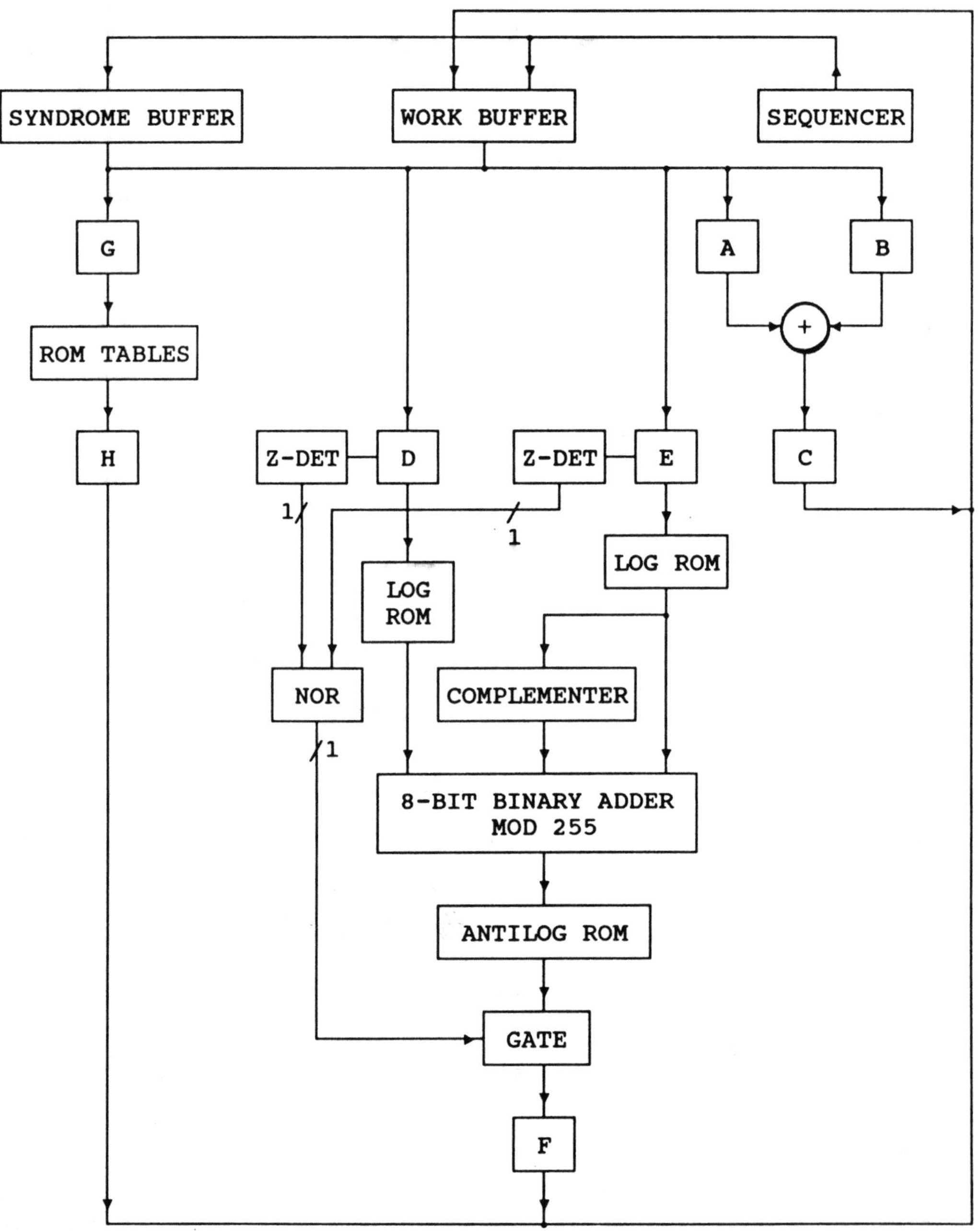

Adding two finite field elements from the work buffer consists of the following steps.

1. Transfer the first element to the A register.

2. Transfer the second element to the B register.

3. XOR the contents of the A and B registers and set the result in the C register.

4. Transfer the C register to the work buffer.

Each of these steps can be a separate instruction or part of a single instruction.

Multiplying finite field elements from the work buffer consists of the following steps:

1. Transfer the first element to the D register.

2. Transfer the second element to the E register.

3. Add logs of the finite field elements and place the antilog of the results in the F register.

4. Transfer the F register to the work buffer.

As in finite field addition, each step can be a separate instruction or part of a single instruction.

If either multiplication operand is zero, the result must be zero. Since the log of zero is undefined, this case must receive special attention. It is handled by the zero--detect circuits connected to the D and E registers and controlling the gate at the input of the F register.

For the processor under consideration, logs must be added modulo 255. Eight-bit binary adders add modulo 256. They can be used to add modulo 255 by connecting "carry out" to "carry in". For the antilog table, the contents of location 255 are the same as location zero.

Finite field division is accomplished with the same steps used for finite field multiplication, except logs are subtracted.

The log operation could be implemented as follows:

1. Load the finite field value in register G.

2. Move the log of the finite field value from the ROM tables to register H.

3. Store register H in the work buffer.

There are many design options available when designing a finite field processor. The options selected depend on the logic family to be used, cost, performance and other design considerations. The options selected for an LSI design would differ from those selected for a discrete design.

A partial list of operations that have been implemented on real world finite-field processors is shown below.

- Finite field addition
- Finite field multiplication
- Finite field division
- Logarithm
- Antilogarithm
- Fetch one root of the equation $y^2 + y + C = 0$
- Take cube root
- Compare finite field values
- Branch unconditional
- Branch conditional

Non-finite-field operations that may be implemented include:

- Binary addition and subtraction
- Logical AND and inclusive-OR operations
- Operations for controlling error-correction hardware.

A finite field processor implementing subfield multiplication is shown in Section 5.4.

2.7 SUBFIELD COMPUTATION

In this section, a large field, $GF(2^{2*n})$, generated by a small field, $GF(2^n)$, is discussed. Techniques are developed to accomplish operations in the large field by performing several operations in the small field.

Let elements of the small field be represented by powers of β. Let elements of the large field be represented by powers of α.

The small field is defined by a specially selected polynomial of degree n over GF(2). The large field is defined by the polynomial:

$$x^2 + x + \beta$$

over the small field.

Each element of the large field, $GF(2^{2*n})$, can be represented by a pair of elements from the small field, $GF(2^n)$. Let x represent an arbitrary element from the large field. Then:

$$x = x_1 \cdot \alpha + x_0$$

where x_1 and x_0 are elements from the small field, $GF(2^n)$. The element x from the large field can be represented by the pair of elements (x_1,x_0) from the small field. This is much like representing an element from the field of Figure 2.5.1 with three elements from GF(2), (x_2,x_1,x_0).

Let α be any primitive root of:

$$x^2 + x + \beta$$

Then:

$$\alpha^2 + \alpha + \beta = 0$$

Therefore:

$$\alpha^2 = \alpha + \beta$$

The elements of the large field $GF(2^{2*n})$, can be defined by the powers of α. For example:

$$0 = 0$$

$$\alpha^0 = \alpha^0$$

$$\alpha^1 = \alpha^1$$

$$\alpha^2 = \alpha + \beta$$

$$\begin{aligned}\alpha^3 &= \alpha \cdot \alpha^2 \\ &= \alpha \cdot (\alpha + \beta) \\ &= \alpha^2 + \alpha \cdot \beta \\ &= \alpha + \beta + \alpha \cdot \beta \\ &= (\beta + 1) \cdot \alpha + \beta\end{aligned}$$

...

This list of elements can be denoted

	α^1	α^0
0	0	0
α^0	0	1
α^1	1	0
α^2	1	β
α^3	$\beta+1$	β
...		

The large field, $GF(2^{2*n})$, can be viewed as being generated by the following shift register. All paths are n bits wide.

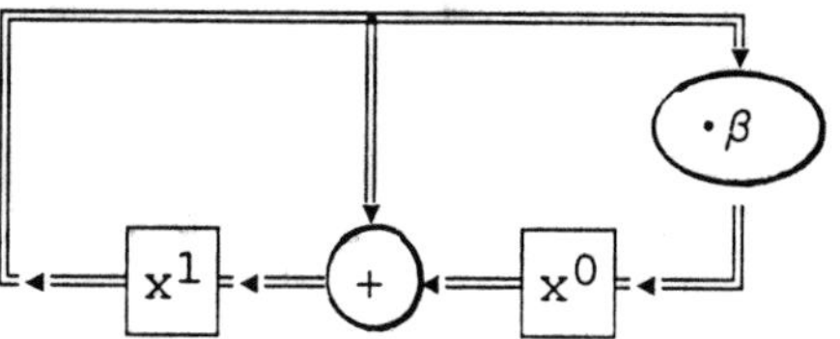

This shift register implements the polynomial $x^2 + x + \beta$ over $GF(2^n)$.

Methods for accomplishing finite field operations in the large field by performing several simpler operations in the small field are developed below.

ADDITION

Let x and w be arbitrary elements from the large field. Then:

$$
\begin{aligned}
y &= x + w \\
&= (x_1 \cdot \alpha + x_0) + (w_1 \cdot \alpha + w_0) \\
&= (x_1 + w_1) \cdot \alpha + (x_0 + w_0)
\end{aligned}
$$

MULTIPLICATION

The multiplication of two elements from the large field can be accomplished with several multiplications and additions in the small field. This is illustrated below:

$$
\begin{aligned}
y &= x \cdot w \\
&= (x_1 \cdot \alpha + x_0) \cdot (w_1 \cdot \alpha + w_0) \\
&= x_1 \cdot w_1 \cdot \alpha^2 + x_1 \cdot w_0 \cdot \alpha + x_0 \cdot w_1 \cdot \alpha + x_0 \cdot w_0
\end{aligned}
$$

But, $\alpha^2 = \alpha + \beta$, so

$$
\begin{aligned}
&= x_1 \cdot w_1 \cdot (\alpha + \beta) + w_0 \cdot x_1 \cdot \alpha + x_0 \cdot w_1 \cdot \alpha + x_0 w_0 \\
&= (x_1 \cdot w_1 + w_0 \cdot x_1 + x_0 \cdot w_1) \cdot \alpha + (x_1 \cdot w_1 \cdot \beta + x_0 \cdot w_0)
\end{aligned}
$$

Methods for accomplishing other operations in the large field can be developed in a similar manner. The method for several additional operations are given below without the details of development.

INVERSION

$$
\begin{aligned}
y &= 1/x \\
&= \frac{x_1}{(x_1)^2 \cdot \beta + x_1 \cdot x_0 + x_0^2} \cdot \alpha + \frac{x_1 + x_0}{(x_1)^2 \cdot \beta + x_1 \cdot x_0 + x_0^2}
\end{aligned}
$$

LOGARITHM

$L = LOG_{\alpha}(x)$

Let,

$J = LOG_{\beta}[(x_1)^2 \cdot \beta + x_1 \cdot x_0 + x_0^2]$

$K = 0$ if $x_1=0$

$= 1$ if $x_1 \neq 0$ and $x_0=0$

$= f_1(x_0/x_1)$ if $x_1 \neq 0$ and $x_0 \neq 0$

Then,

L = {the integer whose residue modulo (2^n-1) is J and whose residue modulo (2^n+1) is K}

This integer can be determined by the application of the Chinese Remainder Method. See Section 1.2 for a discussion of the Chinese Remainder Method.

The function f_1 can be accomplished with a table of 2^n entries which can be generated with the following algorithm.

BEGIN

Set table location $f_1(0)=0$

FOR $I=2$ to 2^n

Calculate the $GF(2^{2*n})$ element $Y = \alpha^I = Y_1 \cdot \alpha + Y_0$

Calculate the $GF(2^n)$ element Y_0/Y_1

Set $f_1(Y_0/Y_1)=I$

NEXT I

END

ANTILOGARITHM

$$X = \text{ANTILOG}_{\alpha}(L)$$

$$= \text{ANTILOG}_{\beta}(\text{INT}(L/(2^n+1))) \quad \text{if } [L \text{ MOD } (2^n+1)]=0$$

$$= [\text{ANTILOG}_{\beta}(\text{INT}(L/(2^n+1)))]\cdot\alpha \quad \text{if } [L \text{ MOD } (2^n+1)]=1$$

$$= x_1\cdot\alpha + x_0 \quad \text{if } [L \text{ MOD } (2^n+1)]>1$$

where x_1 and x_0 are determined as follows. Let

$$a = \text{ANTILOG}_{\beta}[\ L \text{ MOD } (2^n-1)\ ]$$

$$b = f_2[\ (L \text{ mod } (2^n+1))-2\ ]$$

Then,

$$x_1 = \left[\frac{a}{b^2 + b + \beta} \right]^{1/2}$$

$$x_0 = b\cdot x_1$$

The function f_2 can be accomplished with a table of 2^n entries. This table can be generated with the following algorithm.

BEGIN

Set $f_2(2^n-1)=0$

FOR $I=0$ to 2^n-2

Calculate the $GF(2^{2*n})$ element $Y = \alpha^{(I+2)} = Y_1\cdot\alpha + Y_0$

Calculate the $GF(2^n)$ element Y_0/Y_1

Set $f_2(I) = Y_0/Y_1$

NEXT I

END

APPLICATIONS

In this section, techniques were introduced for performing operations in a large field, $GF(2^{2*n})$, by performing several simpler operations in a small field, $GF(2^n)$.

One application of these techniques is for computing in a very large finite field. Assume that it is necessary to perform computation in GF(65536). A multiplication operation might be accomplished by fetching logs from a log table; adding logs modulo 65535; and fetching an antilog. The log and antilog tables would each be 65536 locations of 16 bits each. The total storage space required for these tables would be one quarter million bytes. An alternative is to define GF(65536) as described in this section and to perform operations in GF(65536) by performing several simpler operations in GF(256). These GF(256) operations could be performed with 256 byte log and antilog tables.

Another application is for performing finite field multiplication directly with ROMs for double-bit-memory correction. Instead of using one ROM with 2^{4n} locations, use four ROMs each with 2^{2*n} locations. An example application to multiplier ROMs is shown below.

A GF(256) MULTIPLIER USING A SINGLE ROM

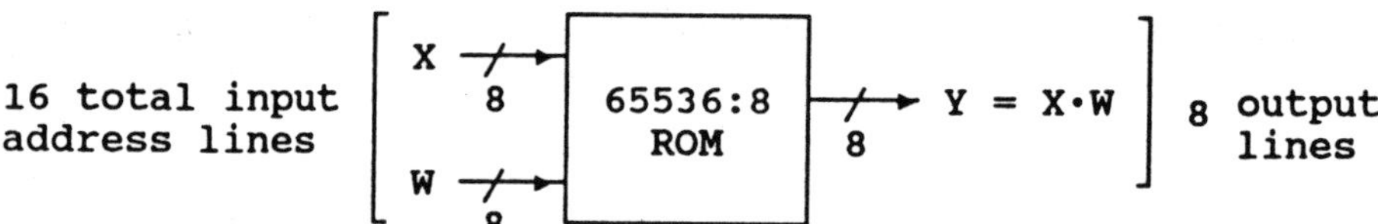

A GF(256) MULTIPLIER USING FOUR SMALLER ROMS

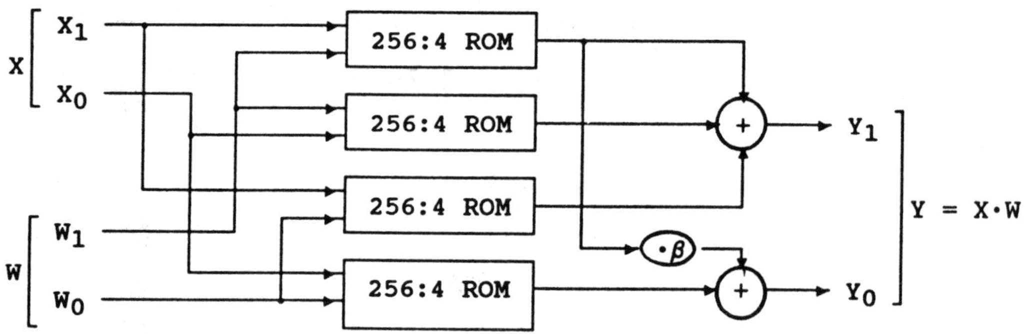

See Section 5.4 for details of a $GF(2^8)$ multiplier constructed from four $GF(2^4)$ multipliers.

CHAPTER 3 - CODES AND CIRCUITS

3.1 FIRE CODES

Fire codes are linear cyclic single-burst-correcting codes defined by generator polynomials of the form:

$$g(x) = c(x) \cdot p(x) = (x^c + 1) \cdot p(x)$$

where

c = Degree of the c(x) factor of g(x)

p(x) is any irreducible polynomial with period e, and e does not divide c.

Let:

z = Degree of the p(x) factor of g(x)

m = Degree of g(x) = total number of check bits = c+z

n = Record length in bits including check bits; $n \leq LCM(e,c)$

b = Guaranteed single-burst correction span in bits

d = Guaranteed single-burst detection span in bits

The maximum record length in bits, including check bits, is equal to the period of g(x), which is the least common multiple of e and c. The guaranteed single-burst correction and detection spans for the Fire codes are subject to the following inequalities:

$$b \leq z$$
$$b \leq d$$
$$b+d \leq c+1$$

These inequalities provide a lower bound for d. When the record length is much less than the period of the polynomial, this bound for d is conservative. In this case, the true detection span should be determined by a computer search.

Given a fixed and limited total number of check bits, selecting the degrees of p(x) and c(x) will be involve a tradeoff. Increasing the degree of p(x) will provide more protection against miscorrection on double-bit errors (less pattern sensitivity), while increasing the degree of c(x) will provide a greater correction span and/or detection span. The degree of c(x) should not be used to adjust the period of a Fire code unless the effects of pattern sensitivity are fully understood.

Overall miscorrection probability for a Fire code for bursts exceeding the guaranteed detection capability is given by the equation below, assuming all errors are possible and equally probable:

$$P_{mc} \approx \frac{n * 2^{b-1}}{2^m}$$

Miscorrection probability for double-bit errors separated by more than the guaranteed detection span, assuming all errors of this type are possible and equally probable, is given by:

$$P_{mcdb} \approx \frac{(b-1)*2}{c} * \frac{n}{c*(2^z-1)}$$

This equation is applicable only when the product of P_{mcdb} and the number of possible double-bit errors is much greater than one. When this is not true, a computer search should be used to determine the actual P_{mcdb}.

An advantage of the Fire Code is simplicity. A disadvantage is pattern sensitivity. The $(x^c + 1)$ factor of the Fire Code generator polynomial causes the code to be susceptible to miscorrection on short double-bursts. The P_{mcdb} equation given above provides a number for this susceptibility for one particular short double-burst (the double-bit error). For more information on the Fire code's pattern sensitivity see Sections 4.4 and 4.6.

The pattern sensitivity of the Fire Code can be reduced to any arbitrary level by adding sufficient redundancy to the p(x) factor.

There are at least five ways to perform the correction step:

1. Clock around the full period of the polynomial.

2. Shorten the code by performing simultaneous multiplication and division of the data polynomial. A computer search may be required to minimize the complexity of feedback logic. The period after shortening can be selected to be precisely the required period.

3. Select a nonprimitive polynomial for p(x). This method yields a less complex feedback structure than method 2. However, it is only possible to select a period that is close to the required period. A computer search is required.

4. Perform the correction function with the reciprocal polynomial. This requires that either a self-reciprocal polynomial be used, or that the feedback terms be modified during correction. In addition, the contents of the shift register must be flipped end-for-end before performing the corrections.

   This method differs from methods 1 through 3 because the maximum number of shifts during correction depends on the record length instead of the polynomial period. Therefore, correction is faster for the case when the record length is shorter than the polynomial period.

5. Decode using the Chinese Remainder Method. This method requires only a fraction of the number of shifts required by the other methods. Thus, significant improvements in decoding speed can be obtained.

Any of the methods above may be implemented in hardware or software. However, for software, methods 4 and 5 are the most applicable. Methods 4 and 5 are more flexible for handling variable record lengths than the other methods.

The Fire Code may be implemented with bit-serial, byte-serial or k-bit-serial logic. See Section 4.7 for k-bit serial techniques.

BIT SERIAL

Fire-code circuit implementations using bit-serial techniques are less complex than those using byte-serial techniques.

Less logic is required for the shift register as well as for detecting the correctable pattern.

Polynomial selection is easier for the bit-serial implementation.

The disadvantage of bit-serial circuit implementations is shift rate limitations.

BYTE SERIAL

Byte-serial circuit implementations have speed as their advantage.

One disadvantage is greater logic complexity compared to bit-serial implementations. More logic is required to implement the shift register and to detect the correctable pattern. Pattern detection is more complex because the pattern is never justified to one end of the shift register. The problem is to determine within one shift (byte time), if a pattern unjustified in several byte-wide registers is of length b bits or less.

Another disadvantage of byte-serial implementations is that a computer search may be required for polynomial selection if the feedback logic is to be minimized.

Both bit-serial and byte-serial logic may be implemented in either hardware or software. Byte-serial implementations in software usually require look-up tables (for effective speed).

DECODING ALTERNATIVES FOR THE FIRE CODE

The Fire code can be decoded with the methods described in Section 2.3. Two examples of real world decoding of the Fire code are discussed in Sections 5.2.2 and 5.2.3.

The internal-XOR or external-XOR forms of shift registers may be used for implementing Fire codes. The decoding methods of Section 5.2 apply to the Fire code as well as to computer-generated codes.

In many cases, logic can be saved by using sequential logic to determine if the shift register is nonzero at the end of a read.

It is possible to use a counter to detect the correctable pattern. The counter counts the number of zeros preceding the error pattern. For the internal-XOR form of shift register the counter can monitor the high order shift register stage. A one clears the counter. A zero bumps the counter. The counter function can also be accomplished by a software routine commanding shifts and monitoring the high order shift register stage.

It is harder to detect the correctable pattern for byte-serial implementations than for bit-serial implementations. The second flowchart of Section 5.3.3 shows a software algorithm for detecting the correctable pattern for a byte-serial software implementation. The following page shows a method for accomplishing this for a byte-serial hardware implementation.

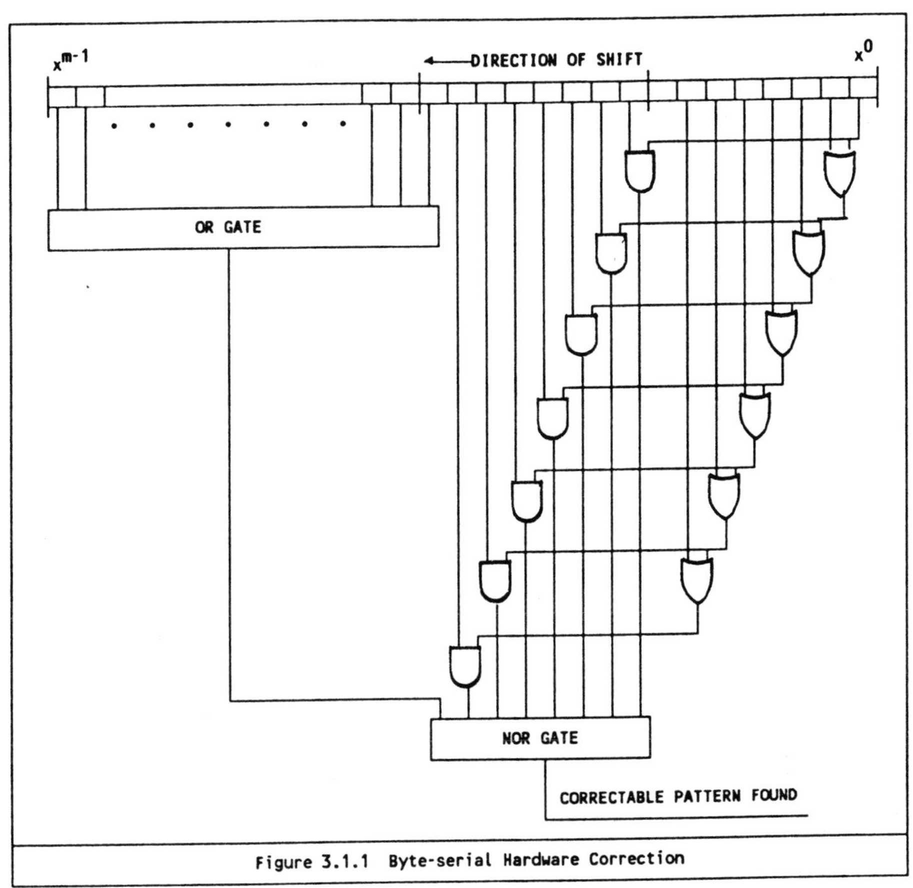

Figure 3.1.1 Byte-serial Hardware Correction

Correction span is assumed to be eight bits. When the correctable pattern first appears in the shift register, at least one bit of the pattern will be in the low order byte.

3.2 COMPUTER-GENERATED CODES

Computer-generated codes are based on the fact that if a large number of polynomials of a particular degree are picked at random, some will meet previously defined specifications, provided the specifications are within certain bounds.

There are equations that predict the probability of success when evaluating polynomials against a particular specification.

For computer-generated codes, correction and detection spans are determined by computer evaluation. Overall miscorrection probability, assuming all errors possible and equally probable, is given by:

$$P_{mc} \approx \frac{n * 2^{b-1}}{2^m}$$

where,

b = Guaranteed single burst correction span in bits
n = Record length in bits including check bits
m = Total number of check bits

In some cases, tens of thousands of computer-generated polynomials have been evaluated in order to find a polynomial with particular characteristics.

Properly selected computer-generated codes do not have the pattern sensitivity of the Fire code. It is possible to select computer-generated codes that have a guaranteed double-burst-detection span. The miscorrecting patterns of these codes are more random than those of the Fire code.

The decoding alternatives for the computer-generated code are the same as those previously described for the Fire code.

COMPUTER SEARCH RUN

This run evaluates polynomials for use with 512-byte records and correction spans to 8 bits. This run is for illustration only. The polynomials below which have a good single-burst detection span may not test well against other criteria.

Polynomial (octal)	Single-burst detection spans for given correction span of: 1	2	3	4	5	6	7	8
40001140741	18	18	18	16	16	16	16	12
41040103211	19	19	19	15	14	14	13	13
42422242001	19	19	19	17	17	12	12	12
42010100127	21	21	16	16	16	15	15	12
42200301203	20	20	19	17	17	15	12	12
40110425041	19	19	17	17	17	17	10	10
40442115001	18	18	18	18	17	16	16	14
44104042501	19	19	16	16	12	12	10	10
40030201415	18	18	18	15	15	13	13	13
40030070211	19	19	18	18	13	11	11	11
40006241441	20	19	18	18	15	15	15	14
40430250401	15	15	15	15	15	13	12	11
44401144041	20	20	20	16	16	14	14	13
41442001203	22	21	20	18	17	16	14	11
44431120001	17	17	17	17	16	15	11	11
40056110021	20	20	15	15	15	9	9	9
40200211701	20	20	20	18	18	9	9	9
40001201163	18	18	18	15	15	14	12	12
40410423003	21	18	17	16	16	16	14	12
42000027421	17	17	17	16	13	13	13	13
40001741005	18	17	17	17	11	11	11	11
42000045065	20	20	17	16	14	14	14	10
41114210201	20	19	19	18	18	16	16	14
44011511001	20	20	18	18	16	13	13	11
41200103203	18	18	15	15	15	15	15	14
43140224001	18	18	18	18	17	7		

COMPUTER SEARCH RUN (CONTINUED)

Polynomial (octal)	Single-burst detection spans for given correction span of: 1	2	3	4	5	6	7	8
40000074461	14	14	14	14	14	13	13	13
40527200001	16	16	16	16	16	16	16	10
40342100221	19	18	18	18	18	16	11	11
40400264411	16	16	16	16	16	13	13	13
44001140305	17	17	17	17	17	13	13	13
41450040051	19	19	18	18	18	16	14	14
40060405013	20	19	19	19	17	14	13	10
41030210031	18	18	18	18	17	17	17	9
40201202131	17	17	17	17	16	16	16	15
41024021025	21	19	19	19	16	12	12	12
40006052403	18	18	18	18	16	15	13	12
40152014401	19	19	18	18	14	14	14	13
46200002341	19	19	19	19	17	14	14	10
44501404011	19	19	16	16	14	14	13	13
40250002053	20	20	18	18	17	17	15	14
43012104011	19	18	18	18	18	17	12	12
42012430201	21	17	17	17	15	15	12	12
42114023001	21	21	20	16	16	11	11	10
43300020241	15	15	15	15	14	14	14	13
40001403207	18	18	18	18	17	16	9	9
40214020503	20	20	20	16	16	16	10	10
40260302005	20	20	19	18	17	7		
40252200241	20	20	20	13	13	13	12	12
40004560111	16	16	16	16	14	14	14	14
40000404347	15	15	15	15	15	15	14	13
42200036011	15	15	15	15	11	11	11	10
42202210241	20	18	18	17	10	10	10	10
40504100431	16	16	15	15	15	12	12	12
42012401111	19	17	17	15	15	14	14	14
43041105001	21	20	17	17	17	14	14	12
40022044225	18	18	18	11	11	11	11	11
40500001465	19	18	18	15	15	15	15	14

SPECTRUM OF DETECTION SPANS FOR COMPUTER SEARCH RUN

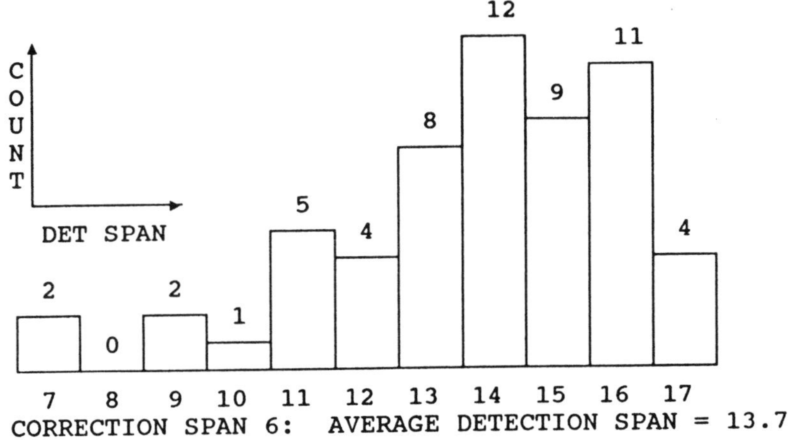

CORRECTION SPAN 6: AVERAGE DETECTION SPAN = 13.7

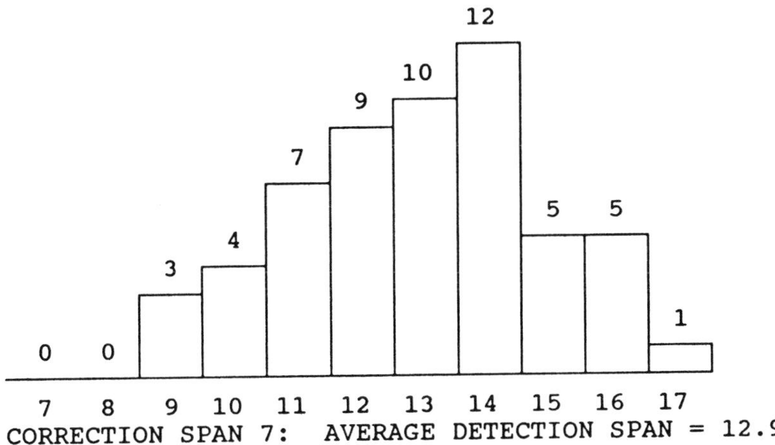

CORRECTION SPAN 7: AVERAGE DETECTION SPAN = 12.9

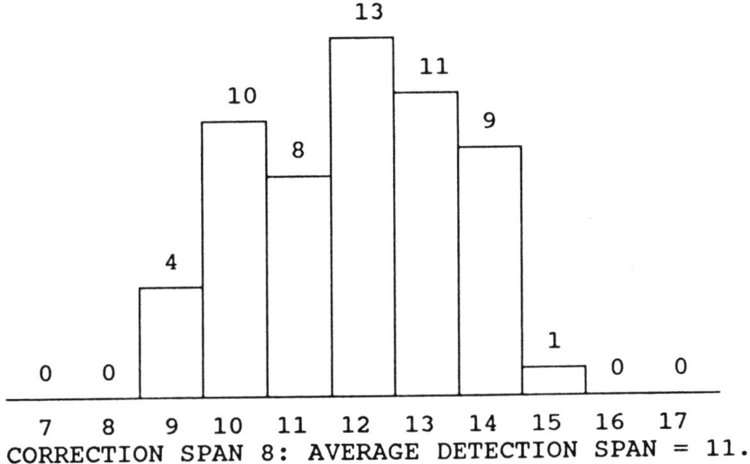

CORRECTION SPAN 8: AVERAGE DETECTION SPAN = 11.9

MOST PROBABLE DETECTION SPAN

The equation below gives an approximation for the most likely single-burst detection span of a single polynomial picked at random.

$$d \approx 0.5287 - \frac{\ln(-\ln(1-(n*2^{b})/2^{m}))}{0.6932} + 1$$

where,

b = Single-burst correction span
d = Single-burst detection span
n = Number of information plus check bits
m = Number of check bits

PROBABILITY OF SUCCESS

The equation below gives an approximation for the probability that a single polynomial picked at random will meet specified criteria.

$$P_s \approx \left[1 - \frac{n*2^{b}}{2^{m}} \right]^{2^{d-1}}$$

where n, m, b, and d are as defined above.

3.3 BINARY BCH CODES

Binary BCH codes correct random bit errors. Coefficients of the data polynomial and check symbols are from GF(2) *i.e.* they are binary '0' or '1', but computation of error locations and values is performed using w-bit symbols in a finite field $GF(2^w)$, where w is greater than one.

BINARY BCH CODE SUMMARY

Let:

w	=	Number of bits required to represent each element of $GF(2^w)$, the field wherein computations are performed.	
n	=	Selected record length in bits, including check bits	
t	=	Number of bits the code is capable of correcting	
d	=	Minimum Hamming distance	
m	=	Degree of code generator polynomial	
	=	Number of check bits	
$m_i(x)$	=	Minimum polynomial in GF(2) of α^i in $GF(2^w)$	
$g(x)$	=	Code generator polynomial	
	=	$LCM[m_1(x), m_3(x), \ldots, m_{2*t-1}(x)]$	
k	=	Number of factors of g(x) [typically t]	
$D(x)$	=	Data polynomial	
$W(x)$	=	Write redundancy polynomial	$= [x^m \cdot D(x)]$ MOD $g(x)$
$C(x)$	=	Transmitted codeword polynomial	$= x^m \cdot D(x) + W(x)$
$E(x)$	=	Error polynomial	$= x^{L1} + x^{L2} + \cdots$
$C'(x)$	=	Received codeword polynomial	$= C(x) + E(x)$

Then the following relationships hold:

$$n \leq 2^w - 1$$

$$d = 2*t+1$$

$$m \leq w*t$$

THE GENERATOR POLYNOMIAL

The generator polynomial for a t-error-correcting binary BCH code is:

$$g(x) = LCM[m_1(x), m_3(x), \ldots, m_{2*t-1}(x)]$$

where $m_i(x)$ is the minimum polynomial in GF(2) of α^i in $GF(2^w)$; see the glossary for the definition of a minimum polynomial. The LCM function above accounts for the fact that if the minimum polynomials of two or more powers of α are identical, only one copy of the polynomial is multiplied into g(x). In most cases no duplicate polynomials exist, and g(x) is the product of them all:

$$g(x) = m_1(x) \cdot m_3(x) \cdot \ldots \cdot m_{2*t-1}(x)$$

ENCODING

Encoding for a binary BCH code can be performed with a bit-serial shift register implementing the generator polynomial of the form shown below. All paths and storage elements are bit-wide. Multipliers comprise either a connection or no connection.

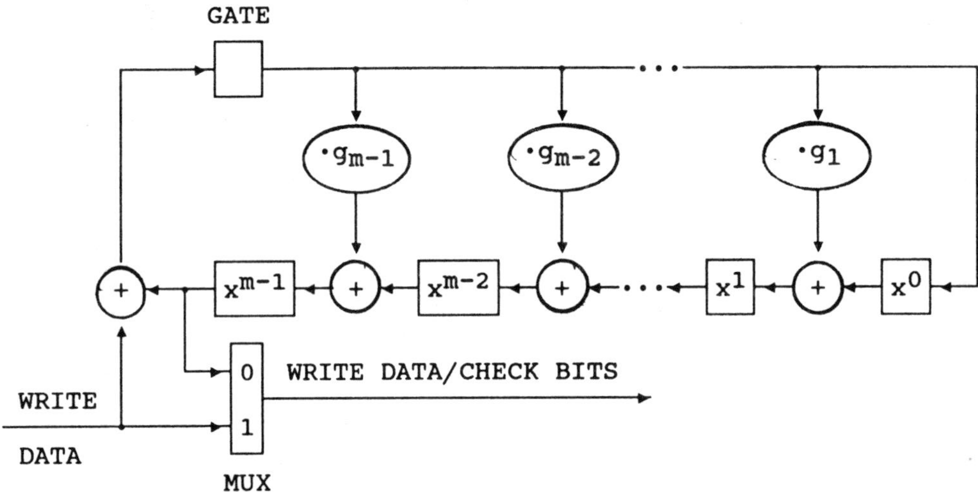

For applications such as error correction on semiconductor memory, an encoder implementing combinatorial logic is preferable to one implementing sequential logic. Such an encoder includes a parity tree for each bit of redundancy. The parity tree for a coefficient W_i of the x^i term of the write redundancy polynomial W(x) includes each data bit D_j for which the coefficient of the x^i term of

$$[x^m \cdot x^j] \text{ MOD } g(x)$$

is one.

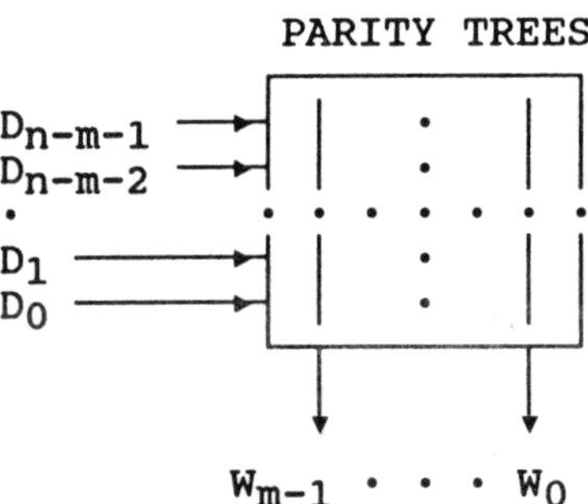

An example of a combinatorial-logic encoder is given in the BINARY BCH CODE EXAMPLE below.

DECODING

Decoding generally requires 5 steps:

1. Generate the syndromes.
2. Calculate the coefficients of an error locator polynomial.
3. Find the roots of the error locator polynomial to determine error location vectors.
4. Calculate logs of error location vectors to obtain error locations.
5. Invert bits in error.

SYNDROME GENERATION

The syndromes contain information about the locations of errors:

$$S_1 = \alpha^{L_1} + \alpha^{L_2} + \cdots$$

$$S_3 = \alpha^{3*L_1} + \alpha^{3*L_2} + \cdots$$

$$S_5 = \alpha^{5*L_1} + \alpha^{5*L_2} + \cdots$$

$$\cdot \quad \cdot \quad \cdot \quad \cdot \quad \cdot \quad \cdot$$

$$S_k = \alpha^{k*L_1} + \alpha^{k*L_2} + \cdots$$

It is possible to compute the syndromes directly from the received codeword polynomial $C'(x)$ with the following equation.

$$S_i = C'(\alpha^i)$$

The above equation can be implemented with either sequential or combinatorial logic.

The syndromes can also be computed by computing the residues of the received codeword when divided by each factor or the generator polynomial. Let:

$$r_i(x) = C'(x) \text{ MOD } m_i(x)$$

then the resulting residues may be used to compute the syndromes:

$$S_i = r_i(\alpha^i)$$

The above equations can be implemented sequentially, combinatorially, or with a mixture of sequential and combinatorial logic.

An example of each of the above methods is shown in the BINARY BCH CODE EXAMPLE below.

COMPUTING COEFFICIENTS OF ERROR LOCATOR POLYNOMIALS

The error locator polynomial has the following form.

$$\prod_{i=1}^{e}(x + \alpha^{L_i}) = x^e + \sigma_1 x^{e-1} + \cdots + \sigma_{e-1} + \sigma_e = 0$$

The coefficients of the error locator polynomial are related to the syndromes by the following system of linear equations, called Newton's identities.

$$\begin{aligned}
&\sigma_1 &&= S_1 \\
&\sigma_1 \cdot S_2 + \sigma_2 \cdot S_1 + \sigma_3 &&= S_3 \\
&\sigma_1 \cdot S_4 + \sigma_2 \cdot S_3 + \sigma_3 \cdot S_2 + \sigma_4 \cdot S_1 + \sigma_5 &&= S_5 \\
&\cdots \qquad \cdots &&\cdots \\
&\sigma_1 \cdot S_{2t-2} + \quad \cdots \quad + \sigma_{2t-2} \cdot S_1 &&= S_{2t-1}
\end{aligned}$$

For error locator polynomials of low degree, the coefficients of the error locator polynomial are computed by solving Newton's identities using determinants. For error locator polynomials of high degree, the coefficients are computed by solving Newton's identities with Berlekamp's iterative algorithm.

FINDING THE ROOTS OF ERROR LOCATOR POLYNOMIALS

The roots of error locator polynomials are error location vectors. The logs of error location vectors are error locations.

The error locator polynomial of degree one is:

$$x + \sigma_1 = 0$$

The single root of this equation is simply:

$$X = \sigma_1$$

The error locator polynomial of degree two is:

$$x^2 + \sigma_1 \cdot x + \sigma_2 = 0$$

This equation can be solved using a precomputed look-up table by first applying a substitution to transform it into following form (see Sections 2.6 and 3.4 for more details):

$$y^2 + y + c = 0$$

There are similar approaches to solving other low degree error locator polynomials. The Chien search is used to solve error locator polynomials of high degree.

BINARY BCH CODE EXAMPLE

Assume a two-error-correcting code over GF(2^4). The generator polynomial is:

$$g(x) = m_1(x) \cdot m_3(x) = (x^4 + x + 1) \cdot (x^4 + x^3 + x^2 + x + 1)$$
$$= x^8 + x^7 + x^6 + x^4 + 1$$

The codeword length is limited to 2^4-1=15 bits, so the code may be used protect a seven-bit data polynomial.

SEQUENTIAL LOGIC ENCODER

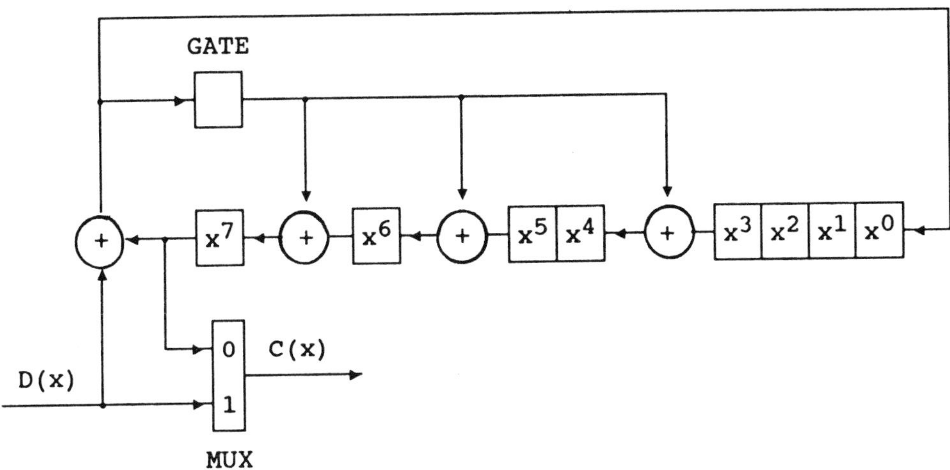

COMBINATORIAL LOGIC ENCODER

The write-redundancy polynomial coefficients are given by the following parity trees. Each coefficient W_j is formed as the XOR sum of those coefficients D_i whose row contains a '1' in W_j's column.

		x^7	x^6	x^5	x^4	x^3	x^2	x^1	x^0
D_6	$[x^8 \cdot x^6]$ MOD g(x)	1	1	1	0	1	0	0	0
D_5	$[x^8 \cdot x^5]$ MOD g(x)	0	1	1	1	0	1	0	0
D_4	$[x^8 \cdot x^4]$ MOD g(x)	0	0	1	1	1	0	1	0
D_3	$[x^8 \cdot x^3]$ MOD g(x)	0	0	0	1	1	1	0	1
D_2	$[x^8 \cdot x^2]$ MOD g(x)	1	1	1	0	0	1	1	0
D_1	$[x^8 \cdot x^1]$ MOD g(x)	0	1	1	1	0	0	1	1
D_0	$[x^8 \cdot x^0]$ MOD g(x)	1	1	0	1	0	0	0	1
		W_7	W_6	W_5	W_4	W_3	W_2	W_1	W_0

SYNDROME GENERATION

SEQUENTIAL CIRCUIT FOR S1

$$S_1 = r_1(\alpha) = [C'(x) \text{ MOD } m_1(x)]\Big|_{\alpha}$$

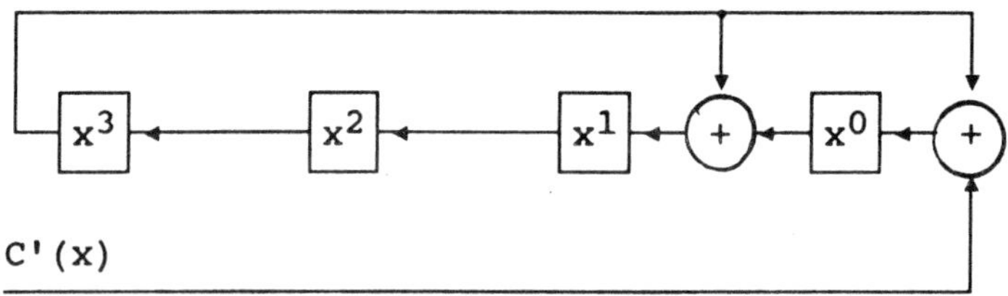

SEQUENTIAL CIRCUIT FOR S3

$$S_3 = r_3(\alpha^3) = [C'(x) \text{ MOD } m_3(x)]\Big|_{\alpha^3}$$

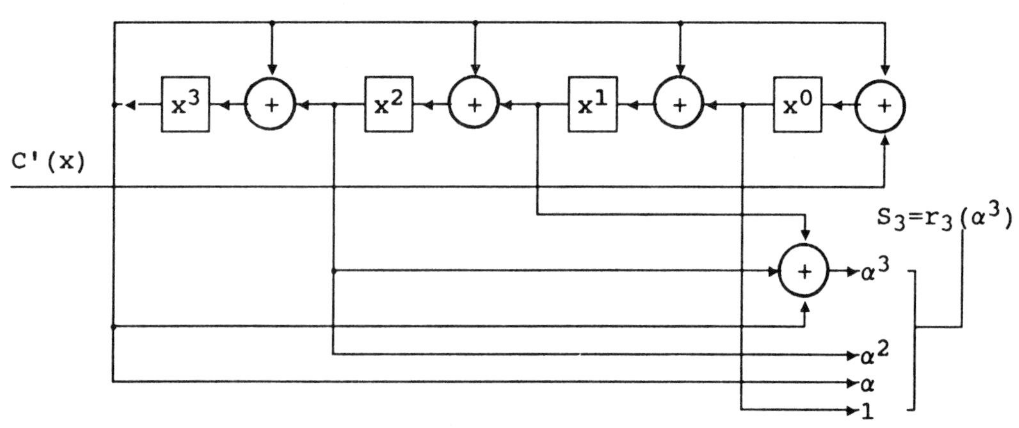

ALTERNATIVE SEQUENTIAL CIRCUIT FOR S3

$$S_3 = C'(\alpha^3)$$

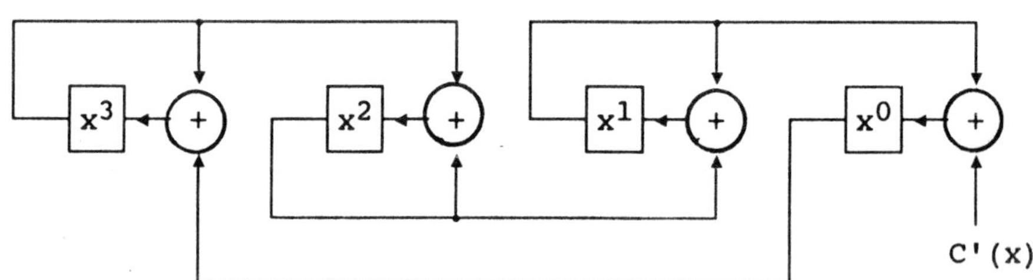

COMBINATORIAL LOGIC SYNDROME CIRCUITS

The parity tree for a coefficient Si_j of the x^j term of syndrome Si includes each received codeword bit C_k' for which the coefficient of the x^j term of

$$[x^{k*i}] \text{ MOD } m_i(x)$$

is one.

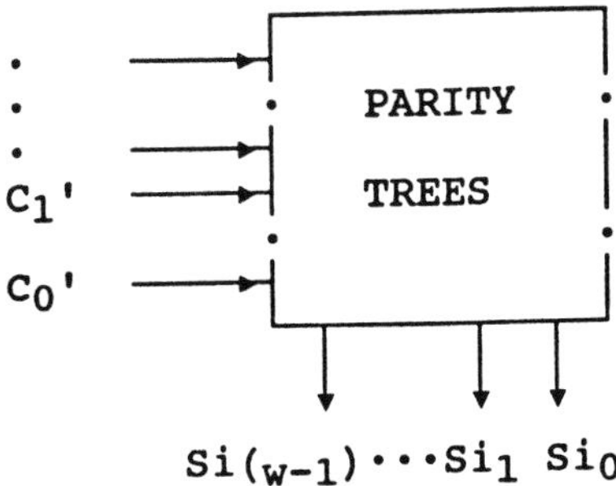

COMPUTING THE COEFFICIENTS OF THE TWO-ERROR LOCATOR POLYNOMIAL

For the two-error case the system of linear equations below must be solved. These equations follow from Newton's identities.

$$(1)\cdot\sigma_1 + (0)\cdot\sigma_2 = S_1$$

$$(S_2)\cdot\sigma_1 + (S_1)\cdot\sigma_2 = S_3$$

$$\sigma_1 = \frac{\begin{vmatrix} S_1 & 0 \\ S_3 & S_1 \end{vmatrix}}{\begin{vmatrix} 1 & 0 \\ S_2 & S_1 \end{vmatrix}} = \frac{(S_1)^2}{S_1} = S_1$$

$$\sigma_2 = \frac{\begin{vmatrix} 1 & S_1 \\ S_2 & S_3 \end{vmatrix}}{\begin{vmatrix} 1 & 0 \\ S_2 & S_1 \end{vmatrix}} = \frac{S_3 + S_1\cdot S_2}{S_1} = \frac{S_3 + S_1\cdot(S_1)^2}{S_1} = \frac{S_3 + (S_1)^3}{S_1}$$

FINDING ROOTS OF THE TWO-ERROR LOCATOR POLYNOMIAL

The algorithm below defines a fast method for finding roots of the error locator polynomial in the two-error case. This algorithm can be performed by a finite field processor. For double bit memory correction is performed by combinatorial logic.

The two-error locator polynomial is

$$x^2 + \sigma_1 \cdot x + \sigma_2 = 0$$

where

$$\sigma_1 = S_1 \quad \text{and} \quad \sigma_2 = \frac{(S_1)^3 + S_3}{S_1}$$

Substitute

$$x = \sigma_1 \cdot y = S_1 \cdot y$$

to obtain

$$y^2 + y + \frac{\sigma_2}{(\sigma_1)^2} = y^2 + y + C = 0$$

where

$$C = \frac{\sigma_2}{(\sigma_1)^2} = \frac{(S_1)^3 + S_3}{(S_1)^3}$$

Fetch Y_1 from TBLA (see Section 2.6) using C as the index. Then form

$$Y_2 = Y_1 + \alpha^0$$

Apply reverse substitution of

$$y = x/\sigma_1$$

to obtain

$$X_1 = \alpha^{L1} = \sigma_1 \cdot Y_1 = S_1 \cdot Y_1 \quad \text{and} \quad X_2 = \alpha^{L2} = \sigma_1 \cdot Y_2 = S_1 \cdot Y_2$$

Finally, calculate the error locations

$$L_1 = \mathrm{LOG}_\alpha(X_1)$$

$$L_2 = \mathrm{LOG}_\alpha(X_2)$$

For a binary BCH code, the error values are by definition equal to '1'.

BCH CODE DOUBLE-BIT MEMORY CORRECTION - EXAMPLE #1

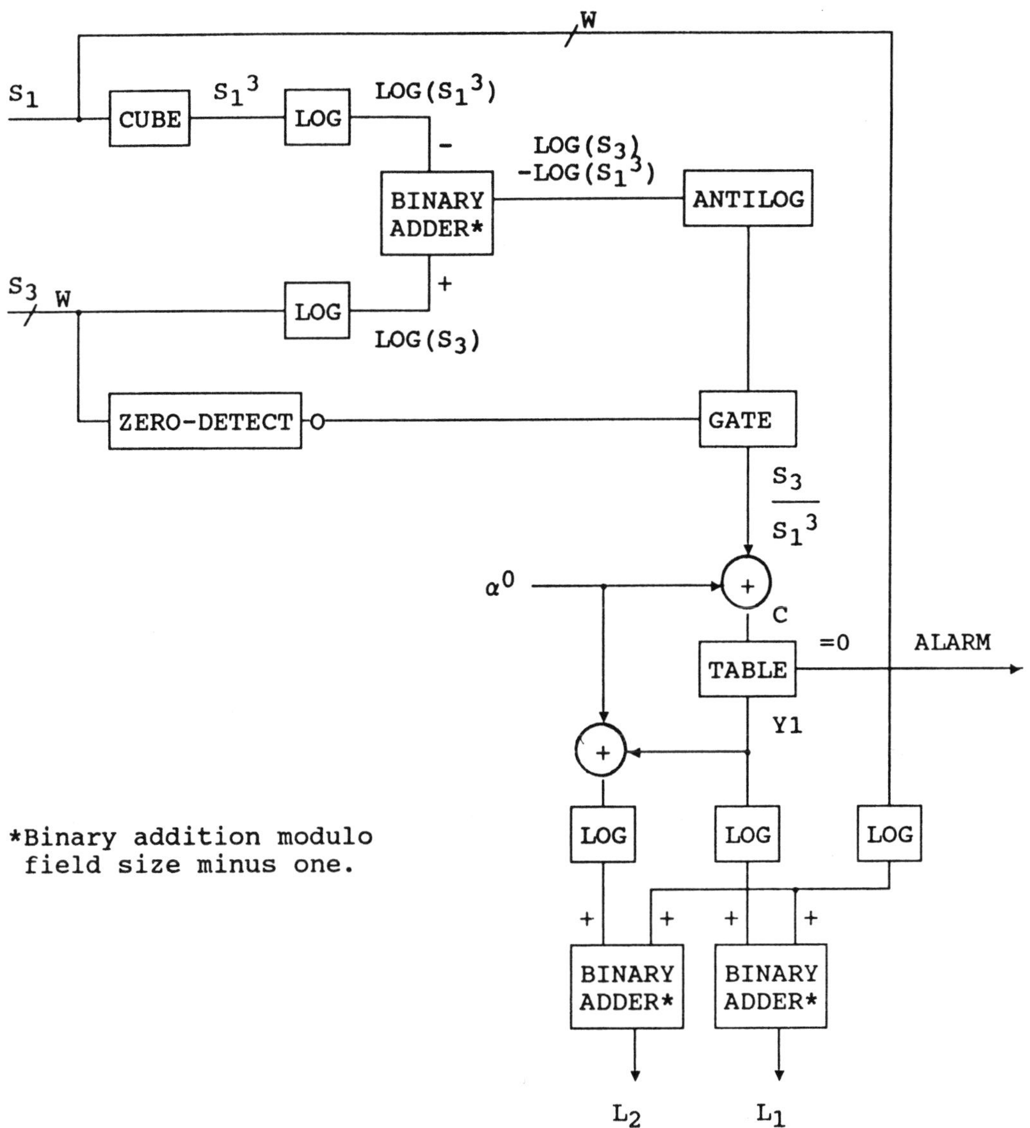

This example is shown in a form that is easier to understand. Example #2 uses the same approach but combines some of the functions.

BCH CODE DOUBLE-BIT MEMORY CORRECTION - EXAMPLE #2

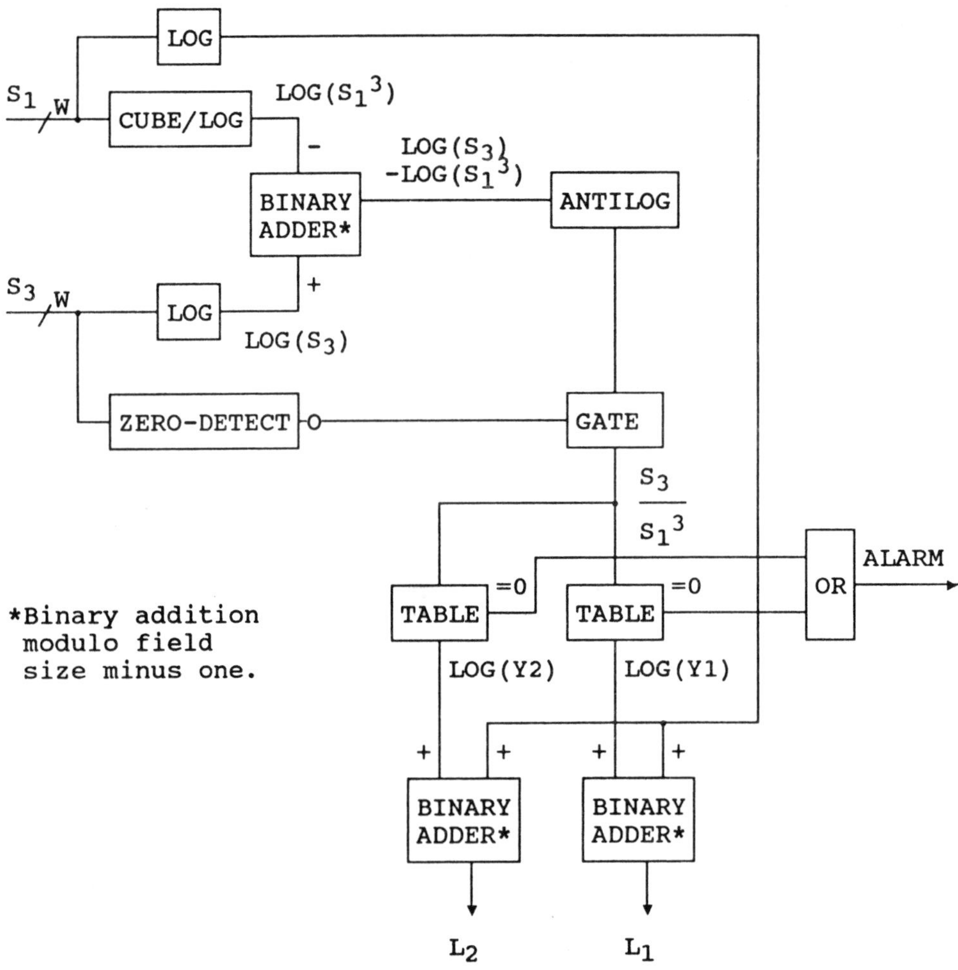

This example uses the same approach as Example #1 but several functions have been combined.

BCH CODE DOUBLE-BIT MEMORY CORRECTION - EXAMPLE #3

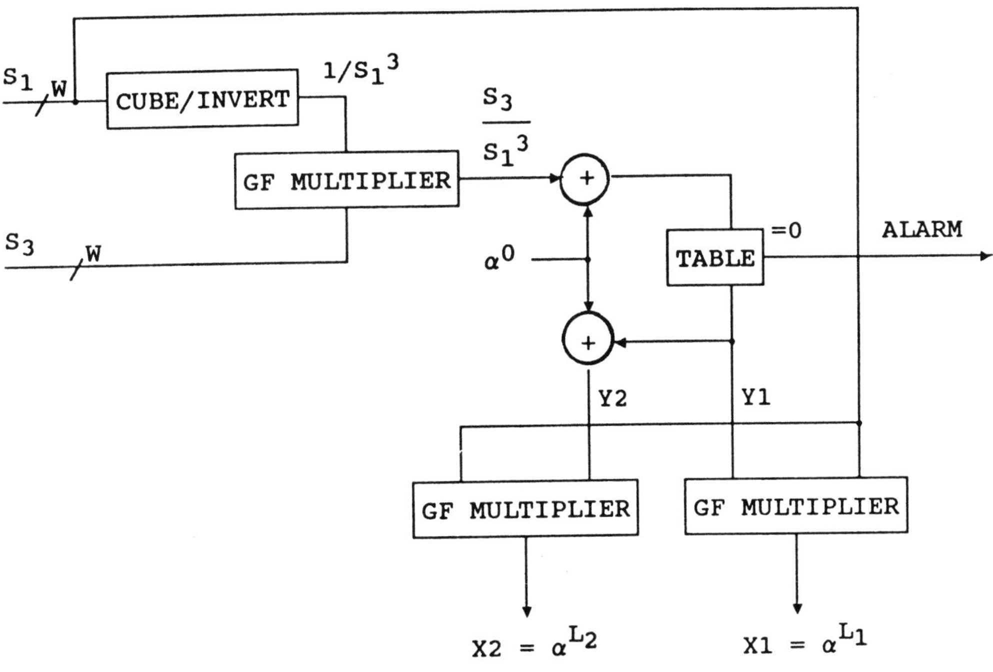

BCH CODE DOUBLE-BIT MEMORY CORRECTION - EXAMPLE #4

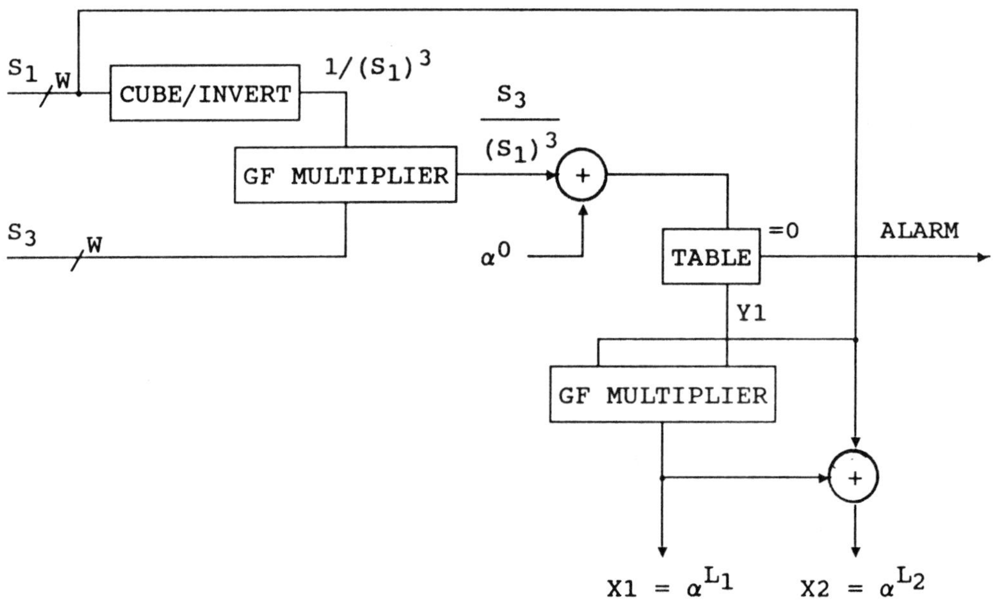

BCH CODE DOUBLE-BIT MEMORY CORRECTION - EXAMPLE #5

The mathematical basis for this example is developed by operating on the error locator polynomial:

$$x^2 + \sigma_1 \cdot x + \sigma_1 = 0$$

First substitute for σ_1 and σy_2 using expressions developed above:

$$x^2 + S_1 \cdot x + \frac{(S_1)^3 + S_3}{S_1} = 0$$

Next multiply by S_1:

$$S_1 \cdot x^2 + (S_1)^2 \cdot x + (S_1)^3 + S_3 = 0$$

Add zero in the form of $(x^3 + x^3)$:

$$(x^3 + \underline{x^3}) + \underline{S_1 \cdot x^2} + \underline{(S_1)^2 \cdot x} + \underline{(S_1)^3} + S_3 = 0$$

Finally, rearrange and combine the underlined terms to obtain a useful relation:

$$\underline{(S_1 + x)^3} + (S_3 + x^3) = 0$$

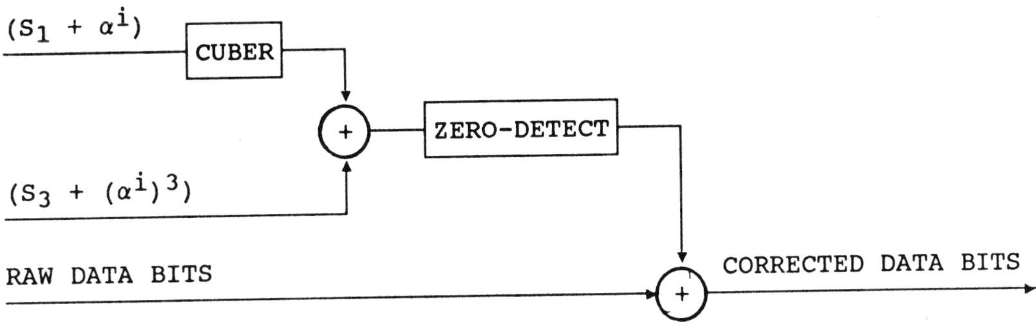

One such circuit is required for each bit of the memory word.

3.4 REED-SOLOMON CODES

Reed-Solomon codes are random single- or multiple-symbol error-correcting codes operating on symbols which are elements of a finite field. The coefficients of the data polynomial and the check symbols are elements of the field, and all encoding, decoding, and correction computations are performed in the field. Reed-Solomon codes are inherently symbol oriented and the circuits implementing them are typically clocked once per data symbol, although bit-serial techniques are also employed.

We shall use the Galois field with eight elements (*i.e.*, GF(8) or GF(2^3)), introduced in Section 2.5 in illustrating the properties and implementation of Reed-Solomon codes.

REED-SOLOMON CODE SUMMARY

Let

w	=	Number of bits per symbol; each symbol ϵ GF(2^w)	
m	=	Degree of generator polynomial = number of check symbols	
n	=	Selected record length in symbols, including check symbols	
d	=	Minimum Hamming distance of the code	
t	=	Number of symbol errors correctable by the code	
e_c	=	Selected number of symbol errors to be corrected	
e_d	=	Number of symbol errors which the code is capable of detecting beyond the number selected for correction	
b	=	Burst length in bits	
c	=	Number of bursts correctable by the code	
A(x)	=	Any polynomial in the field	
G(x)	=	The code generator polynomial	
$g_i(x)$	=	Any of the m factors of G(x)	$= (x + \alpha^i)$ when m0=0
D(x)	=	Data polynomial	
W(x)	=	Write redundancy polynomial	$= [x^m \cdot D(x)]$ MOD G(x)
C(x)	=	Transmitted codeword polynomial	$= x^m \cdot D(x) + W(x)$
E(x)	=	Error polynomial	$= e_1 \cdot x^{L1} + e_2 \cdot x^{L2} + \cdots$
C'(x)	=	Received codeword polynomial	= C(x) + E(x)
R(x)	=	Remainder polynomial	= C'(x) MOD G(x)
S_i	=	ith syndrome	= C'(x) MOD $g_i(x)$

Then the following relationships hold:

$$n \le 2^w - 1$$

$$d = m + 1$$

$$e_c \le t = \text{INT}[(d-1)/2] = \text{INT}[m/2]$$

$$e_d = d_{min} - 2*e_c - 1 = m - 2*e_c$$

$$b \le (e_c - 1)*w + 1$$

$$c = e_c/(1 + \text{INT}[(b+w-2)/w])$$

^R
^R
^R
^R
^R
REED-SOLOMON CODE SUMMARY (CONT.)

$A(x) \text{ MOD } g_i(x) = [A(x) \text{ MOD } G(x)] \text{ MOD } g_i(x)$ (1)

$A(x) \text{ MOD } g_i(x) = A(x)\Big|_{\alpha^i} = A(\alpha^i)$ (2)

$C(x) \text{ MOD } G(x) = 0$ (3)

$C(x) \text{ MOD } g_i(x) = 0$ (4)

$R(x) = C'(x) \text{ MOD } G(x)$

$= [C(x) + E(x)] \text{ MOD } G(x)$ {by definition of C'}

$= E(x) \text{ MOD } G(x)$ {by equation (3)} (5)

$S_i = C'(x) \text{ MOD } g_i(x)$

$= [C(x) + E(x)] \text{ MOD } g_i(x)$ {by definition of C'}

$= E(x) \text{ MOD } g_i(x)$ {by equation (4)}

$= E_1 \cdot \alpha^{i*L_1} + E_2 \cdot \alpha^{i*L_2} + \cdots$ {by equation (2)}

$= [E(x) \text{ MOD } G(x)] \text{ MOD } g_i(x)$ {by equation (1)}

$= R(x) \text{ MOD } g_i(x)$ {by equation (5)}

CONSTRUCTING THE CODE GENERATOR POLYNOMIAL

The generator polynomial of a Reed-Solomon code is given by:

$$G(x) = \prod_{i=0}^{m-1} (x + \alpha^{m_0+i})$$

where m is the number of check symbols and m_0 is an offset, often zero or one. In the interest of simplicity, we take m_0 equal to zero for the remainder of the discussion. Note that many expressions derived below must be modified for cases where m0 is not zero. Let m=4; the code will be capable of correcting:

$$t = \mathrm{INT}(m/2) = 2$$

symbol errors in a codeword. The generator polynomial is:

$$\begin{aligned} G(x) &= \prod_{i=0}^{3} (x + \alpha^{i}) \\ &= (x + \alpha^0)(x + \alpha^1)(x + \alpha^2)(x + \alpha^3) \\ &= x^4 + (\alpha^0 + \alpha^1 + \alpha^2 + \alpha^3)\cdot x^3 \\ &\quad + (\alpha^0\alpha^1 + \alpha^0\alpha^2 + \alpha^0\alpha^3 + \alpha^1\alpha^2 + \alpha^1\alpha^3 + \alpha^2\alpha^3)\cdot x^2 \\ &\quad + (\alpha^0\alpha^1\alpha^2 + \alpha^0\alpha^1\alpha^3 + \alpha^0\alpha^2\alpha^3 + \alpha^1\alpha^2\alpha^3)\cdot x + (\alpha^0\alpha^1\alpha^2\alpha^3) \\ &= x^4 + \alpha^2\cdot x^3 + \alpha^5\cdot x^2 + \alpha^5\cdot x + \alpha^6 \end{aligned}$$

VERIFYING THE CODE GENERATOR POLYNOMIAL

The code generator polynomial evaluates to zero at each of its roots. This fact can be used to prove the computations used in generating it:

$$G(x)\Big|_{\alpha^0} = \alpha^0 + \alpha^2 + \alpha^5 + \alpha^5 + \alpha^6 = 0$$

$$G(x)\Big|_{\alpha^1} = \alpha^0\cdot(\alpha^1)^4 + \alpha^2\cdot(\alpha^1)^3 + \alpha^5\cdot(\alpha^1)^2 + \alpha^5\cdot\alpha^1 + \alpha^6 = 0$$

$$G(x)\Big|_{\alpha^2} = \alpha^0\cdot(\alpha^2)^4 + \alpha^2\cdot(\alpha^2)^3 + \alpha^5\cdot(\alpha^2)^2 + \alpha^5\cdot\alpha^2 + \alpha^6 = 0$$

$$G(x)\Big|_{\alpha^3} = \alpha^0\cdot(\alpha^3)^4 + \alpha^2\cdot(\alpha^3)^3 + \alpha^5\cdot(\alpha^3)^2 + \alpha^5\cdot\alpha^3 + \alpha^6 = 0$$

FINITE FIELD CONSTANT MULTIPLIERS

To design a constant multiplier to implement

$$y=\alpha^{n}\cdot x$$

in $GF(2^{w})$, fill in the diagram below with the binary representations of

$$\alpha^{n},\alpha^{n+1},\ \ldots\ ,\alpha^{n+w-1}.$$

x			⋮			
x_{w-1}			⋮			α^{n+w-1}
x_{w-2}			⋮			α^{n+w-2}
⋅	⋅	⋅	⋅	⋅	⋅	⋅
x_{1}			⋮			α^{n+1}
x_{0}			⋮			α^{n}
	y_{w-1}	y_{w-2}		y_{1}	y_{0}	

$$y = \alpha^{n}\cdot x$$

Then construct parity trees down columns. The parity tree for a given y bit includes each x bit with a '1' at the intersection of the corresponding column and row.

Example: Using the field of Figure 2.5.1, construct a constant multiplier to compute:

$$y = \alpha^{3}\cdot x$$

x				
x_{2}	1	1	1	α^{5}
x_{1}	1	1	0	α^{4}
x_{0}	0	1	1	α^{3}
	y_{2}	y_{1}	y_{0}	

$$y = \alpha^{3}\cdot x$$

$$\begin{aligned} y_{2} &= \phantom{x_{0} +{}} x_{1} + x_{2} \\ y_{1} &= x_{0} + x_{1} + x_{2} \\ y_{0} &= x_{0} \phantom{{}+x_{1}} + x_{2} \end{aligned}$$

ENCODING OF REED-SOLOMON CODES

Encoding is typically, but not always, performed using an internal-XOR shift register with symbol-wide data paths, implementing the form of generator polynomial shown above. Other encoding alternatives will be discussed later in this section.

The following circuit computes C(x) for our example field and code in symbol-serial fashion:

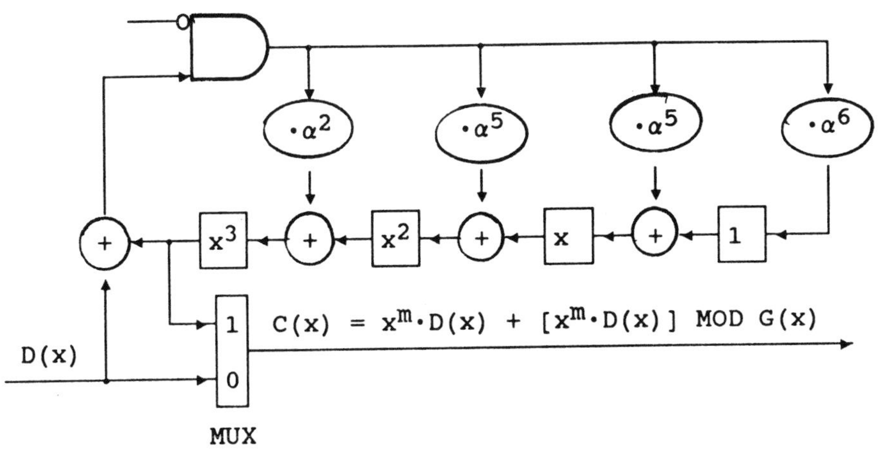

The circuit above multiplies the data polynomial D(x) by x^m and divides by G(x). All paths are symbol-wide (three bits for this example). The AND gate and the MUX are fed by a signal which is low during data time and high during redundancy time.

The circuit below performs the same function.

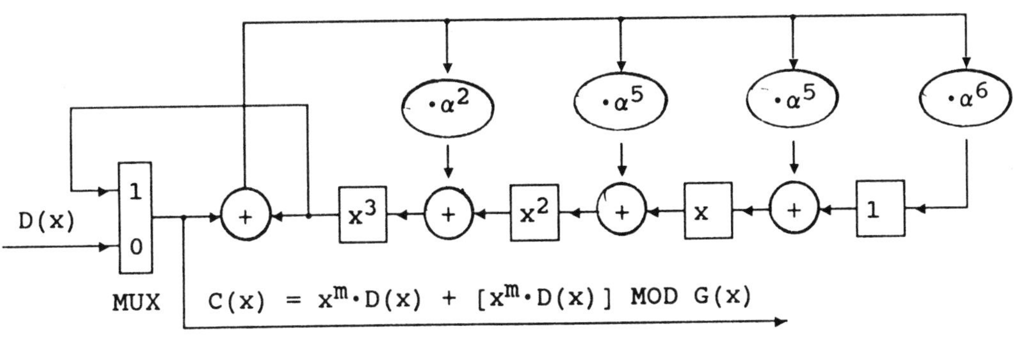

DECODING OF REED-SOLOMON CODES

Decoding generally requires five steps:

1. Compute syndromes.
2. Calculate the coefficients of the error locator polynomial.
3. Find the roots of the error locator polynomial. The logs of the roots are the error locations.
4. Calculate the error values.

The following circuit computes the syndromes for our example field and code in symbol-serial fashion:

$$S_i = C'(x) \text{ MOD } g_i(x) = C'(x) \text{ MOD } (x + \alpha^i)$$

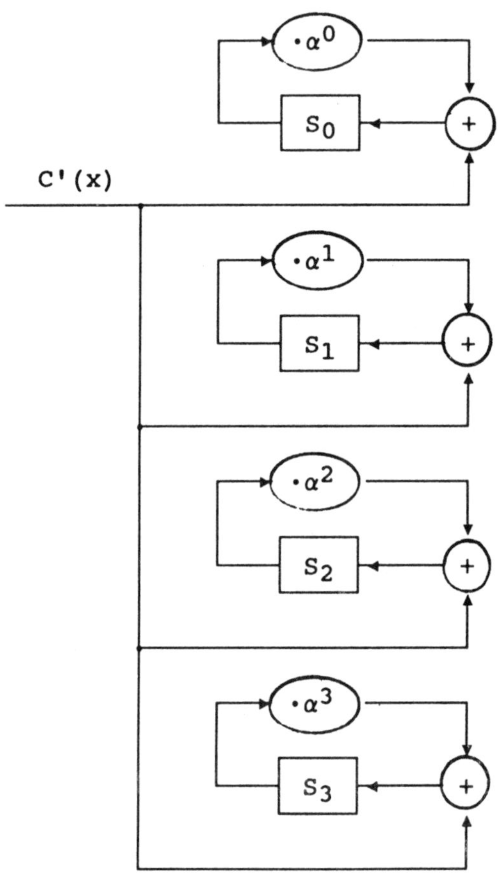

This circuit computes the syndromes by dividing the received codeword C'(x) by the factors of G(x). All paths are symbol-wide (three bits for this example). After all data and redundancy symbols have been clocked, the registers contain the syndromes S_i.

COMPUTING COEFFICIENTS OF ERROR LOCATOR POLYNOMIALS

Recall the syndrome equations derived above:

$$S_i = E_1 \cdot \alpha^{i*L_1} + E_2 \cdot \alpha^{i*L_2} + \cdots$$

These form a system of nonlinear equations with error values and error location vectors as unknowns. More easily solved is the error locator polynomial, which contains only error location information. Error locator polynomials have the following form:

$$\prod_{i=1}^{e} (x + \alpha^{L_i}) = x^e + \sigma_1 \cdot x^{e-1} + \cdots + \sigma_{e-1} \cdot x + \sigma_e = 0$$

where e is the number of errors. The coefficients of the error locator polynomial are related to the syndromes by the following system of linear equations, called Newton's generalized identities:

$$\begin{array}{ccccccccc} S_0 \cdot \sigma_e & + & S_1 \cdot \sigma_{e-1} & + & \cdots & + & S_{e-1} \cdot \sigma_1 & = & S_e \\ S_1 \cdot \sigma_e & + & S_2 \cdot \sigma_{e-1} & + & \cdots & + & S_e \cdot \sigma_1 & = & S_{e+1} \\ \cdot & & \cdot & & \cdots & & \cdot & & \cdot \\ S_{m-e-1} \cdot \sigma_e & + & S_{m-e} \cdot \sigma_{e-1} & + & \cdots & + & S_{m-2} \cdot \sigma_1 & = & S_{m-1} \end{array}$$

where m is the number of syndromes.

When computation of the error location polynomial is begun, the number of errors, and thus the degree of the error locator polynomial, is unknown. One method of computing coefficients of the error locator polynomial first assumes a single error. If this assumption is found to be incorrect, the number of assumed errors is increased to two, and so on. This method is fastest for the least number of errors. This is desirable because in most cases few errors are more likely than many.

For error locator polynomials of low degree, the coefficients σ_i are computed by directly solving the above system of equations using determinants. Examples are worked out below.

For error locator polynomials of high degree, the coefficients σ_i are computed by solving the system of equations above using Berlekamp's iterative algorithm. One version of the iterative algorithm is outlined, and an example is worked out, below.

ITERATIVE ALGORITHM

(0) Initialize a table as shown below; the parenthesized superscript on $\sigma(x)$ is a counter and not an exponent.

n	$\sigma^{(n)}(x)$	d_n	L_n	$n-L_n$
-1	1	1	0	-1
0	1	S_0	0	0
1	.	.	.	.
.	.	.	.	.

The table will be completed in the steps below.
Initialize n to zero.

(1) If $d_n=0$ then set $L_{n+1} = L_n$, set $\sigma^{(n+1)}(x) = \sigma^{(n)}(x)$, and go to Step (3).

(2) Find a row k where $k<n$ and $d_k \neq 0$, such that $k-L_k$ (the last column of row k in the table) has the largest value. Compute:

$$L_{n+1} = \text{MAX}[L_n, L_k+(n-k)]$$

$$\sigma^{(n+1)}(x) = x^{L_{n+1}-L_n} \cdot \sigma^{(n)}(x) - (d_n/d_k) \cdot \sigma^{(k)}(x)$$

(3) If $n+1=m$ (or $n+1=t+L_{n+1}$)* then exit; $\sigma^{(n+1)}(x)$ is the desired error locator polynomial.

* $t+L_n$ iterations are required to satisfy the basic guarantees of the code; terminating on this second criterion is sufficient for generating the proper error locator polynomial for correctable error cases, but may sacrifice some protection against miscorrection of uncorrectable error cases.

(4) Compute:

$$d_{n+1} = \sum_{i=0}^{L_{n+1}} \sigma_i^{(n+1)} \cdot S_{n+1-i}$$

then set $n = n+1$ and go to Step (1).

FINDING ERROR LOCATIONS USING THE LOCATOR POLYNOMIAL

Methods for solving the error locator polynomial for cases of one and two errors are developed below. A method for solving the three-error case is given in Section 5.4. Methods for directly solving the four-error case are also known, but we shall not discuss them.

For cases of more than four errors, the Chien search is typically used to find the roots of the error locator polynomial. The Chien search could be used to find the roots of the error locator polynomial for cases of fewer errors, but it is slower and in most cases requires more logic.

COMPUTING ERROR VALUES

Once error location vectors are known, the syndrome equations become a system of linear equations with the error values as unknowns. Determinants can be used to solve the syndrome equations when the number of errors is low. The following method can be used to solve the syndrome equations when the number of errors is high:

$$E_i = \frac{z(1/\alpha^{L_i})\cdot\alpha^{L_i}}{\prod_{\substack{j=1\\ j\neq i}}^{e}\left[1 + (\alpha^{L_j}/\alpha^{L_i})\right]}$$

where,

e = number of errors

$i = 1,2,\cdots,e$

L_i = error locations

α^{L_i} = error location vectors

σ_i = coefficients of the error locator polynomial.

$$\begin{aligned} z(1/\alpha^{L_i}) = {} & 1 && \\ & + (S_0 + \sigma_1)\cdot(1/\alpha^{L_i}) && \text{(e=1: through this term)} \\ & + (S_1 + \sigma_1\cdot S_0 + \sigma_2)\cdot(1/\alpha^{L_i})^2 && \text{(e=2: through this term)} \\ & + (S_2 + \sigma_1\cdot S1 + \sigma_2\cdot S_0 + \sigma_3)\cdot(1/\alpha^{L_i})^3 && \text{(e=3: through this term)} \\ & + \cdots \end{aligned}$$

THE REED-SOLOMON SINGLE ERROR CASE

ERROR LOCATOR POLYNOMIAL

$$\sigma(x) = \prod_{i=1}^{1} (x + \alpha^{L_i}) = x + \sigma \qquad (1)$$

SYNDROME EQUATIONS

$S_0 = E$ (2)

$S_1 = E \cdot \alpha^{L}$ (3)

$S_2 = E \cdot \alpha^{2L}$ (4)

$S_3 = E \cdot \alpha^{3L}$ (5)

NEWTONS IDENTITIES

$S_0 \cdot \sigma = S_1$ (6)

$S_1 \cdot \sigma = S_2$ (7)

$S_2 \cdot \sigma = S_3$ (8)

Solving equation (6) gives a simple expression for σ:

$$\sigma = S_1/S_0$$

σ may then be substituted into equations (7) and (8) for verification. The location L is given as the log of σ from equation (1) and the value E is given as S_0 from equation (2).

THE REED-SOLOMON TWO ERROR CASE

ERROR LOCATOR POLYNOMIAL

$$\sigma(x) = \prod_{i=1}^{2} (x + \alpha^{L_i}) = x^2 + \sigma_1 \cdot x + \sigma_2 \qquad (1)$$

SYNDROME EQUATIONS

$S_0 = E_1 + E_2$ (2)

$S_1 = E_1 \cdot \alpha^{L_1} + E_2 \cdot \alpha^{L_2}$ (3)

$S_2 = E_1 \cdot \alpha^{2L_1} + E_2 \cdot \alpha^{2L_2}$ (4)

$S_3 = E_1 \cdot \alpha^{3L_1} + E_2 \cdot \alpha^{3L_2}$ (5)

NEWTONS IDENTITIES

$S_0 \cdot \sigma_2 + S_1 \cdot \sigma_1 = S_2$ (6)

$S_1 \cdot \sigma_2 + S_2 \cdot \sigma_1 = S_3$ (7)

The syndrome equations are a set of simultaneous non-linear equations which are difficult to solve. Newton's identities are a set of simultaneous linear equations which can be solved by determinants for σ_1 and σ_2 in terms of the syndromes. Once we have computed σ_1 and σ_2, we must solve (1) for L_1 and L_2. From (1) we have:

$$\sigma_1 = \alpha^{L_1} + \alpha^{L_2}$$

$$\sigma_2 = \alpha^{(L_1+L_2)}$$

FINDING ROOTS OF THE TWO-ERROR LOCATOR POLYNOMIAL

One method for finding the roots of the two-error locator polynomial:

$$x^2 + \sigma_1 \cdot x + \sigma_2 = 0 \qquad (1)$$

is to employ the substitution:

$$x = \sigma_1 \cdot y, \quad C = \sigma_2/(\sigma_1)^2 \qquad (8)$$

to transform equation (1) into the form:

$$y^2 + y + C = 0 \qquad (9)$$

Equation (9) can be solved by using C as an index into a table of roots:

C	Y_1	Y_2
0	0	α^0
α^0	--	--
α^1	α^2	α^6
α^2	α^4	α^5
α^3	--	--
α^4	α^1	α^3
α^5	--	--
α^6	--	--

Once roots Y_1 and Y_2 of (9) have been found, roots X_1 and X_2 of (1) can be computed by reverse substitution of equation (8). Then L_1 and L_2 may be computed as the logs of X_1 and X_2.

DETERMINING ERROR VALUES FOR THE TWO-ERROR CASE

When L_1 and L_2 are known, the syndrome equations become a set of linear simultaneous equations in E_1 and E_2 and we can solve:

$$S_0 = E_1 + E_2 \qquad (2)$$

$$S_1 = \alpha^{L1} \cdot E_1 + \alpha^{L2} \cdot E_2 \qquad (3)$$

by determinants to obtain E_1 and E_2.

ONE- AND TWO-ERROR CORRECTION ALGORITHM

A) Assume a single error exists.

1) If S0=0 or S1=0, go to the two error case.

2) Compute $\sigma = S_1/S_0$.

3) Verify $S_2 = S_1 \cdot \sigma$ and $S_3 = S_2 \cdot \sigma$. If either test fails, go to the two error case.

4) Compute $L = LOG(\sigma)$ and $E = S_0$.

5) If the symbol at L is data, XOR value E at location L.

6) Exit.

B) Assume two errors exist.

1) Compute

$$\sigma_1 = \frac{S_0 \cdot S_3 + S_1 \cdot S_2}{S_0 \cdot S_2 + (S_1)^2} \quad \text{and} \quad \sigma_2 = \frac{(S_2)^2 + S_1 \cdot S_3}{S_0 \cdot S_2 + (S_1)^2}$$

If the denominator or either numerator is zero, post an uncorrectable error flag and exit.

2) Compute $C = \sigma_2/(\sigma_1)^2$ and fetch Y_1 and Y_2 from the root table. If C does not correspond to a valid pair of roots, post an uncorrectable error flag and exit.

4) Compute $X_1 = \sigma_1 \cdot Y_1$ and $X_2 = \sigma_1 \cdot Y_2$

5) Compute $L_1 = LOG(X_1)$ and $L_2 = LOG(X_2)$

6) Compute

$$E_1 = \frac{\alpha^{L_2} \cdot S_0 + S_1}{\alpha^{L_1} + \alpha^{L_2}} \quad \text{and} \quad E_2 = S_0 + E_1$$

7) if the symbol at L_1 is data, XOR value E_1 at location L_1; if the symbol at L_2 is data, XOR value E_2 at location L_2.

8) Exit.

CODEWORD EXAMPLE

Data Symbols | Redundant Symbols

Assume the data symbols are (in order of transmission) α^2, α^1, and α^5. Then the data polynomial is:

$$D(x) = \alpha^2 \cdot x^2 + \alpha^1 \cdot x + \alpha^5$$

The redundant symbols can be computed using one of the encoder circuits shown above. A trace of the contents of the registers is shown below:

data	x^3	x^2	x	1
init	0	0	0	0
α^2	α^4	α^0	α^0	α^1
α^1	α^5	0	α^3	α^1
α^5	0	α^3	α^1	0

The transmitted codeword is:

$$C(x) = \alpha^2 \cdot x^6 + \alpha^1 \cdot x^5 + \alpha^5 \cdot x^4 + 0 \cdot x^3 + \alpha^3 \cdot x^2 + \alpha^1 \cdot x + 0$$

SINGLE ERROR EXAMPLE

$$C(x) = \alpha^2 \cdot x^6 + \alpha^1 \cdot x^5 + \alpha^5 \cdot x^4 + 0 \cdot x^3 + \alpha^3 \cdot x^2 + \alpha^1 \cdot x + 0$$

$$E(x) = \alpha^2 \cdot x^4$$

$$C'(x) = \alpha^2 \cdot x^6 + \alpha^1 \cdot x^5 + \alpha^3 \cdot x^4 + 0 \cdot x^3 + \alpha^3 \cdot x^2 + \alpha^1 \cdot x + 0$$

COMPUTE SYNDROMES

C'(x)	S_0	S_1	S_2	S_3
INIT	0	0	0	0
α^2	α^2	α^2	α^2	α^2
α^1	α^4	α^0	α^2	α^6
α^3	α^6	α^0	α^6	α^5
0	α^6	α^1	α^1	α^1
α^3	α^4	α^5	0	α^6
α^1	α^2	α^5	α^1	α^4
0	α^2	α^6	α^3	α^0

COMPUTE σ

$$\sigma = S_1/S_0 = \alpha^6/\alpha^2 = \alpha^4$$

VERIFY NEWTONS IDENTITIES

$$S_1 \cdot \sigma \overset{?}{=} S_2$$
$$\alpha^6 \cdot \alpha^4 \overset{?}{=} \alpha^3$$
$$\alpha^{(6+4 \text{ MOD } 7)} \overset{?}{=} \alpha^3$$
$$\alpha^3 = \alpha^3$$

$$S_2 \cdot \sigma \overset{?}{=} S_3$$
$$\alpha^3 \cdot \alpha^4 \overset{?}{=} \alpha^0$$
$$\alpha^{(3+4 \text{ MOD } 7)} \overset{?}{=} \alpha^0$$
$$\alpha^0 = \alpha^0$$

COMPUTE ERROR LOCATION AND VALUE

$$L = \text{LOG}(\sigma) = \text{LOG}(\alpha^4) = 4$$

$$E = S_0 = \alpha^2$$

TWO ERROR EXAMPLE

$$C(x) = \alpha^2 \cdot x^6 + \alpha^1 \cdot x^5 + \alpha^5 \cdot x^4 + 0 \cdot x^3 + \alpha^3 \cdot x^2 + \alpha^1 \cdot x + 0$$

$$E(x) = \alpha^2 \cdot x^5 + \alpha^1 \cdot x^2$$

$$C'(x) = \alpha^2 \cdot x^6 + \alpha^4 \cdot x^5 + \alpha^5 \cdot x^4 + 0 \cdot x^3 + \alpha^0 \cdot x^2 + \alpha^1 \cdot x + 0$$

COMPUTE SYNDROMES

C'(x)	S_0	S_1	S_2	S_3
INIT	0	0	0	0
α^2	α^2	α^2	α^2	α^2
α^4	α^1	α^6	0	α^0
α^5	α^6	α^4	α^5	α^2
0	α^6	α^5	α^0	α^5
α^0	α^2	α^2	α^6	α^3
α^1	α^4	α^0	0	α^5
0	α^4	α^1	0	α^1

COMPUTE σ

$$\sigma = S_1/S_0 = \alpha^1/\alpha^4 = \alpha^4$$

VERIFY NEWTONS IDENTITIES

$$S_1 \cdot \sigma \stackrel{?}{=} S_2$$

$$\alpha^4 \cdot \alpha^4 \stackrel{?}{=} \alpha^3$$

$$\alpha^{(4+4 \text{ MOD } 7)} \stackrel{?}{=} \alpha^3$$

$$\alpha^1 \neq \alpha^3 \Rightarrow \text{TWO ERRORS}$$

COMPUTE ERROR LOCATIONS

$$\sigma_1 = \frac{S_0 \cdot S_3 + S_1 \cdot S_2}{S_0 \cdot S_2 + (S_1)^2} = \alpha^3 \qquad \sigma_2 = \frac{(S_2)^2 + S_1 \cdot S_3}{S_0 \cdot S_2 + (S_1)^2} = \alpha^0$$

$$C = \sigma_2/(\sigma_1)^2 = \alpha^1$$

$Y_1 = \alpha^2$
$X_1 = \sigma_1 \cdot Y_1 = \alpha^3 \cdot \alpha^2 = \alpha^5$
$L_1 = \text{LOG}(X_1) = 5$

$Y_2 = \alpha^6$
$X_2 = \sigma_1 \cdot Y_2 = \alpha^3 \cdot \alpha^6 = \alpha^2$
$L_2 = \text{LOG}(X_2) = 2$

COMPUTE ERROR VALUES

$$E_1 = \frac{\alpha^{L_2} \cdot S_0 + S_1}{\alpha^{L_1} + \alpha^{L_2}} = \alpha^2 \qquad E_2 = S_0 + E_1 = \alpha^1$$

ITERATIVE ALGORITHM EXAMPLE

Use the iterative algorithm to generate $\sigma(x)$ for the case above.

S_0	S_1	S_2	S_3
α^4	α^1	0	α^1

TABLE GENERATED BY ITERATIVE ALGORITHM

n	$\sigma^{(n)}(x)$	d_n	L_n	$n-L_n$
-1	1	1	0	-1
0	1	α^4	0	0
1	$x + \alpha^4$	0	1	0
2	$x + \alpha^4$	α^5	1	1
3	$x^2 + \alpha^4 \cdot x + \alpha^1$	α^4	2	1
4	$x^2 + \alpha^3 \cdot x + \alpha^0$			

TRACE OF ITERATIVE ALGORITHM

n=0 (1) $d_0=\alpha^4$<>0 => Go to (2).

(2) k = -1. $d_0/d_{-1} = \alpha^4/1 = \alpha^4$. L_1 = MAX[0,0+0-(-1)] = 1.

$\sigma^{(1)}(x) = x^1 \cdot (1) + \alpha^4 \cdot (1) = x + \alpha^4$.

(3) (n+1):m => 1<4 => Continue.

(4) $d_1 = \sigma_0 \cdot S_1 + \sigma_1 \cdot S_0 = 1 \cdot \alpha^1 + \alpha^4 \cdot \alpha^4 = 0$.

n = 0+1 = 1. Go to (1).

n=1 (1) $d_1=0$ => $L_2 = L_1 = 1$. $\sigma^{(2)}(x) = \sigma^{(1)}(x) = x + \alpha^4$.

Go to (3).

(3) (n+1):m => 2<4 => Continue.

(4) $d_2 = \sigma_0 \cdot S_2 + \sigma_1 \cdot S_1 = 1 \cdot 0 + \alpha^4 \cdot \alpha^1 = \alpha^5$.

n = 1+1 = 2. Go to (1).

n=2 (1) $d_2=\alpha^5$<>0 => Go to (2).

(2) k = 0. $d_2/d_0 = \alpha^5/\alpha^4 = \alpha^1$. L_3 = MAX[1,0+2-0] = 2.

$\sigma^{(3)}(x) = x^1 \cdot (x + \alpha^4) + \alpha^1 \cdot (1) = x^2 + \alpha^4 \cdot x + \alpha^1$.

(3) (n+1):m => 3<4 => Continue.

(4) $d_3 = \sigma_0 \cdot S_3 + \sigma_1 \cdot S_2 + \sigma_2 \cdot S_1 = 1 \cdot \alpha^1 + \alpha^4 \cdot 0 + \alpha^1 \cdot \alpha^1 = \alpha^4$.

n = 2+1 = 3. Go to (1).

n=3 (1) $d_3=\alpha^4$<>0 => Go to (2).

(2) k = 2. $d_3/d_2 = \alpha^4/\alpha^5 = \alpha^6$. L_4 = MAX[2,1+3-2] = 2.

$\sigma^{(4)}(x) = x^0 \cdot (x^2 + \alpha^4 \cdot x + \alpha^1) + \alpha^6 \cdot (x + \alpha^4)$

$= x^2 + \alpha^3 \cdot x + \alpha^0$.

(3) (n+1):m => 4=4 => Stop

n=4 $\sigma(x) = \sigma^{(4)}(x) = x^2 + \alpha^3 \cdot x + \alpha^0$; same as case above.

UNCORRECTABLE ERROR EXAMPLE

$$C(x) = \alpha^2 \cdot x^6 + \alpha^1 \cdot x^5 + \alpha^5 \cdot x^4 + 0 \cdot x^3 + \alpha^3 \cdot x^2 + \alpha^1 \cdot x + 0$$

$$E(x) = \alpha^2 \cdot x^5 + \alpha^2 \cdot x^4 + \alpha^1 \cdot x^2$$

$$C'(x) = \alpha^2 \cdot x^6 + \alpha^4 \cdot x^5 + \alpha^3 \cdot x^4 + 0 \cdot x^3 + \alpha^0 \cdot x^2 + \alpha^1 \cdot x + 0$$

COMPUTE SYNDROMES

C'(x)	S_0	S_1	S_2	S_3
INIT	0	0	0	0
α^2	α^2	α^2	α^2	α^2
α^4	α^1	α^6	0	α^0
α^3	α^0	α^1	α^3	0
0	α^0	α^2	α^5	0
α^0	0	α^1	0	α^0
α^1	α^1	α^4	α^1	α^0
0	α^1	α^5	α^3	α^3

COMPUTE σ

$$\sigma = S_1/S_0 = \alpha^5/\alpha^1 = \alpha^4$$

VERIFY NEWTONS IDENTITIES

$$S_1 \cdot \sigma \overset{?}{=} S_2$$

$$\alpha^5 \cdot \alpha^4 \overset{?}{=} \alpha^3$$

$$\alpha^{(5+4 \text{ MOD } 7)} \overset{?}{=} \alpha^3$$

$$\alpha^2 \neq \alpha^3 \Rightarrow \text{TWO ERRORS}$$

COMPUTE ERROR LOCATIONS

$$\sigma_1 = \frac{S_0 \cdot S_3 + S_1 \cdot S_2}{S_0 \cdot S_2 + (S_1)^2} = \alpha^3 \qquad \sigma_2 = \frac{(S_2)^2 + S_1 \cdot S_3}{S_0 \cdot S_2 + (S_1)^2} = \alpha^6$$

$$C = \sigma_2/(\sigma_1)^2 = \alpha^0 \Rightarrow \text{UNCORRECTABLE ERROR}$$

MISCORRECTION EXAMPLE

$$C(x) = \alpha^2 \cdot x^6 + \alpha^1 \cdot x^5 + \alpha^5 \cdot x^4 + 0 \cdot x^3 + \alpha^3 \cdot x^2 + \alpha^1 \cdot x + 0$$

$$E(x) = \alpha^2 \cdot x^5 + \alpha^1 \cdot x^2 + \alpha^4$$

$$C'(x) = \alpha^2 \cdot x^6 + \alpha^4 \cdot x^5 + \alpha^5 \cdot x^4 + 0 \cdot x^3 + \alpha^0 \cdot x^2 + \alpha^1 \cdot x + \alpha^4$$

COMPUTE SYNDROMES

C'(x)	S_0	S_1	S_2	S_3
INIT	0	0	0	0
α^2	α^2	α^2	α^2	α^2
α^4	α^1	α^6	0	α^0
α^5	α^6	α^4	α^5	α^2
0	α^6	α^5	α^0	α^5
α^0	α^2	α^2	α^6	α^3
α^1	α^4	α^0	0	α^5
α^4	0	α^2	α^4	α^2

COMPUTE σ

$S_0 = 0$ => TWO ERRORS

COMPUTE ERROR LOCATIONS

$$\sigma_1 = \frac{S_0 \cdot S_3 + S_1 \cdot S_2}{S_0 \cdot S_2 + (S_1)^2} = \alpha^2 \qquad \sigma_2 = \frac{(S_2)^2 + S_1 \cdot S_3}{S_0 \cdot S_2 + (S_1)^2} = \alpha^5$$

$$C = \sigma_2/(\sigma_1)^2 = \alpha^1$$

$Y_1 = \alpha^2$
$X_1 = \sigma_1 \cdot Y_1 = \alpha^2 \cdot \alpha^2 = \alpha^4$
$L_1 = LOG(X_1) = 4$

$Y_2 = \alpha^6$
$X_2 = \sigma_1 \cdot Y_2 = \alpha^2 \cdot \alpha^6 = \alpha^1$
$L_2 = LOG(X_2) = 1$

COMPUTE ERROR VALUES

$$E_1 = \frac{\alpha^{L_2} \cdot S_0 + S_1}{\alpha^{L_1} + \alpha^{L_2}} = \alpha^0 \qquad E_2 = S_0 + E_1 = \alpha^0$$

REFERENCE TABLES

FINITE FIELD

	α^2	α^1	α^0
0	0	0	0
α^0	0	0	1
α^1	0	1	0
α^2	1	0	0
α^3	0	1	1
α^4	1	1	0
α^5	1	1	1
α^6	1	0	1

ADDITION TABLE

	0	α^0	α^1	α^2	α^3	α^4	α^5	α^6
0	0	α^0	α^1	α^2	α^3	α^4	α^5	α^6
α^0	α^0	0	α^3	α^6	α^1	α^5	α^4	α^2
α^1	α^1	α^3	0	α^4	α^0	α^2	α^6	α^5
α^2	α^2	α^6	α^4	0	α^5	α^1	α^3	α^0
α^3	α^3	α^1	α^0	α^5	0	α^6	α^2	α^4
α^4	α^4	α^5	α^2	α^1	α^6	0	α^0	α^3
α^5	α^5	α^4	α^6	α^3	α^2	α^0	0	α^1
α^6	α^6	α^2	α^5	α^0	α^4	α^3	α^1	0

MULTIPLICATION TABLE

	0	α^0	α^1	α^2	α^3	α^4	α^5	α^6
0	0	0	0	0	0	0	0	0
α^0	0	α^0	α^1	α^2	α^3	α^4	α^5	α^6
α^1	0	α^1	α^2	α^3	α^4	α^5	α^6	α^0
α^2	0	α^2	α^3	α^4	α^5	α^6	α^0	α^1
α^3	0	α^3	α^4	α^5	α^6	α^0	α^1	α^2
α^4	0	α^4	α^5	α^6	α^0	α^1	α^2	α^3
α^5	0	α^5	α^6	α^0	α^1	α^2	α^3	α^4
α^6	0	α^6	α^0	α^1	α^2	α^3	α^4	α^5

ROOT TABLE

C	Y_1	Y_2
0	0	α^0
α^0	--	--
α^1	α^2	α^6
α^2	α^4	α^5
α^3	--	--
α^4	α^1	α^3
α^5	--	--
α^6	--	--

AN INTUITIVE DISCUSSION OF THE SINGLE-ERROR CASE

The following discussion provides an intuitive description of how the Reed-Solomon code single-error case works. A particular example is used in order to make the discussion more understandable. Finite field theory is intentionally avoided.

Consider a single-error correcting Reed-Solomon code operating on 8-bit symbols and employing, on read, the binary polynomials:

$$P0 = (x^8 + 1)$$

$$P1 = (x^8 + x^6 + x^5 + x^4 + 1)$$

The correction algorithm requires residues of a function of the data, f(data), modulo P0 and P1 where:

$$\text{for } P_0,\ f(\text{DATA}) = \sum_{i=0}^{m-1} D_i(x)$$

$$\text{for } P_1,\ f(\text{DATA}) = \sum_{i=0}^{m-1} x^i D_i(x)$$

and m is the number of data bytes. Di represents the individual data byte polynomials. D0 is the lowest order data byte (last byte to be transmitted or received).

The residues are computed by hardware implementing the logical circuits shown below. These logical circuits are clocked once per byte.

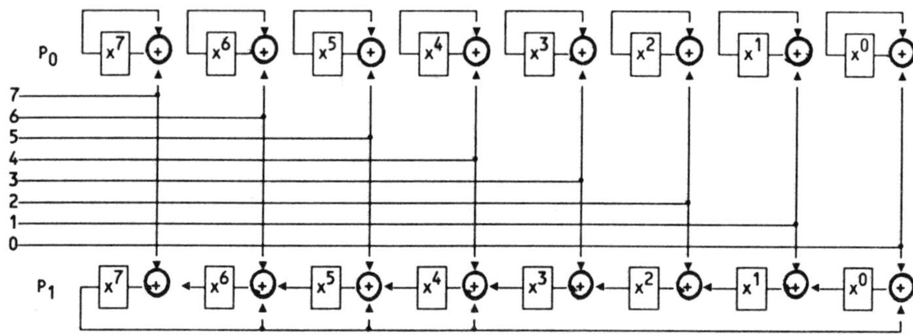

The shift register for P0 computes an XOR sum of all data bytes including the check bytes. Since P1 is primitive, its shift register generates a maximum length sequence (255 states). When the P1 shift register is non-zero, but its input is zero, each clock sets it to the next state of its sequence.

CORRECTION ALGORITHM

Consider what happens when the data record is all zeros and a byte in error is received.

Both shift registers will remain zero until the byte in error arrives. The error byte is XOR'd into the P0 and P1 shift registers. Since the P0 shift register preserves its current value as long as zeros are received, the error pattern remains in it until the end of record. XOR'ing the error byte into the P1 shift register places the shift register at a particular state in its sequence. As each new byte of zeros is received the P1 shift register is clocked along its sequence, one state per byte.

The terminal states of the P0 and P1 shift registers are sufficient for determining displacement. To find displacement, it is necessary to determine the number of shifts of the P1 shift register that occurred between the occurrence of the error byte and the end of record.

To better understand the correction algorithm, consider a sequence of 255 states as represented by the circle in the drawing on the following page. Let S1 be the ending state of the P1 shift register and let S0 be the ending state of the P0 shift register (S0 is also the initial state of the P1 shift register). Let Sr be the reference state '0000 0001'. The number of states between S0 and S1 must be determined. There are several ways to do this. In this description a table method is used.

Refer again to the diagram on the following page. What we need to know is the number of states between S0 and S1. We construct a table. The table is addressed by S0 and S1, and contains the distance along the P1 sequence between the reference state and any arbitrary state Sx.

First, S0 is used to address the table to fetch distance d1. Next, S1 is used to address the table to fetch distance d2. The desired distance (d), distance between S0 and S1 is computed as follows:

$$d = d2\text{-}d1;\ \text{if } d<0 \text{ then } d = d+255$$

The distance d is the reverse displacement from the end of the record. The forward displacement can be computed by subtracting the reverse displacement from the record length minus one. The error pattern is simply the terminal state of P0, which is S0.

Consider the case when the data is not all zeros. The check bytes would have been selected on write, such that on read, when the entire record (including check bytes) is processed by the P0 and P1 shift registers, residues of zero result.

When an error occurs, the operation differs from the all-zeros data case only while residues are being generated. A given error condition results in the same residues, regardless of data content. Once residues have been generated, the operation is the same as previously described for the all-zeros data case.

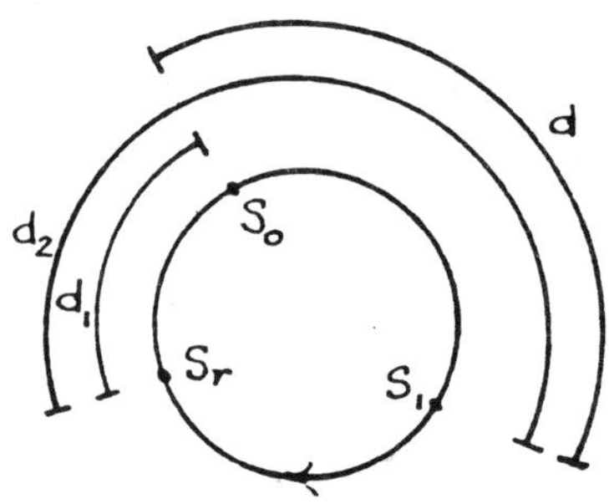

THE P1 SEQUENCE

ALTERNATIVE ENCODING AND DECODING METHODS
FOR THE REED-SOLOMON CODE

There are many encoding and decoding alternatives for the Reed-Solomon code. The best alternative for a given application depends on such factors as:

- Cost requirements
- Speed requirements
- Space requirements
- Sharing of circuits and resources

Some of these alternatives are described below.

ENCODING ALTERNATIVES

Encoding can be accomplished with the external-XOR form of shift register as well as the internal-XOR form. An encoder circuit example using the external-XOR form of shift register is shown below:

$$g(x) = (x + \alpha^0)\cdot(x + \alpha^1) = x^2 + \alpha^a\cdot x + \alpha^b$$

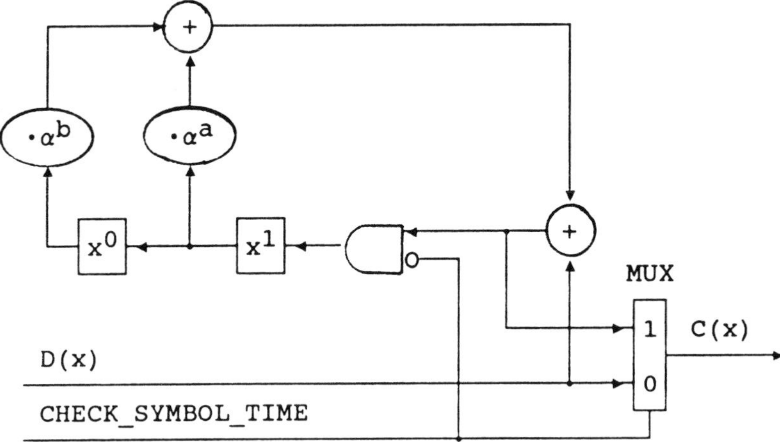

Another encoding alternative is illustrated by the following example.

STANDARD ENCODER

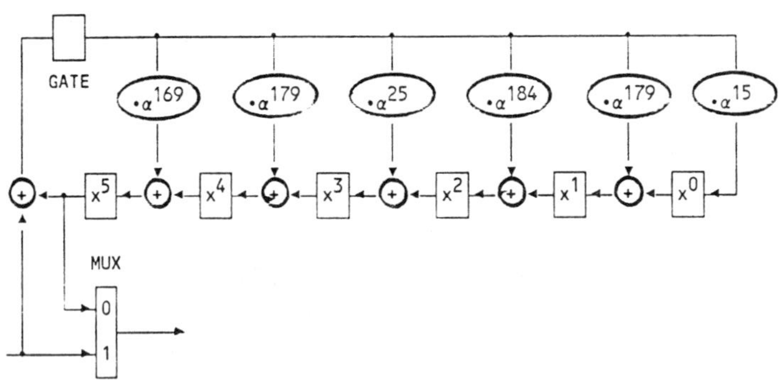

AN EQUIVALENT ALTERNATIVE ENCODER

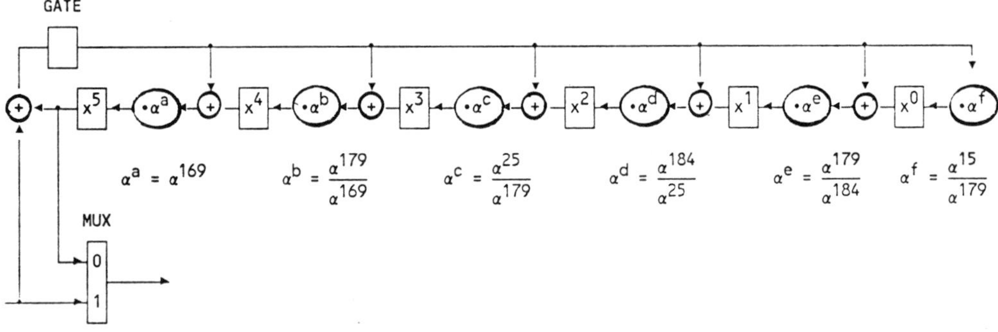

BIT-SERIAL ENCODING

Encoding can be accomplished with bit-serial techniques. We illustrate using the encoder implementing the external-XOR form of shift register introduced above. Rearranging to place low-order to the right, we have:

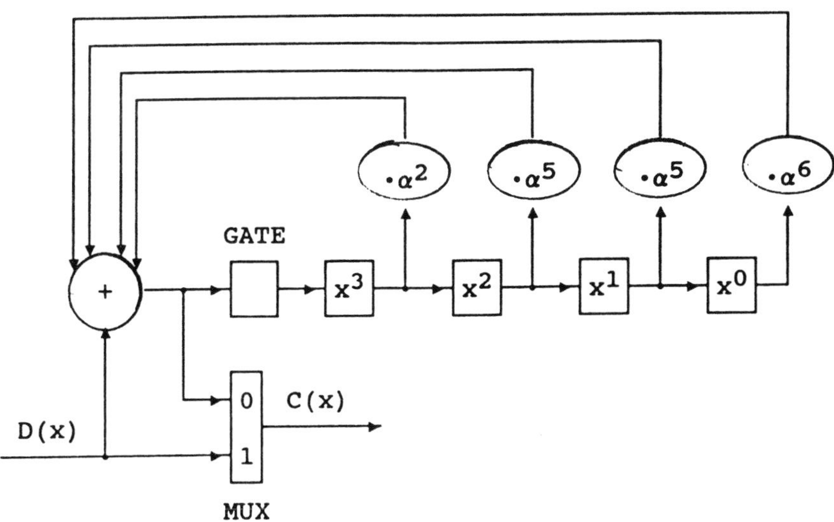

All paths are symbol-wide (three bits for this example) and the GATE and MUX are controlled by a signal which is asserted during clocks for data symbols. The field GF(8) is generated by the polynomial:

$$x^3 + x + 1$$

over GF(2). The code generator polynomial over GF(2^3) is:

$$G(x) = (x + \alpha^0)\cdot(x + \alpha^1)\cdot(x + \alpha^2)\cdot(x + \alpha^3)$$

$$= x^4 + \alpha^2\cdot x^3 + \alpha^5\cdot x^2 + \alpha^5\cdot x + \alpha^6$$

The external-XOR form of shift register requires the computation of the sum of four products of variable field elements with fixed field elements. Bit-serial multipliers were introduced in Section 2.6. The circuit below shows a bit-serial implementation of the encoder.

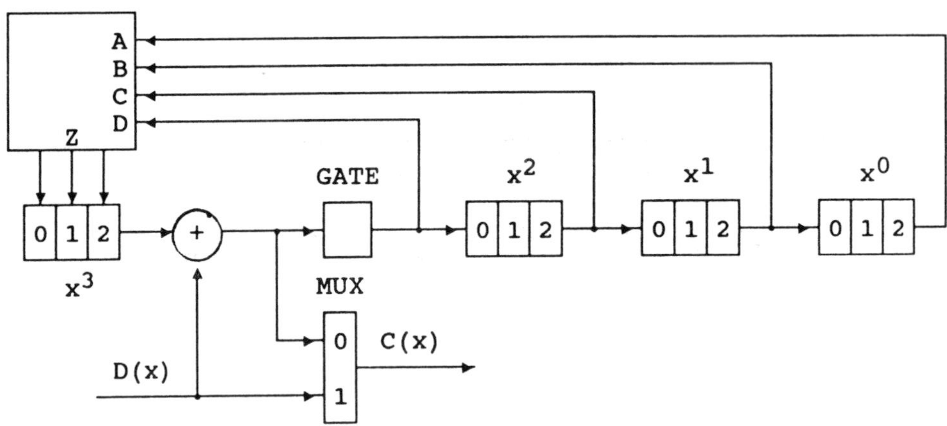

The bit-serial multiplier circuit accomplishes in three clocks what the four multipliers of the symbol-serial encoder accomplish in one clock. The Z register is initially cleared, then on every third clock it is again cleared and what would have been shifted in is clocked into the x^3 register.

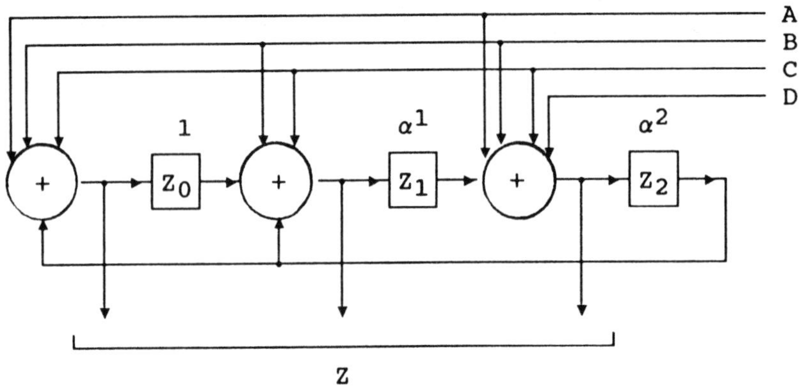

DECODING ALTERNATIVES

The standard form of syndrome circuit is:

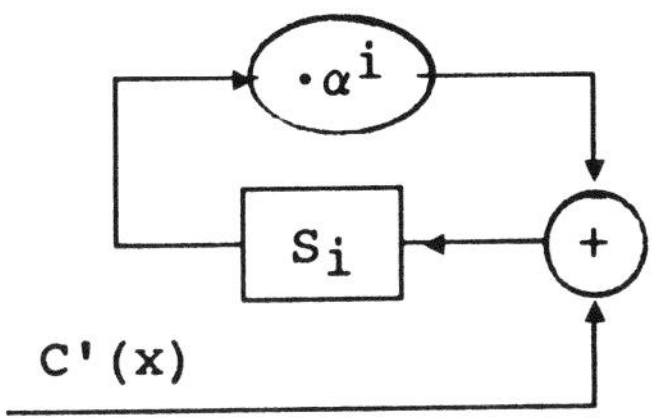

This circuit computes the syndromes:

$$S_i = C'(x) \text{ MOD } (x + \alpha^i)$$

It is also possible to use the circuit form below:

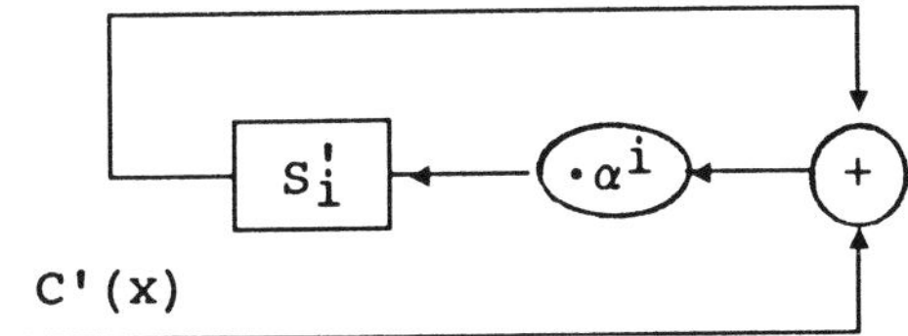

This circuit computes modified syndromes:

$$S_i' = \alpha^i \cdot S_i = \alpha^i \cdot [C'2(x) \text{ MOD } (x + \alpha^i)]$$

When this circuit form is used, the correction algorithm must be adjusted accordingly.

Decoding can also be accomplished with bit-serial multiply-and-sum techniques like those discussed in Section 2.6 and implemented above for encoding.

SHARING CIRCUITRY BETWEEN ENCODER AND DECODER

It is possible to share circuitry between the encoder and the decoder in several different ways. Recall the general case of the generator polynomial, write redundancy polynomial, and syndromes of a Reed-Solomon code of degree m:

$$G(x) = \prod_{j=0}^{m-1} g_j(x) = \prod_{j=0}^{m-1} (x + \alpha^{m_0+j})$$

$$W(x) = x^m \cdot D(x) \text{ MOD } G(x)$$

$$S_j = C'(x) \text{ MOD } g_j(x)$$

As we have seen, the processes of generating W(x) and generating S_j each require a different circuit configuration and a different set of finite field multipliers. Cost motivates us to find some means for reducing hardware by sharing circuitry in performing these two functions.

One method for sharing circuitry is to use the encoder on read to assist with syndrome generation by feeding it the received codeword to generate the composite read remainder:

$$R(x) = C'(x) \text{ MOD } G(x)$$

The individual remainders (syndromes) can then be generated by dividing the composite remainder by each factor of the generator polynomial. This second step can be accomplished with sequential logic, combinatorial logic, or software. In many cases, more time can be allotted for the processing of each symbol during the second step than during the first step due to the difference in degrees between the composite remainder and the full received codeword.

Another method for sharing circuitry is to use the syndrome circuits for encoding. The validity of the following approach is guaranteed by the Chinese Remainder Theorem for polynomials.

Consider the set of parameters:

$$P_j = D(x) \text{ MOD } g_j(x) = D(x) \text{ MOD } (x + \alpha^{m_0+j})$$

which are the contents of registers of a set of circuits for j=0 to m-1 like the one shown below, after clocking in a data polynomial D(x). We use P_j here to distinguish from the syndromes S_j, which are produced by similar circuits but have a received codeword C'(x) polynomial as input.

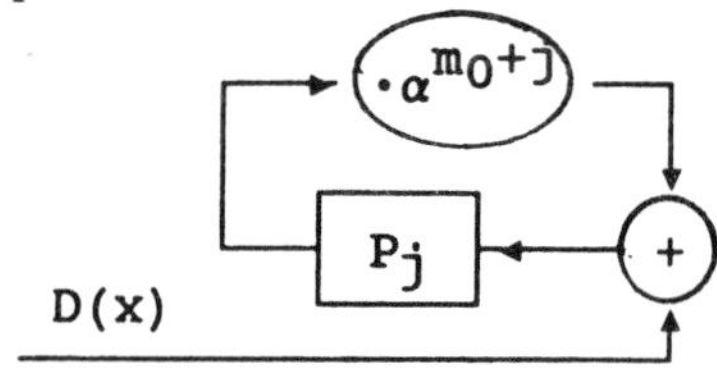

Now observe that:

$$P_j = \alpha^{-m*(m_0+j)} \cdot [x^m \cdot D(x) \text{ MOD } g_j(x)]$$

Since $g_j(x)$ is a factor of $G(x)$, we know that:

$$x^m \cdot D(x) \text{ MOD } g_j(x) = [x^m \cdot D(x) \text{ MOD } G(x)] \text{ MOD } g_j(x)$$

and so by definition of $W(x)$, we have:

$$P_j = \alpha^{-m*(m_0+j)} \cdot [W(x) \text{ MOD } g_j(x)]$$

By noting that:

$$W(x) \text{ MOD } g_j(x) = W(x) \Big|_{\alpha^{m_0+j}} = W(\alpha^{m_0+j})$$

we may expand in terms of the coefficients W_i of $W(x)$ to obtain:

$$P_j = \sum_{i=0}^{m-1} \alpha^{(i-m)*(m_0+j)} \cdot W_i$$

These equations give the parameters P_j in terms of the write redundancy coefficients W_i and a matrix of constants which are powers of α that depend only on i, j, m and m_0. Inverting this matrix gives the write redundancy coefficients W_i in terms of the parameters P_j and a set of transform constants $K_{i,j}$ which also depend only on i, j, m, and m_0:

$$W_i = \sum_{j=0}^{m-1} K_{i,j} \cdot P_j$$

To aid in understanding the implementation of the above equations, we first discuss the following circuit, which is equivalent to the conventional form of encoder circuit discussed previously.

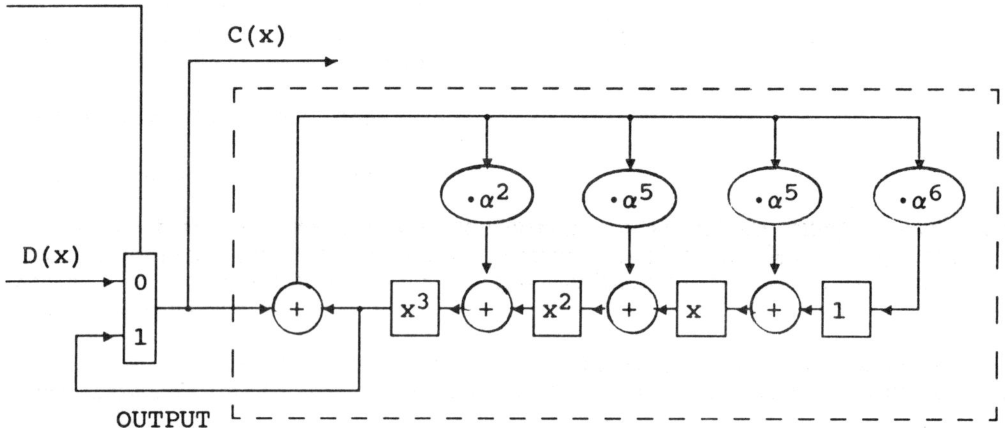

When the CHECK_SYMBOL_TIME signal is asserted, the OUTPUT bus is fed back into the dashed portion of the circuit by the MUX. The input to the multipliers is then zero, so the contents of the registers, the write redundancy polynomial W(x), are not altered as they are shifted out and appended to the data polynomial D(x) to form the codeword C(x).

The circuit below illustrates the method for sharing circuitry.

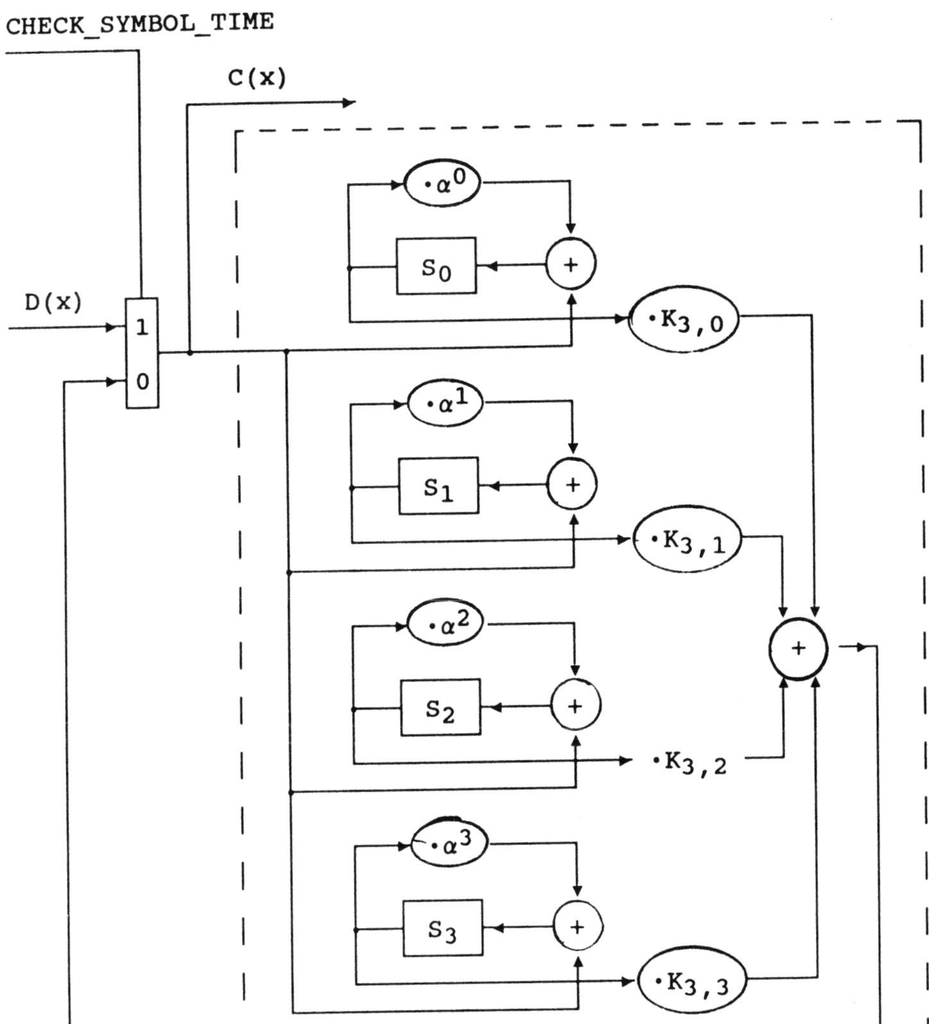

Note that in practice it is necessary to implement only the multipliers corresponding to $K_{m-1,j}$. To understand this, observe that from the development above it is clear that given the same input data polynomial D(x), the dashed portions of both the conventional and shared-circuitry methods will produce the same OUTPUT for W_{m-1}, the coefficient of the high-order term of the write redundancy polynomial W(x). Since the portion outside the dashed box of each circuit is the same, and W_{m-1} is fed back in the same manner for each circuit, they will produce the same output for W_{m-1}, etc. Mathematical proof is left as an exercise for the reader.

The registers are labeled S_j because this same circuitry can be used to generate the syndromes on read by presenting $C'(x)$ at the input and not asserting the CHECK-_SYMBOL_TIME signal. The OUTPUT bus is simply ignored, and the syndromes may be loaded from the registers after the last symbol of the received codeword has been clocked in.

It is possible to take the input to the $K_{m-1,j}$ multipliers from the input to the S_j registers instead of from their output. If this is done, a register must be inserted in the OUTPUT path before the MUX preserve clocking. It is also possible to take the multipliers from the output of the α^j multipliers. If this is done, the values of the $K_{m-1,j}$ multipliers are changed to:

$$K'_{m-1,j} = \frac{K_{m-1,j}}{\alpha^{m_0+j}}$$

to remove the extra factor of α^{m_0+j}.

Finally, it is also possible to implement the shared circuitry method using the modified form of the syndrome circuit introduced above:

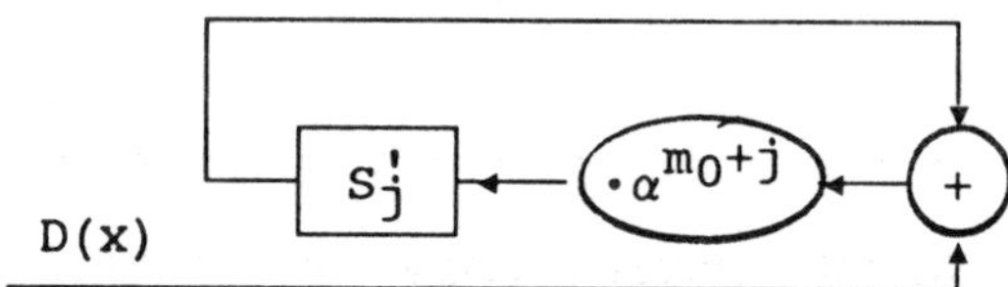

The appropriate set ($K_{m-1,j}$ or $K'_{m-1,j}$) of multiplier values is used on write, depending on where their inputs are taken. For the general case where $m_0 \neq 0$, using this form of syndrome circuit on read produces modified syndromes:

$$S_j' = \alpha^{m_0+j} \cdot S_j$$

and correction algorithms must be modified accordingly.

EXTENDED REED-SOLOMON CODES

It is possible to extend certain Reed-Solomon codes by one or two symbols. The additional symbols may be used as data in which case the minimum distance of the extended code is the same as that of the original code, or as additional redundancy in which case the minimum distance, and hence the correction power, of the extended code is greater than that of the original code. When a Reed-Solomon code over w-bit symbols is extended by two symbols, the maximum codeword size is 2^w+1 symbols. We illustrate using a basic code of degree m=2t=2. Let us use our example field, GF(8) generated by $(x^3 + x + 1)$, and let the generator polynomial of the basic code be

$$G(x) = \prod_{i=1}^{2} (x + \alpha^i)$$

$$= (x + \alpha^1)\cdot(x + \alpha^2)$$

$$= x^2 + (\alpha^1 + \alpha^2)\cdot x + \alpha^1\cdot\alpha^2$$

$$= x^2 + \alpha^4\cdot x + \alpha^3$$

ENCODING OF EXTENDED REED-SOLOMON CODE

Proceed in the usual fashion for the basic code:

$$W(x) = [x^2\cdot D(x) \text{ MOD } G(x)] = W_1x + W_0$$

$$C(x) = x^2\cdot D(x) + W(x)$$

and form two extension symbols:

$$X_3 = C(x) \text{ MOD } (x + \alpha^3)$$

$$X_0 = C(x) \text{ MOD } (x + \alpha^0)$$

Notice that the extension polynomials are identical to polynomials we would use to expand G(x) by one factor on each end of its set of factors.

which may be transmitted following C(x):

DATA SYMBOLS					REDUNDANT SYMBOLS		EXTENSION SYMBOLS	
D_4	D_3	D_2	D_1	D_0	W_1	W_0	X_3	X_0

The unextended degree four code discussed above and this extended degree two code each have four redundant symbols per codeword and each can correct two symbols in error per codeword, but the former has three data symbols per seven-symbol codeword while the latter has five data symbols per nine-symbol extended codeword.

ENCODING CIRCUITRY FOR EXTENDED REED-SOLOMON CODE

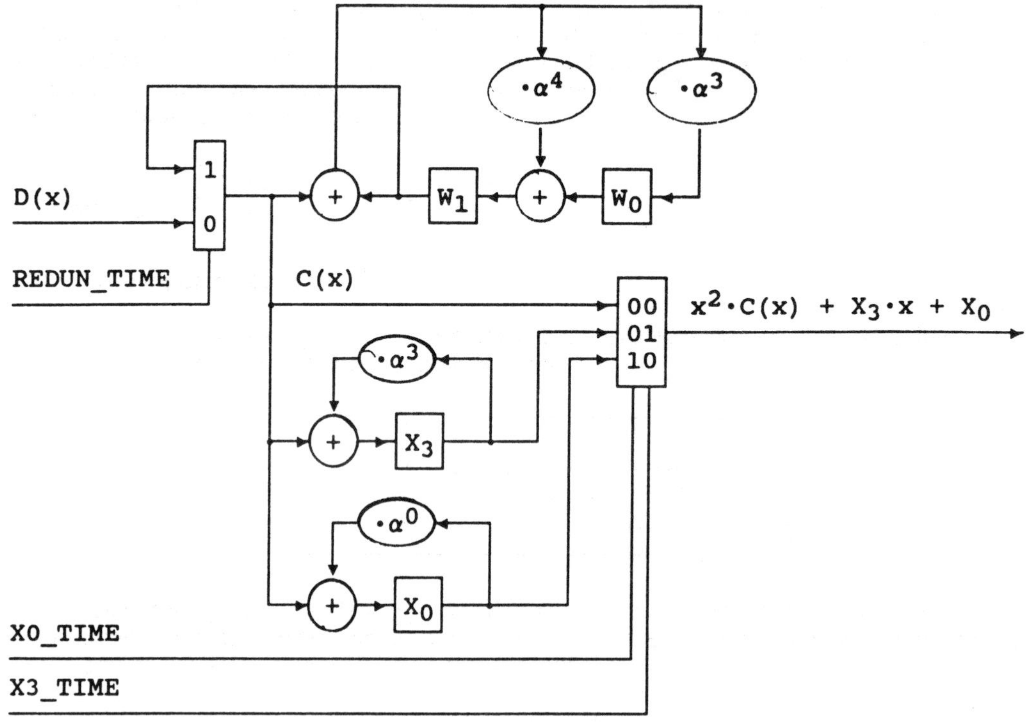

SIGNAL DEFINITIONS

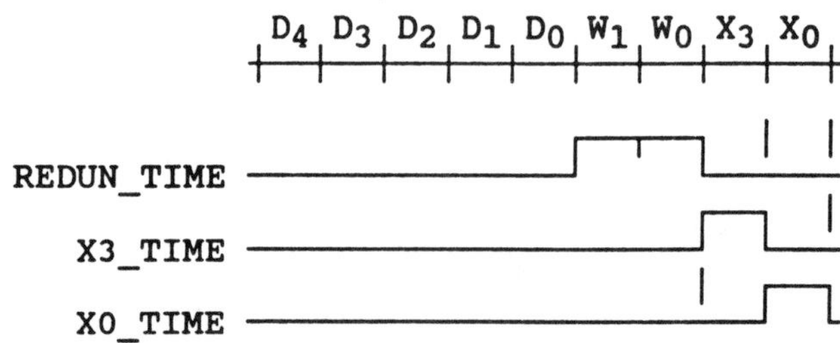

DECODING CIRCUITRY FOR EXTENDED REED-SOLOMON CODE

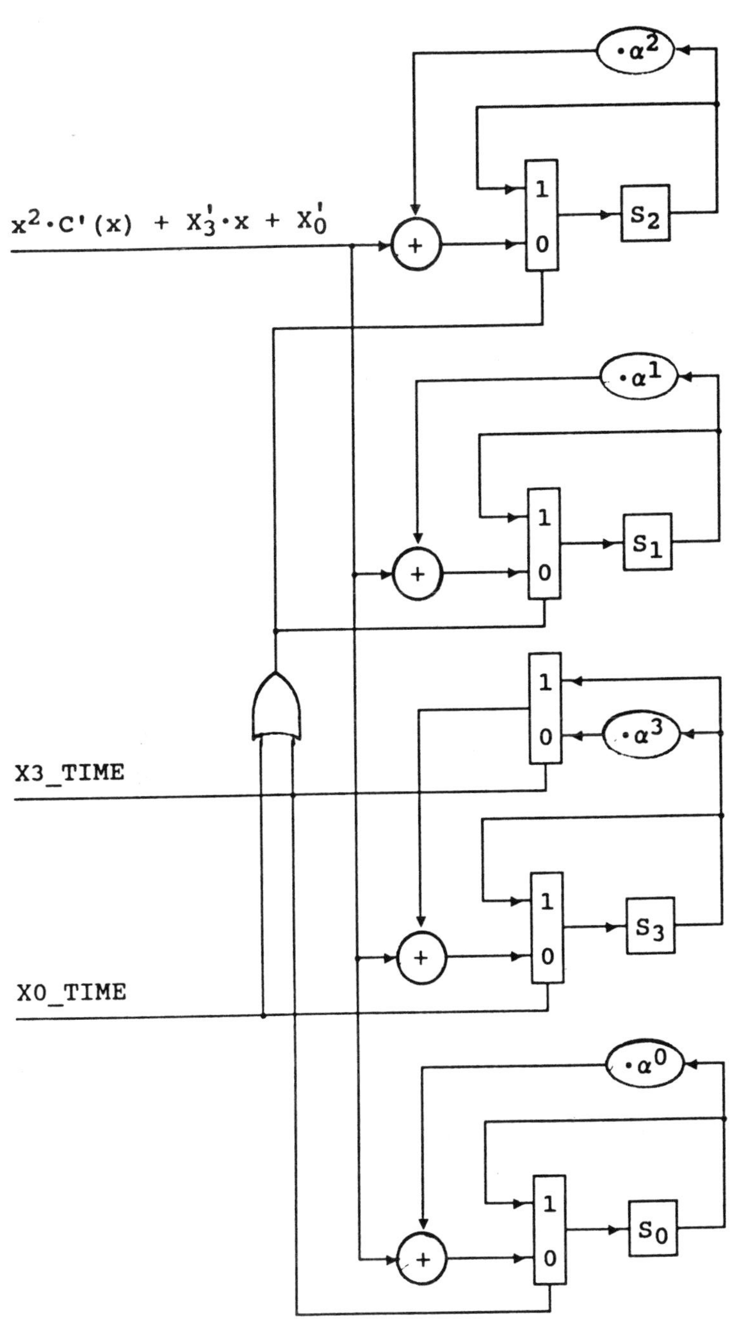

DECODING OF EXTENDED REED-SOLOMON CODE

Compute the syndromes for the basic code over the received codeword $C'(x)$ in the usual fashion:

$$S_2 = C'(x) \text{ MOD } (x + \alpha^2)$$

$$S_1 = C'(x) \text{ MOD } (x + \alpha^1)$$

and compute two more syndromes using the received extension symbols X_3' and X_0':

$$S_3 = X_3' + [C'(x) \text{ MOD } (x + \alpha^3)]$$

$$S_0 = X_0' + [C'(x) \text{ MOD } (x + \alpha^0)]$$

CORRECTION OF EXTENDED REED-SOLOMON CODE

The four syndromes of the extended code allow correction of up to two symbols in error. When no error falls into either extension symbol, the decoding process produces the same syndromes as for the degree four code shown above, and correction proceeds in exactly the same manner.

When at least one error falls into an extension symbol, we have two cases: those in which one or two errors occur and all errors fall into the extension symbols, and those in which two errors occur and one of the errors falls into a data symbol or one of the redundant symbols of the basic code and one of the errors falls into an extension symbol.

When errors occur only in the extension symbols, S_1 and S_2 will both be zero. This cannot occur for any correctable error pattern, so we know within the power of the code that no error in the basic codeword exists.

When one error falls in a data symbol or one of the redundant symbols of the basic code and one error falls into an extension symbol, both S_1 and S_2 and either S_0 or S_3 will be the same as for the degree four code shown above. We may solve for the location and value of the first error using S_1 and S_2 by a process similar to that used above, and the fact that either S_0 or S_3 satisfies Newton's identities is sufficient to confirm within the power of the code that the computed location and value of the single error in the basic codeword are valid.

EXTENDED DECODING OF REED-SOLOMON CODES

Extended decoding refers to techniques that allow successful correction of many error situations which exceed the basic guaranteed correction capability of a Reed-Solomon code. This is distinct from and not to be confused with the concept of an extended Reed-Solomon code introduced above. Several extended decoding techniques are discussed below.

ERASURE CORRECTION WITH EXTERNAL OR INTERNAL POINTERS

A Reed-Solomon code can correct higher raw error rates if error pointer information is available from some external source. External pointer sources include modulation-code run-length violations, marginal timing, and marginal amplitude. If signal drop-out is the predominant type of error and if the burst length distribution shows a high probability of long defects, modulation-code run-length violations can be an excellent pointer source. When a block-modulation code is used with byte or nibble boundaries, run-length violation pointers will accurately identify bytes in error. When a (2,7)-like code is used, a run-length violation pointer may flag a byte adjacent to the byte in error. This error location uncertainty can be overcome to some extent in the decoding algorithms.

A simple method for transferring pointers from a storage device to its controller is to implement a special read command that places pointer flags on the data line (or bus). These flags replace data field bytes (data and redundancy) only; all track format information bytes (sync, resync, etc.) are transferred as for a normal read command. When the correction algorithm encounters an uncorrectable sector, it returns to the calling routine with a flag requesting that pointers be read. The calling routine executes the special read command for the required sector and pointer flags are placed on the data line (or bus) and buffered at the controller. The calling routine returns control to the correction routine, which uses the pointers to assist correction. No support hardware is required at the controller to support this technique. If modulation-code run-length violations are the only pointer source, the only support hardware required at the drive is a multiplexer to switch between the data line and the invalid-decode line from the modulation decoder.

The capability of an interleaved Reed-Solomon code can be extended by using error locations from adjacent correctable interleaves as erasure pointers for an interleave that is uncorrectable without the use of erasure pointers.

Described below is a well-known algorithm for erasure correction. More efficient algorithms do exist, but this one was chosen for inclusion for its instructional value.

1) Generate an erasure-locator polynomial from the known (or suspected) erasure locations:

$$\Gamma(x) = \prod_{i=1}^{n} (x + \alpha^{P_i})$$

$$= x^n + \Gamma_1 \cdot x^{n-1} + \cdots + \Gamma_{n-1} \cdot x + \Gamma_n$$

where
n = the number of available erasure pointers
P_i = the location specified by erasure pointer number i

2) Generate (m-n) modified syndromes T_i from the m raw syndromes S_i and the coefficients of the erasure-locator polynomial:

$$T_{i-n} = \sum_{j=0}^{n} \Gamma_j \cdot S_{i-j}$$

for $i=n$ to m-1, where m is the degree of the code's generator polynomial.

3) Generate the coefficients of the error locator $\sigma(x)$ from the modified syndromes T_i.

4) Find error locations using the error locator polynomial.

5) Compute error values using the raw syndromes S_i and the erasure pointers and error locations.

The error value for a false erasure pointer will be zero, so a false erasure pointer will not necessarily cause miscorrection, but each false erasure pointer decreases the remaining correction capability, and increases the chance of miscorrection, by decreasing by one the number of available modified syndromes.

ERASURE CORRECTION EXAMPLE

[This example uses the same field and polynomials as the uncorrectable error example shown above.]

$$C(x) = \alpha^2 \cdot x^6 + \alpha^1 \cdot x^5 + \alpha^5 \cdot x^4 + 0 \cdot x^3 + \alpha^3 \cdot x^2 + \alpha^1 \cdot x + 0$$

$$E(x) = \alpha^2 \cdot x^5 + \alpha^2 \cdot x^4 + \alpha^1 \cdot x^2$$

$$C'(x) = \alpha^2 \cdot x^6 + \alpha^4 \cdot x^5 + \alpha^5 \cdot x^4 + 0 \cdot x^3 + \alpha^0 \cdot x^2 + \alpha^1 \cdot x + 0$$

COMPUTE SYNDROMES

$C'(x)$	S_0	S_1	S_2	S_3
INIT	0	0	0	0
α^2	α^2	α^2	α^2	α^2
α^4	α^2	α^2	α^2	α^2
α^5	0	α^5	α^1	α^3
0	0	α^6	α^3	α^6
α^0	α^1	α^3	α^6	α^4
α^1	α^1	α^4	α^1	α^0
0	α^1	α^5	α^3	α^3

POINTERS

$$n = 2$$
$$P_1 = 4$$
$$P_2 = 5$$

COMPUTE ERASURE LOCATOR $\tilde{a}(x)$

$$\Gamma(x) = \prod_{i=1}^{n} (x + \alpha^{P_i})$$

$$= (x + \alpha^4) \cdot (x + \alpha^5)$$

$$= x^2 + (\alpha^4 + \alpha^5) \cdot x + \alpha^4 \cdot \alpha^5$$

$$= \alpha^0 \cdot x^2 + \alpha^0 \cdot x + \alpha^2$$

$$= \Gamma_0 \cdot x^2 + \Gamma_1 \cdot x + \Gamma_2$$

GENERATE MODIFIED SYNDROMES

$$T_{i-n} = \sum_{j=0}^{n} \Gamma_j \cdot S_{i-j} \quad \text{for } i = n \text{ to } m-1$$

$$\begin{aligned} T_0 &= \Gamma_0 \cdot S_2 + \Gamma_1 \cdot S_1 + \Gamma_2 \cdot S_0 \\ &= \alpha^0 \cdot \alpha^3 + \alpha^0 \cdot \alpha^5 + \alpha^2 \cdot \alpha^1 = \alpha^5 \end{aligned}$$

$$\begin{aligned} T_1 &= \Gamma_0 \cdot S_3 + \Gamma_1 \cdot S_2 + \Gamma_2 \cdot S_1 \\ &= \alpha^0 \cdot \alpha^3 + \alpha^0 \cdot \alpha^3 + \alpha^2 \cdot \alpha^5 = \alpha^0 \end{aligned}$$

COMPUTE ERROR LOCATOR Í(x)

$$\sigma = T_1/T_0 = \alpha^0/\alpha^5 = \alpha^2$$

COMPUTE ERROR LOCATION

$$L = \log(\sigma) = 2$$

COMPUTE ERRATA VALUES

(The following equations are from Section 5.4.)

$$X_1 = \alpha^L = \alpha^2$$

$$X_2 = \alpha^{P_1} = \alpha^4$$

$$X_3 = \alpha^{P_2} = \alpha^5$$

$$E_1 = \frac{S_2+S_1 \cdot (X_2+X_3)+S_0 \cdot X_2 \cdot X_3}{(X_1+X_2) \cdot (X_1+X_3)} = \alpha^1$$

$$E_2 = \frac{S_0 \cdot X_3+S_1+E_1 \cdot (X_1+X_3)}{X_2+X_3} = \alpha^2$$

$$E_3 = S_0+E_1+E_2 = \alpha^2$$

EXTENDED CORRECTION ALGORITHMS

It is possible to extend the correction capability of a Reed-Solomon code by using algorithms that decode beyond the basic code guarantees without using erasure correction. Examples of error situations which, though not guaranteed to be handled by extended decoding techniques, have a certain probability of being handled include:

(a) A single long burst where the number of bytes in error in a codeword exceeds the basic guarantees of the code.

(b) Multiple long bursts, or a long burst in combination with random byte-errors, where the total number of bytes in error in a codeword exceeds the basic guarantees of the code.

(c) A number of random byte-errors in a codeword which exceeds the basic guarantees of the code.

AN EXAMPLE OF EXTENDED DECODING

Assume a code with generator polynomial G(x) over $GF(2^8)$ of degree 16, distance d=17, guaranteed to correct t=8 symbols in error in a codeword. Recall some definitions for Reed-Solomon codes:

G(x)	=	The generator polynomial of a Reed-Solomon code over $GF(2^w)$.
n	=	The length of a codeword; $n \le 2^w-1$.
m	=	The degree of the generator polynomial G(x).
d	=	The minimum distance of a Reed-Solomon code with generator polynomial of degree m; d=m+1.
t	=	The maximum number of symbols in error guaranteed correctable by a Reed-Solomon code with generator polynomial of degree m; t=INT[m/2].

We first illustrate decoding beyond code guarantees without erasure pointers with a method for case (a) above. Consider a single error burst which is thirteen bytes in length and affects the last thirteen bytes of the received codeword. The error polynomial is:

$$E(x) = E_{12} \cdot x^{12} + E_{11} \cdot x^{11} + \cdots + E_1 \cdot x^1 + E_0$$

Clearly, the sixteen-byte remainder:

$$R(x) = E(x) \text{ MOD } G(x) = E(x)$$

contains three consecutive high-order bytes that are all zeros followed by thirteen low-order bytes that constitute the error pattern.

Now consider a single thirteen-byte error burst that ends J bytes prior to the end of the received codeword. The error polynomial is:

$$E(x) = E_{J+12} \cdot x^{J+12} + \cdots + E_{J+1} \cdot x^{J+1} + E_J \cdot x^J$$

and nothing can be guaranteed about the zero/nonzero status of the coefficients of the sixteen byte remainder:

$$R(x) = E(x) \text{ MOD } G(x)$$

However, if we premultiply R(x) by x^{-J} and form a new remainder:

$$\begin{aligned} R_J(x) &= x^{-J} \cdot R(x) \text{ MOD } G(x) \\ &= x^{-J} \cdot [E(x) \text{ MOD } G(x)] \text{ MOD } G(x) \\ &= x^{-J} \cdot E(x) \text{ MOD } G(x) \\ &= x^{-J} \cdot E(x) \\ &= E_{J+12} \cdot x^{12} + \cdots + E_{J+1} \cdot x + E_J \end{aligned}$$

we again obtain a remainder which contains three consecutive high-order bytes that are all zeros followed by thirteen low-order bytes that constitute the error pattern.

The equation above is the basis for the decoding method. We count and record the number of consecutive high-order zero coefficients in the initial remainder, recording the low-order coefficients if the number of consecutive high-order zero coefficients is sufficiently high. Then we compute:

$$R_1(x) = x^{-1} \cdot R(x) \text{ MOD } G(x)$$

and repeat the counting/recording process. This process is performed n-1 times, where n is the length of the codeword, to compute $R_1(x)$ through $R_{n-1}(x)$ and account for all possible ending locations of the long burst. The pattern containing the highest number of consecutive high-order zero coefficients will be that of the long burst itself, which will have been segregated at the low-order end of the remainder.

The detection of some minimum number of consecutive high-order zero bytes (three for the given code operating on a full-length codeword, as shown below) can be used to flag the existence of a single long burst. The necessary number of consecutive high-order zero coefficients is established by the required miscorrection probability.

MISCORRECTION

For a codeword of length n, the miscorrection probability (units: miscorrected codewords per uncorrectable codeword) for a conventional decoding method against all combinations of random errors which exceeds the capability of the code is:

$$P_{mc1} = \sum_{i=0}^{t} \frac{\begin{bmatrix} n \\ i \end{bmatrix} 255^i}{256^{t+i}}$$

where,

$$\begin{bmatrix} n \\ r \end{bmatrix} = \frac{n!}{r!(n-r)!}$$

The miscorrection probability (units: miscorrected codewords per uncorrectable codeword) of the extended decoding method outlined above when used to decode a single burst of up to L bytes is roughly:

$$P_{mc2} = 1-[1-256^{-(m-L)}]^n$$

For a full-length (n=255) codeword with t=8, we have:

$$P_{mc1} = \sum_{i=0}^{8} \frac{\begin{bmatrix} 255 \\ i \end{bmatrix} 255^i}{256^{8+i}} \approx 2.1E-5$$

while for n=255, m=16, and L=13, we have:

$$P_{mc2} = 1-[1-256^{-3}]^{255} \approx 1.5E-5$$

Thus the extended decoding method outlined above could be used to correct a single burst of up to thirteen bytes in a full-length codeword with a miscorrection probability comparable to that of a conventional decoding method against all combinations of random errors.

It is important to note that for high-performance ECC applications, an auxiliary error detecting code is usually implemented to improve data accuracy. In some cases, the dedicated error detection code may provide most of the protection against the transfer of undetected erroneous data.

INTERLEAVING

When interleaving is used, the maximum length of a decodable single burst is multiplied by the number of interleaves. Consider the same code described above but implemented with ten-way interleaving in sectors of 1040 data bytes; each interleave contains n=(1040/10+16)=120 bytes. The conventional miscorrection probability (units: miscorrected codewords per uncorrectable codeword) against all combinations of random errors is:

$$P_{mc1} = \sum_{i=0}^{8} \frac{\begin{bmatrix}120\\i\end{bmatrix} 255^{i}}{256^{8+i}} \approx 4.4E-8$$

while for I=10, m=16, and L=12, the miscorrection probability for this extended decoding method is:

$$P_{mc2} = 1-[1-256^{-4}]^{120} \approx 2.8E-8$$

Thus the method outlined above will allow successful decoding of a single burst of up to I*L=120 bytes in a ten-way interleaved sector of 1040 data bytes with a miscorrection probability comparable to that achieved using a conventional decoding method in decoding all combinations of random errors.

A SECOND EXAMPLE

We next illustrate decoding beyond code guarantees without erasure pointers with a method for case (b) above, also for a ten-way interleaved sector of 1040 data bytes. Consider an error burst which is 100 bytes in length (ten consecutive bytes in error in each of the ten interleaves) that ends J bytes prior to the end of an interleave, together with other error burst(s) or random byte error(s) which affect no more than one byte in any one interleave. The error polynomial for an interleave is:

$$E(x) = E_A \cdot x^A + E_{J+9} \cdot x^{J+9} + \cdots + E_J \cdot x^J$$

where A is the location of the single byte in error, which may either precede or follow the long burst. If we premultiply by x^{-J} then the sixteen byte remainder is:

$$R_J(x) = x^{-J} \cdot E(x) \text{ MOD } G(x)$$

$$= [E_A \cdot x^{A-J} \text{ MOD } G(x)] + [E_{J+9} \cdot x^9 + \cdots + E_J]$$

All of the sixteen coefficients of the first term are nonzero, while the six high-order coefficients of the second term are equal to zero. Methods are known for decoding directly from the remainder without computing the conventional syndromes, and it is possible to solve for the location and value of a single error using two of the six high-order remainder coefficients, leaving four for verification. Once the location and value of the single byte in error have been computed and verified, its contribution to the ten low-order coefficients of the remainder can be removed, leaving just the error pattern of the long burst.

The decoding method for case (b) is similar to that for case (a) above. We attempt to decode some restricted number of bytes in error (one for this particular example) using the first few high-order coefficients of the initial remainder, count and record the number of consecutive high-order coefficients which are consistent, and record the low-order coefficients if the number of consecutive consistent high-order coefficients is sufficiently high (six for this particular example). Then we compute:

$$R_1(x) = R(x) \text{ MOD } G(x)$$

and repeat the decoding/counting/recording process. This process is repeated n-1 times, where n is the length of the codeword, to compute $R_1(x)$ through $R_{n-1}(x)$ and account for all possible ending locations of the long burst. The low-order coefficients of the remainder containing the highest number of consecutive consistent high-order coefficients can be adjusted to remove the contribution of the decoded and verified errors, leaving the pattern of the long burst, which will have been segregated at the low-order end of the remainder.

The detection of some minimum number of consecutive consistent high-order coefficients can be used to flag the existence of a single long burst together with up to some maximum number of other bytes in error in a codeword. The necessary number of consecutive high-order zero coefficients is again established by the required miscorrection probability.

The miscorrection probability for each remainder $R_J(x)$ for this method when used on n-byte codewords to decode a long burst contributing L consecutive bytes in error together with up to K other bytes in error per codeword, where K<INT[(m-L)/2], is:

$$P_{mc3} = \sum_{i=0}^{K} \frac{\binom{n}{i} 255^i}{256^{m-L}}$$

and the total miscorrection probability (units: miscorrected codewords per uncorrectable codeword) for all n values of J is:

$$P_{mc4} = 1-[1-P_{mc3}]^n$$

For m=16, L=10, K=1 in a ten-way interleaved sector of 1040 data bytes, the miscorrection probability at each value of J is:

$$P_{mc3} = \sum_{i=0}^{1} \frac{\binom{120}{i} 255^i}{256^6} \approx 1.09E-10$$

and the total miscorrection probability is:

$$P_{mc4} = 1-[1-1.09E-10]^{120} = 1.3E-8$$

Thus this method would allow successful decoding of a long burst of up to I*L=100 bytes in combination with up to one other byte in error per interleave with a miscorrection probability comparable to that achieved using a conventional decoding method in decoding all combinations of random errors.

Note that a logical extension of the decoding method for both cases (a) and (b) for an interleaved code is to require consistency across interleaves in the decoded location of the long burst.

CONCLUDING REMARKS

The techniques discussed above were selected for ease of understanding and are by no means the best or only methods which exist for extending the correction power of long-distance Reed-Solomon codes. It is possible, with or without erasure pointers, to efficiently decode multiple long-burst errors and combinations of long-burst errors and random byte-errors which exceed the basic guarantees of a code. Long-distance Reed-Solomon codes possess much greater correction power against both long-burst and random byte-errors than has traditionally been understood.

3.5 b-ADJACENT CODES

The b-Adjacent codes are parity check codes constructed with symbols from $GF(2^b)$, $b>1$. A subset of these codes is similar to the Reed-Solomon codes, but in many cases encoding for a b-adjacent code is less complex than encoding for a Reed-Solomon code with an equivalent capability.

Check symbols are generated on write and appended to data. On read, check symbols are generated and compared with the write check symbols. The XOR differences between the read check symbols and write check symbols determine the syndromes. The syndromes are used to compute error pattern and displacement information. Errors within the check bytes must be detected with special tests.

The IBM 3370, 3375, and 3380 magnetic disk drives employ b-Adjacent code techniques. Several of these techniques are described below.

EXAMPLE #1 - *A CODE TO CORRECT A SINGLE WORD ERROR*

Consider a b-Adjacent code using two 16-bit shift registers, P_0 and P_1, defined by the polynomials below:

$$P_0 = (x^{16} + 1)$$

$$P_1 = (x^{16} + x^{12} + x^3 + x + 1) \quad \text{[Primitive]}$$

The properties of these polynomials enable the code to correct a single word (16 bits) in error in a 65,535 word record.

The write and read check words (C0 and C1) are generated by taking residues of a function of the data, f(data), modulo P0 and P1, where:

$$\text{for } P_0,\ f(\text{DATA}) = \sum_{i=0}^{m-1} D_i(x)$$

$$\text{for } P_1,\ f(\text{DATA}) = \sum_{i=0}^{m-1} x^i \cdot D_i(x)$$

and m is the number of data words. $D_i(x)$ are the individual data word polynomials. D_0 is the lowest order data word (last data word to be transmitted and received).

The residues are computed by hardware implementing the logical circuits shown in figure 3.4.1 below. These logical circuits are clocked once per word. The P_0 shift register computes an XOR sum of all data words. The P_1 shift register computes a cyclic XOR sum of all data words. Since P_1 is primitive, its shift register generates a maximum length sequence (65,535 states). When the P_1 shift register is nonzero, but its input is zero, each word clock sets it to the next state of its sequence.

On read, the check words read from media are XOR-ed with the computed check words to obtain syndromes S0 and S1.

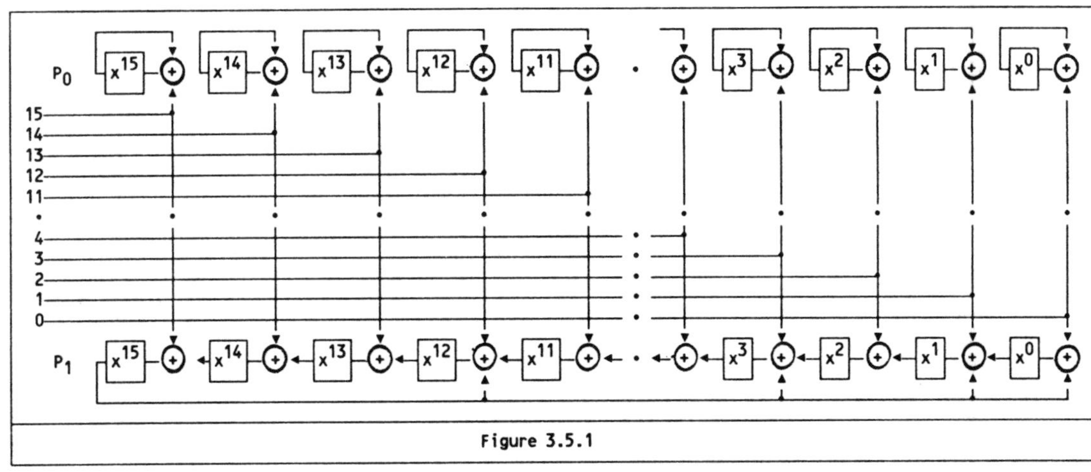

Figure 3.5.1

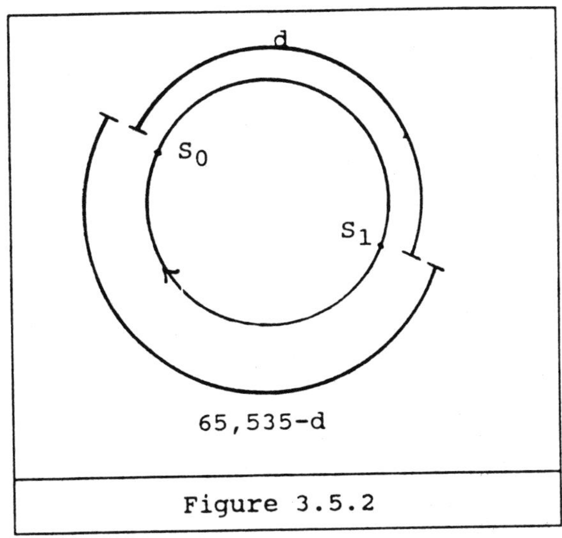

Figure 3.5.2

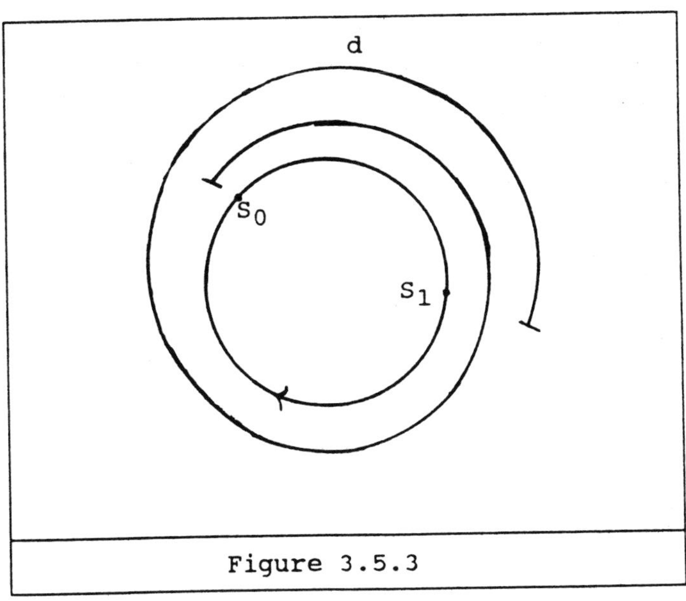

Figure 3.5.3

CORRECTION ALGORITHM

Consider what happens when the data record is all zeros and a word in error is received.

Both shift registers will remain zero until the word in error arrives. The error word is XOR-ed into the P_0 and P_1 shift registers. Since P_0 preserves its current value as long as zeros are received, the error pattern remains until the end of record. XOR-ing the error word into P_1, places it to a particular state in its sequence. This state will be referred to as the initial state. As each new word of zeros is received the P_1 shift register is clocked along its sequence, one state per word.

The terminal state of P_0 is the error pattern. The terminal states of P_0 and P_1 together establish error displacement.

To find displacement, it is necessary to determine the number of shifts of the P_1 shift register that occurs between the occurrence of the error word and the end of record.

To better understand the correction algorithm, consider a sequence of 65,535 states as represented by the circle of Figure 3.5.2.

Let S_1 be the terminal state of the P_1 shift register and let S_0 be the terminal state of the P_0 shift register. S_0 is also the initial state of the P_1 shift register.

The number of states between S_0 and S_1 must be determined. There are several ways to do this. For this simple example an implementation is assumed that clocks S_1 forward along the P_1 sequence until a match is found with S_0. The number of clocks subtracted from 65,535 is the displacement from the end of data counting the last data word as one.

Consider the case when the data is not all-zeros. The check words are selected on write such that residues of zero result on read, when the entire record is processed by the P_0 and P_1 shift registers. When an error occurs, the operation differs from the all-zeros data case only while residues are being computed. A given error condition results in the same residues, regardless of data values. Once residues have been computed, the operation is the same as previously described for the all-zeros data case.

If there is a single word in error in the record and it is check word C_0, then S_1 will be zero and S0 will be nonzero. However, if check word C_1 is the word in error, S_0 will be zero, and S_1 will be nonzero.

EXAMPLE #2 - *SINGLE-WORD ERROR CORRECTION IN TWO INTERLEAVES*

The code of example #1 could be implemented in multiple interleaves.

Consider a code with two interleaves. Assume four shift registers P_0, P_1, P_2 and P_3.

The P_0 shift register computes an XOR sum of all even data words. P_1 computes an XOR sum of all odd data words. P_2 and P_3 compute cyclic XOR sums of even and odd data words respectively.

P_0 and P_2 determine the pattern and displacement for the even interleave. P_1 and P_3 determine the pattern and displacement for the odd interleave.

This code can be used to correct a single word error in an even interleave and a single word error in an odd interleave. The error words need not be adjacent. However, correction can be restricted to double word adjacent errors by requiring a particular relationship between interleave displacements.

If the record length is even, then the odd interleave displacement (from the end of the record) must be either equal to, or one greater than the even interleave displacement.

A double adjacent word error starting on an even word will cause interleave displacements to be equal. A double adjacent word error starting on an odd word will cause the odd interleave displacement to be one greater than the even interleave displacement.

EXAMPLE #3 - *SINGLE-WORD ERROR CORRECTION USING A NONPRIMITIVE POLYNOMIAL*

The polynomial P_1 of example #1 is primitive. Therefore, it generates two sequences; a sequence of length one when initialized to zero; a sequence of length 65,535 when initialized to any nonzero state.

Consider another code where P_1 is degree 16 and irreducible but nonprimitive. Assume that P_1 has a period of 257. Then it would have 256 sequences, the zero sequence of length one and 255 sequences of length 257. The operation of the code and displacement computation would be identical to the code of example #1 except that the record length, including check words would be limited to 257.

The operation of the code is unaffected by the fact that P_1 has multiple sequences. However, it is very important that all sequences of P_1 are of an equal length, excepting the zero sequence. This condition is met by all irreducible polynomials. The condition is also met by some composite polynomials, but not all.

EXAMPLE #4 - *A CODE TO COMPUTE DISPLACEMENT MODULO SOME INTEGER*

The code of Example #3 could be part of a larger code. For example, instead of computing error displacement for a 257-word record, displacement modulo 257 could be computed for a larger record.

In this case, if the data record is all-zeros and an error is received, the P_1 shift register could traverse its sequence many times before the end of record is reached. See Figure 3.5.3.

Another part of the overall code might compute displacement modulo some other integer that is relatively prime to 257. The overall displacement then would be computed using the Chinese Remainder Method.

EXAMPLE #5 - *A CODE TO CORRECT DOUBLE-WORD ADJACENT ERRORS*

The interleave code of Example #2 uses four shift registers. Its capability includes the correction of double-word adjacent errors.

Consider a code using only three shift registers (P_0, P_1, P_2) that corrects most double-word adjacent errors.

The P_0 shift register computes an XOR sum of all even data words. The P_1 shift register computes an XOR sum of all odd data words. The P_2 shift register processes all data words (odd and even). Its definition and operation are identical to that of the P_1 shift register in the previous examples.

Assume the data to be all zeros. Assume that a double word adjacent error occurs. The two adjacent words in error will be XOR-ed into the P_0 and P_1 shift registers. Which shift register receives the first word in error depends on whether the error starts on an odd or even word. When the first error word is received, it is XOR-ed into the P_2 shift register, after which P_2 is advanced one state along its sequence. Next, the second error word is XOR-ed into P_2. P_2 is again advanced one state along its sequence.

P_2 continues to be advanced along its sequence once per data word until the end of record is reached.

The final states of shift registers P_0, P_1, P_2 are syndromes S_0, S_1, S_2.

S_0 and S_1 are the error pattern. Assume that it is known from another part of an overall code, that the error started in an even word. Then, the error displacement can be found by advancing S_2 along the P_2 sequence until a k'th state is found, such that, zero results when S_0 is XOR-ed with the k'th state and the result is advanced one state along the P_2 sequence and XOR-ed with S_1. The procedure for finding displacement would be slightly different if the error started on an odd word.

This code would not allow correction of all double word adjacent errors. If the second word in error is equal to the first word in error shifted once along the P_2 sequence, the error is not detected at all and correction cannot be accomplished.

Using two codes of this type will overcome the problem, providing the P_2 polynomials of the two codes are different and satisfy a particular criteria.

USING FINITE FIELD MATH WITH THE b-ADJACENT CODE

Let powers of α represent the elements of a field. Let reverse displacement mean the displacement from the last data word to the first word in error, counting the last data word as one.

In example #1, displacement is computed by shifting S1 forward along the P_1 sequence until a match is found with S0. In terms of finite field math, j must be determined, where:

$$S_1 \cdot \alpha^j + S_0 = 0$$

The reverse displacement is then (-j) MOD 65,535.

For example #5, j must be determined where if the double-word error starts in an even word:

$$(S_2 \cdot \alpha^j + S_0) \cdot \alpha = S_1$$

and if the double-word error starts in an odd word:

$$(S_2 \cdot \alpha^j + S_1) \cdot \alpha = S_0$$

The reverse displacement is then (-j) MOD 65,535.

CHAPTER 4 - APPLICATION CONSIDERATIONS

4.1 RAW ERROR RATES AND NATURE OF ERROR

Error rates and the nature of error must be characterized before designing and testing a real-world error-control system. The error characteristics should be determined by a combination of measurement and estimation. The estimation should be based on experiences with similar products and technologies. Data typically required is listed below.

1. Defect distribution (number of defects per media of each defect length).

2. Soft-error distribution (number of soft errors versus total bits transferred for each error burst length).

3. Methods of defect identification at the time of manufacture.

4. Percentage of defects and percentage of soft errors that result in loss of sync.

5. Probability that a loss of sync results in the phase lock loop (PLL) staying off frequency.

6. Probability of sync framing error.

7. Probability of false sync detection.

8. Change in defect rate versus media usage and storage time.

9. Change in soft error rate versus media usage and storage time.

10. Information on any clustering of defects or soft errors, such as:

    a. High probability of multiple bursts.
    b. High probability of long bursts.
    c. Higher error rate at particular tracks.
    d. Periodic misregistration.
    e. Interference from another function.
    f. Weak areas of media.
    g. Media deformity.
    h. Contamination.

11. Other recovery means that may be used and their effectiveness. Some recovery techniques used on prior storage products are listed below.

    a. Head offset.
    b. Detection window shift.
    c. VFO bandwidth change.
    d. Detector threshold change.

12. Relationship between decoded bits in error and encoded bits in error for the read/write modulation method used.

13. Available pointer information that can be used for erasure correction. Some sources of pointer information on prior storage products are listed below.

    a. Excessive phase shift.
    b. Excessive amplitude deviations.
    c. Invalid code found by the modulation method.
    d. Error locations from adjacent interleaves.

14. Information on usage. For example, expected bits read per day and expected accesses per day.

15. Record sizes.

4.2 DECODED ERROR RATES

Error correction is used in communication systems to improve channel throughput. It is used in storage device subsystems to improve data recoverability. Part of the design of every error control system is determining code performance. The block (or decoded) error rate for a specified raw error rate is one measure of performance.

The equations and tables below and on the following pages can be used to determine the block error rate when raw error rate and the number of errors corrected per block are known. A block error exists if, after performing error correction, the data is erroneous. The block error rate is the ratio of block occurrences to blocks transferred. Raw error rate for the equations is the ratio of raw error occurrences to a unit of data transfer. The unit of data transfer is specified in each case. The raw error rate for the tables is the ratio of raw error occurrences to bits transferred. An error may be a bit, symbol, or burst error. Errors are assumed to be random; the equations and tables give erroneous results if they are not.

In the equations, the following notation represents the number of ways to chose r out of n without regard to order.

$$\begin{bmatrix} n \\ r \end{bmatrix} = \frac{n!}{r!*(n-r)!} = \prod_{j=0}^{r-1} \frac{(n-j)}{(r-j)}$$

Some of the probability equations given on the following pages can be reduced in complexity by using the following relationships when applicable.

$$(1+P_e)^{n-r} \approx 1 \quad \text{if } P_e<<1$$

$$\begin{bmatrix} n \\ r \end{bmatrix} \approx \frac{n^r}{r!} \quad \text{if } n>>r$$

BIT-ERROR PROBABILITIES

Let P_e be the raw-bit-error rate. Let the raw-bit-error rate be defined as the ratio of bit error occurrences to total bits transferred; that is, bit errors per bit. The equations below give probabilities for various numbers of bit errors occurring in a block of n bits.

PROBABILITY OF EXACTLY r BIT ERRORS IN A BLOCK OF n BITS

$$P_r = \begin{bmatrix} n \\ r \end{bmatrix} * (P_e)^r * (1-P_e)^{n-r}$$

PROBABILITY OF ZERO BIT ERRORS IN A BLOCK OF n BITS

$$P_0 = \begin{bmatrix} n \\ 0 \end{bmatrix} * (P_e)^0 * (1-P_e)^n = (1-P_e)^n$$

PROBABILITY OF ONE BIT ERROR IN A BLOCK OF n BITS

$$P_1 = \begin{bmatrix} n \\ 1 \end{bmatrix} * (P_e)^1 * (1-P_e)^{n-1}$$

PROBABILITY OF AT LEAST ONE BIT ERROR IN A BLOCK OF n BITS

$$\sum_{r>0}^{n} P_r = P_1+P_2+\cdots+P_n = 1-P_0$$

PROBABILITY OF TWO OR MORE BIT ERRORS IN A BLOCK OF n BITS

$$\sum_{r>1}^{n} P_r = P_2+P_3+\cdots+P_n = 1-P_0-P_1$$

DECODED ERROR PROBABILITIES FOR A BIT-CORRECTING CODE

n = Block length in bits

e = Number of bits corrected per block
= 0 for an error-detection-only code

P_e = Raw bit error probability (units: bit errors per bit)

$$\frac{\text{BLOCK ERRORS}}{\text{BLOCK}} \approx \sum_{i>e}^{n} \begin{bmatrix} n \\ i \end{bmatrix} * (P_e)^{i} * (1-P_e)^{n-i}$$

$$\frac{\text{BLOCK ERRORS}}{\text{BIT}} \approx \frac{1}{n} * \sum_{i>e}^{n} \begin{bmatrix} n \\ i \end{bmatrix} * (P_e)^{i} * (1-P_e)^{n-i}$$

$$\frac{\text{BIT ERRORS}}{\text{BLOCK}} \approx \sum_{i>e}^{n} (i+e) * \begin{bmatrix} n \\ i \end{bmatrix} * (P_e)^{i} * (1-P_e)^{n-i}$$

$$\frac{\text{BIT ERRORS}}{\text{BIT}} \approx \frac{1}{n} * \sum_{i>e}^{n} (i+e) * \begin{bmatrix} n \\ i \end{bmatrix} * (P_e)^{i} * (1-P_e)^{n-i}$$

BURST-ERROR PROBABILITIES

Let P_e be the raw burst-error rate, defined as the ratio of burst error occurrences to total bits transferred, with units of burst errors per bit. The equations below give the probabilities for various numbers of burst errors occurring in a block of n bits. It is assumed that burst length is short compared to block length.

PROBABILITY OF EXACTLY r BURST ERRORS IN A BLOCK OF n BITS

$$P_r = \begin{bmatrix} n \\ r \end{bmatrix} * (P_e)^r * (1-P_e)^{n-r}$$

PROBABILITY OF ZERO BURST ERRORS IN A BLOCK OF n BITS

$$P_0 = \begin{bmatrix} n \\ 0 \end{bmatrix} * (P_e)^0 * (1-P_e)^n = (1-P_e)^n$$

PROBABILITY OF ONE BURST ERROR IN A BLOCK OF n BITS

$$P_1 = \begin{bmatrix} n \\ 1 \end{bmatrix} * (P_e)^1 * (1-P_e)^{n-1}$$

PROBABILITY OF AT LEAST ONE BURST ERROR IN A BLOCK OF n BITS

$$\sum_{r>0}^{n} P_r = P_1+P_2+\cdots+P_n = 1-P_0$$

PROBABILITY OF TWO OR MORE BURST ERRORS IN A BLOCK OF n BITS

$$\sum_{r>1}^{n} P_r = P_2+P_3+\cdots+P_n = 1-P_0-P_1$$

DECODED ERROR PROBABILITIES FOR A BURST-CORRECTING CODE

n = Block length in bits

e = Number of bursts corrected per block
= 0 for an error-detection-only code

P_e = Raw burst error probability
(units: burst errors per bit)

$$\frac{\text{BLOCK ERRORS}}{\text{BLOCK}} \approx \sum_{i>e}^{n} \begin{bmatrix} n \\ i \end{bmatrix} * (P_e)^{i} * (1-P_e)^{n-i}$$

$$\frac{\text{BLOCK ERRORS}}{\text{BIT}} \approx \frac{1}{n} * \sum_{i>e}^{n} \begin{bmatrix} n \\ i \end{bmatrix} * (P_e)^{i} * (1-P_e)^{n-i}$$

$$\frac{\text{BURST ERRORS}}{\text{BLOCK}} \approx \sum_{i>e}^{n} (i+e) * \begin{bmatrix} n \\ i \end{bmatrix} * (P_e)^{i} * (1-P_e)^{n-i}$$

$$\frac{\text{BURST ERRORS}}{\text{BIT}} \approx \frac{1}{n} * \sum_{i>e}^{n} (i+e) * \begin{bmatrix} n \\ i \end{bmatrix} * (P_e)^{i} * (1-P_e)^{n-i}$$

SYMBOL-ERROR PROBABILITIES

Let P_e be the raw-symbol-error rate, defined as the ratio of symbol error occurrences to total symbols transferred, with units of symbol errors per symbol. The equations below give probabilities for various numbers of symbol errors occurring in a block of n symbols.

PROBABILITY OF EXACTLY r SYMBOL ERRORS IN A BLOCK OF n SYMBOLS

$$Pr = \begin{bmatrix} n \\ r \end{bmatrix} * (P_e)^r * (1-P_e)^{n-r}$$

PROBABILITY OF ZERO SYMBOL ERRORS IN A BLOCK OF n SYMBOLS

$$P_0 = \begin{bmatrix} n \\ 0 \end{bmatrix} * (P_e)^0 * (1-P_e)^n = (1-P_e)^n$$

PROBABILITY OF ONE SYMBOL ERROR IN A BLOCK OF n SYMBOLS

$$P_1 = \begin{bmatrix} n \\ 1 \end{bmatrix} * (P_e)^1 * (1-P_e)^{n-1}$$

PROBABILITY OF AT LEAST ONE SYMBOL ERROR IN A BLOCK OF n SYMBOLS

$$\sum_{r>0}^{n} Pr = P_1+P_2+\cdots+P_n = 1-P_0$$

PROBABILITY OF TWO OR MORE SYMBOL ERRORS IN A BLOCK OF n SYMBOLS

$$\sum_{r>1}^{n} Pr = P_2+P_3+\cdots+P_n = 1-P_0-P_1$$

DECODED ERROR PROBABILITIES FOR A SYMBOL-CORRECTING CODE

n = Block length in symbols

e = Number of bits corrected per block
= 0 for an error-detection-only code

P_e = Raw symbol error probability
(units: symbol errors per symbol)

w = Symbol width in bits

$$\frac{\text{BLOCK ERRORS}}{\text{BLOCK}} \approx \sum_{i>e}^{n} \binom{n}{i} * (P_e)^i * (1-P_e)^{n-i}$$

$$\frac{\text{BLOCK ERRORS}}{\text{SYMBOL}} \approx \frac{1}{n} * \sum_{i>e}^{n} \binom{n}{i} * (P_e)^i * (1-P_e)^{n-i}$$

$$\frac{\text{BLOCK ERRORS}}{\text{BIT}} \approx \frac{1}{w*n} * \sum_{i>e}^{n} \binom{n}{i} * (P_e)^i * (1-P_e)^{n-i}$$

$$\frac{\text{SYMBOL ERRORS}}{\text{BLOCK}} \approx \sum_{i>e}^{n} (i+e) * \binom{n}{i} * (P_e)^i * (1-P_e)^{n-i}$$

$$\frac{\text{SYMBOL ERRORS}}{\text{SYMBOL}} \approx \frac{1}{n} * \sum_{i>e}^{n} (i+e) * \binom{n}{i} * (P_e)^i * (1-P_e)^{n-i}$$

$$\frac{\text{SYMBOL ERRORS}}{\text{BIT}} \approx \frac{1}{w*n} * \sum_{i>e}^{n} (i+e) * \binom{n}{i} * (P_e)^i * (1-P_e)^{n-i}$$

$$* \quad \frac{\text{BIT ERRORS}}{\text{BIT}} \approx \frac{1}{2*n} * \sum_{i>e}^{n} (i+e) * \binom{n}{i} * (P_e)^i * (1-P_e)^{n-i}$$

* Assuming a symbol error results in k/2 bit errors.

DECODED ERROR PROBABILITIES FOR A SYMBOL-CORRECTING CODE WHEN ERASURE POINTERS ARE AVAILABLE FOR SYMBOL ERRORS

n = Block length in symbols

e = Number of bits corrected per block
= 0 for an error-detection-only code

P_e = Raw symbol error probability
(units: symbol errors per symbol)

w = Symbol width in bits

$$\frac{\text{BLOCK ERRORS}}{\text{BLOCK}} \approx \sum_{i>e}^{n} \begin{bmatrix} n \\ i \end{bmatrix} * (P_e)^{i} * (1-P_e)^{n-i}$$

$$\frac{\text{BLOCK ERRORS}}{\text{SYMBOL}} \approx \frac{1}{n} * \sum_{i>e}^{n} \begin{bmatrix} n \\ i \end{bmatrix} * (P_e)^{i} * (1-P_e)^{n-i}$$

$$\frac{\text{BLOCK ERRORS}}{\text{BIT}} \approx \frac{1}{w*n} * \sum_{i>e}^{n} \begin{bmatrix} n \\ i \end{bmatrix} * (P_e)^{i} * (1-P_e)^{n-i}$$

$$\frac{\text{SYMBOL ERRORS}}{\text{BLOCK}} \approx \sum_{i>e}^{n} i * \begin{bmatrix} n \\ i \end{bmatrix} * (P_e)^{i} * (1-P_e)^{n-i}$$

$$\frac{\text{SYMBOL ERRORS}}{\text{SYMBOL}} \approx \frac{1}{n} * \sum_{i>e}^{n} i * \begin{bmatrix} n \\ i \end{bmatrix} * (P_e)^{i} * (1-P_e)^{n-i}$$

$$\frac{\text{SYMBOL ERRORS}}{\text{BIT}} \approx \frac{1}{w*n} * \sum_{i>e}^{n} i * \begin{bmatrix} n \\ i \end{bmatrix} * (P_e)^{i} * (1-P_e)^{n-i}$$

$$* \quad \frac{\text{BIT ERRORS}}{\text{BIT}} \approx \frac{1}{2*n} * \sum_{i>e}^{n} (i) * \begin{bmatrix} n \\ i \end{bmatrix} * (P_e)^{i} * (1-P_e)^{n-i}$$

* Assuming a symbol error results in k/2 bit errors.

4.3 DATA RECOVERABILITY

Error correction is used in storage device subsystems to improve data recoverability. There are other techniques that improve data recoverability as well. Some of these techniques are discussed in this section. System manufacturers may want to include data recovery techniques on their list of criteria for comparing subsystems.

DATA RECOVERY TECHNIQUES

Some storage device subsystems attempt data recovery with the techniques below when ECC is unsuccessful.

a. Head offset.
b. Detection window shift.
c. VFO bandwidth change.
d. Detector threshold change.
e. Rezero and reread.
f. Remove and reinsert media then reread.
g. Move media to another device and reread.

DATA SEPARATOR

The design of the data separator will have a significant influence on data recoverability. Some devices have built-in data separators. Other devices require a data separator in the controller.

Controller manufacturers should consult their device vendors for recommendations when designing a controller for devices which require external data separators.

Circuit layout and parts selection are very important for data separators. Even if one has a circuit recommended by a drive vendor, it may be advisable to use a highly experienced read/write consultant for the detailed design and layout.

WRITE VERIFY

Another technique that can improve the probability of data recovery is write verify (read back after write). Write verify can be very effective for devices using magnetic media due to the nature of defects in this media. One may write/read over a defect hundreds of times without an error. An error will result only when the write occurs with the proper phasing across the defect. Once the error occurs, it may then have a high incidence rate until the record is rewritten. Hundreds of writes may be required before the error occurs again.

When an error is detected by write verify, the record is rewritten or retired or defect skipping is applied. This reserves error correction for errors that develop with time or usage. Since it affects performance, write verify should be optional.

DEFECT SKIPPING

Defect-skipping techniques include alternate-sector assignment, header move functions, and defect skipping within a data field. These techniques are used to handle media defects detected during formatting and persistent errors detected on read.

Under alternate-sector assignment, a defective sector may be retired and logically replaced with a sector physically located elsewhere. Space for alternate sector(s) may be reserved on each track or cylinder, or one or more tracks or cylinders may be reserved exclusively for alternate sectors. The header contains an alternate-sector assignment field; when a sector is retired, this field in its header is written to point to the alternate sector which is to logically replace it. An assigned alternate sector typically has a field which points back to the retired sector that it is replacing.

When a header-move function is implemented, a defect falling in a header is avoided by moving the header further along the track. Space may be allotted in the track format to allow a normal-length data field to follow a moved header, or the moved header may contain a field pointing to an assigned alternate sector. In the latter case, since the data field following a moved header is not used, it need not be of normal length; it may or may not actually be written, depending on implementation alternatives.

Defect skipping within a data field is used in some high-capacity magnetic disk subsystems employing variable-length records as a means of handling known defects. Each record has a count field which records information on the locations of defects within the track. Writing is interrupted when the current byte displacement from the index corresponds to the starting offset of a skip as recorded in the count field. When the recording head passes beyond the known length of the defect, a preamble pattern and sync mark are written, then writing of data re-commences. Some IBM devices allow up to seven defects per track to be skipped in this manner.

Defect skipping within a data field is also used on magnetic devices employing fixed-length records. In this case, each sector header records displacement information for defects in that sector. Some implementations write a preamble pattern and sync mark at the end of a skip as discussed above for variable-length records while others do not. The former practice handles defects which can cause loss of sync. If a preamble pattern and sync mark are not written, some other method must be used to map out defects which can cause loss of sync.

Devices employing defect skipping within a data field must allocate extra media area for each sector, track, or cylinder, depending on whether or not embedded servoing is used and on other implementation choices. In devices using embedded servoing, the space allotted for each sector must allow room for the maximum-length defect(s) which may be skipped. In devices not using embedded servo techniques, the track format need accommodate only some maximum number of skips per track, which may be much less than one per sector.

When defect-skipping techniques are used and skip or alternate-sector information is stored in headers, care must be taken to make sure that the storage of information in headers other than track and sector number does not weaken the error tolerance of the headers. A different method for alternate-sector assignment, which avoids this complication, is sector slipping. Each track or cylinder contains enough extra area to write one or more extra sectors. When a sector must be retired, it and each succeeding sector are slipped one sector-length along the track or cylinder. This method has the additional advantage that sectors remain consecutive and no additional seek time is required to find an alternate sector at the end of the track or cylinder, or on a different track or track or cylinder. This method is discussed in more detail under A HEADER STRATEGY EXAMPLE below.

ERROR-TOLERANT TRACK FORMATS

Achieving error tolerance in the track format is a major consideration when architecting a storage device and controller for high error rate media. All special fields and all special bytes of the track format must be error-tolerant. This includes but is not limited to preambles, sync marks, header fields, sector marks, and index marks.

Experience shows that designing an error-tolerant track format (one that does not dominate the uncorrectable sector event rate) to support high defect densities can be even more difficult than selecting a high performance ECC code.

SYNCHRONIZATION

For high defect rate devices, it is essential that the device/controller architectures include a high degree of tolerance to defects that fall within sync marks. There are several synchronization strategies that achieve this. The selection will be influenced by the nature of the device and the nature of defects (e.g., length distribution, growth rate, etc.). Both false detection and detection failure probabilities must be considered. Synchronization is discussed in detail in Section 4.8.1; some high points are briefly covered below.

One method for achieving tolerance to defects that fall within sync marks is to employ error-tolerant sync marks. Error-tolerant sync marks have been used in the past that can be detected at the proper time even if several small error bursts or one large error burst occurs within the mark. See Section 4.8.1 for a more in-depth discussion of synchronization codes.

Another strategy is to replicate sync marks with some number of bytes between. The number of bytes between replications would be determined by the maximum defect length to be accommodated. A different code is used for each replication so that the detected code identifies the true start of data. The number of replications required is selected to achieve a high probability of synchronization for the given rate and nature of defects. Mark lengths, codes, and detection qualification criteria are selected to achieve an acceptable rate of false sync mark detection.

If synchronization consists of several steps, each must be error-tolerant. If sector marks (also called address marks) and preambles precede sync marks they must also be error tolerant. Today, in some implementations correct synchronization will not be achieved if an error occurs in the last bit or last few bits of a preamble. Such sensitivities must be avoided. Section 4.8.1 discusses how error tolerance can be achieved in the clock-phasing step of synchronization as well as in the byte-synchronization step.

MAINTAINING SYNCHRONIZATION THROUGH LARGE DEFECTS

Obviously, it is desirable to maximize the defect length that the PLL can flywheel through without losing synchronization. Engineers responsible for defect handling strategy will want to influence the device's rotational speed stability and PLL flywheeling characteristics. One technique that has been used to extend the length of bursts the PLL can flywheel through is to coast the PLL through defects by using some criteria (run-length violation, loss of signal amplitude, etc.) to temporarily shut off updating of the PLL's frequency and phase memory.

FALSE SYNC MARK DETECTION

The false detection of a sync mark can result in synchronization failure. The probability of false mark detection must be kept low by careful selection of mark lengths, codes, and qualification criteria.

In some architectures, once data acquisition has been achieved, sync mark detection is qualified with a timing window in order to minimize the probability of false detection. In such an architecture, it is desirable to generate the timing window from the reference clock; if the timing window is generated from the data clock and the PLL loses sync while clocking over a large defect in a known defective sector, the following good sector may be missed due to the subsequent mispositioning of the timing window.

HEADERS

For high error-rate devices, header strategy is influenced by defect event rates, growth rates, length distributions, performance requirements, and write prerequisites.

One header strategy requires replication. A number of contiguous headers with CRC are written, then on read one copy must be read error-free. Another strategy is to allow a data field to be recovered even if its header is in error. This requires that headers consist solely of address information such as track and sector number. If a header is in error, such information can be generated from known track orientation. Some devices combine this strategy with header replication in order to minimize the frequency at which address information is generated rather than read. In any case, devices using high error-rate media must be insensitive to defects falling into the headers of several consecutive sectors. When address information is generated rather than read, the data field can be further qualified by subsequent headers.

Using error correction on the header field as well as the data field will increase the probability of recovering data. However, one must either be able to store and correct both a header and the associated data field, or provide a way to space over a defective header in order to recover the associated data field on a succeeding revolution.

An alternative to correcting the header is to keep only address information in the header and to provide a way to space over a defective header. When a defective header is detected, record address is computed from track orientation. A disadvantage of this method is that it does not allow flags to be part of the header field.

Some devices also include address information within the highly protected data field to use as a final check that the proper data field was recovered. This check must take place after error correction. The best time to perform it may be just before releasing the sector for transfer to the host.

A HEADER STRATEGY EXAMPLE

A typical error-tolerant header and sector-retirement strategy might be: Store in the header only track and sector address information. Reserve K sectors at the end of each cylinder for spare sectors. When a sector must be retired, slip all data sectors down the cylinder by one sector position and write a special "sector-retired" flag in place of the sector number in the header of the retired sector. On searches if a header is read error-free and the "sector-retired" flag is found instead of a sector number, adjust the sector number in the known orientation and continue searching.

If a header-in-error is encountered during a search then it is either the header of a sector that had been previously retired or it is a header containing a temporary error or a new hard defect. The sector number sequence encountered in continuing the search can be used to determine which is the case. If the header-in-error was that of an already-retired sector, the sector number sequence should be adjusted and the search continued. Otherwise the search should still be continued unless the header-in-error was that of the desired sector, in which case the search should be interrupted and a re-read attempted. If the error is not present on re-read, assume it was a temporary error and proceed to read the data field. If the error persists on re-read, assume a new hard defect: orient on the preceding sector, skip the header-in-error, and read the desired data field. A sector whose header contains a new hard defect should be retired as soon as possible.

Note that the error-tolerant header strategy outlined above will not work if it is necessary to store control data, such as location information for defect skipping, within headers.

SERVO SYSTEMS

In many devices, the ability to handle large defects is limited by the servo system(s). Engineers responsible for defect handling strategy must understand the limits of the servo system(s) relative to defect tolerance. In particular, any testing of defect handling capabilities should include the servo system(s).

MODULATION CODES

The modulation code selected will affect EDAC performance by influencing noise--generated error rates, the extension of error bursts, the ability to acquire synchronization, the ability to hold synchronization through defects, the ability to generate erasure pointers, and the resolution of erasure pointers.

The following summarizes the results of an analysis of the error propagation performance of the (2,7) code described in U.S. Patent #4,115,768, inventors Eggenberger and Hodges, assignee IBM (1978). Analysis was confined to cases of single-bit errors defined below:

```
Drop-in:  A code-bit '1' where '0' was encoded
Drop-out: A code-bit '0' where '1' was encoded
Shift:    A code-bit '1' where '0' was encoded, coincident
          with an adjacent code-bit '0' where '1' was encoded
```

Error propagation length is defined as the inclusive number of data-bits between the first data-bit in error and the last data-bit in error caused by a given code-bit error case.

Random fifteen-data-bit sequences were generated and encoded using the encoder described in the patent. Drop-in, drop-out, and shift errors were created in turn in the twelfth through the eighteenth bits of the resulting code-bit sequences. The corrupted code-bit sequences were decoded using the decoder described in the patent, the resulting data-bit sequences were analyzed, and the error propagation lengths recorded. Results of 2000 trials are shown below:

ERROR TYPE		ERROR PROPAGATION LENGTH 0	1	2	3	4	5	TOTAL
DROP-IN	#	2674	5009	1220	195	127	0	9225
	%	29	54	13	2	1	0	
DROP-OUT	#	201	1258	776	448	92	0	2775
	%	7	45	28	16	3	0	
SHIFT	#	150	1955	1496	1242	508	125	5476
	%	3	36	27	23	9	2	
TOTAL	#	3025	8222	3492	1885	727	125	17476
	%	17	47	20	11	4	1	

4.4 DATA ACCURACY

Data accuracy is one of the most important considerations in error correction system design. The following discussion on data accuracy is concerned primarily with magnetic disk applications. However, the concepts are extendable to many other error correction applications.

The transfer of undetected erroneous data can be one of the most catastrophic failures of a data storage system; consider the consequences of an undetected error in the money field of a financial instrument or the control status of a nuclear reactor. Most users of disk subsystems consider data accuracy even more important than data recoverability. Nevertheless, many disk subsystem designers are unaware of the factors determining data accuracy.

Some causes of undetected erroneous data transfer are listed below.

- Miscorrection by an error-correcting code.
- Misdetection by an error-detecting or error-correcting code.
- Synchronization framing errors in an implementation without synchronization framing error protection.
- Occasional failure on an unprotected data path on write or read.
- Occasional failure on an unprotected RAM buffer within the data path on write or read.
- A software error resulting in the transfer of the wrong sector.
- A broken error latch which never flags an error; other broken hardware.

Some other factors impacting data accuracy are discussed below.

POLYNOMIAL SELECTION

In disk subsystems, the error-correction polynomial has a significant influence on data accuracy. Fire code polynomials, for example, have been widely used on disk controllers, yet they provide less accuracy than carefully selected computer-generated codes.

Many disk controller manufacturers have employed one of the following Fire code polynomials:

$$(x^{21} + 1) \bullet (x^{11} + x^2 + 1) \text{ or } (x^{21} + 1) \bullet (x^{11} + x^9 + 1)$$

The natural period of each polynomial is 42,987. Burst correction and detection spans are both eleven bits for record lengths, including check bits, no greater than the natural period. These codes are frequently used to correct eleven-bit bursts on record lengths of 512 bytes.

When used for correction of eleven-bit bursts on a 512-byte record, these codes miscorrect ten percent of all possible double bursts where each burst is a single bit in error. With the same correction span and record length, the miscorrection probability for all possible error bursts is one in one thousand. The short double burst, with each burst a single bit in error, has a miscorrection probability two orders of magnitude greater.

Such codes have a high miscorrection probability on other short double bursts as well. Double bursts are not as common as single bursts. However, due to error clustering, they occur frequently enough to be a problem.

The data accuracy provided by the above Fire codes for all possible error bursts is comparable to that provided by a ten-bit CRC code. The data accuracy for all possible double-bit errors is comparable to that provided by a three-bit or four-bit CRC code.

Fire codes are defined by generator polynomials of the form:

$$g(x) = c(x) \bullet p(x) = (x^c + 1) \bullet p(x)$$

where $p(x)$ is any irreducible polynomial of degree z and period e, and e does not divide c.

The period of the generator polynomial $g(x)$ is the least common multiple of c and e. For record lengths (including check bits) not exceeding the period of $g(x)$, these codes are guaranteed to correct single bursts of length b bits and detect single bursts of length d bits where $d \geq b$, provided $z \geq b$ and $c \geq (d+b-1)$.

The composite form of the generator polynomial ($g(x)$) is used for encoding. Decoding can be performed with a shift register implementing the composite generator polynomial ($g(x)$) or by two shift registers implementing the factors of the generator polynomial ($c(x)$ and $p(x)$). Code performance is the same in either case.

The $p(x)$ factor of the Fire code generator polynomial carries error displacement information. The $c(x)$ factor carries error pattern information. It is this factor that is responsible for the Fire code's pattern sensitivity. To understand the pattern sensitivity, assume that decoding is performed with shift registers implementing the individual factors of the generator polynomial. For a particular error burst to result in miscorrection, it must leave in the c(x) shift register a pattern that qualifies as a correctable error pattern. A high percentage of short double bursts do exactly that. For example, two bits in error, (c+1) bits apart, would leave the same pattern in the c(x) shift register as an error burst of length two. The same would be true of two bits in error separated by any multiple of (c+1) bits.

If p(x) has more redundancy than required by the Fire code formulas, the excess redundancy reduces the miscorrection probability for short double bursts, as well as the miscorrection probability for all possible error bursts.

The overall miscorrection probability (P_{mc}) for a Fire code is given by the following equation, assuming all errors are possible and equally probable.

$$P_{mc} \approx \frac{n*2^{(b-1)}}{2^m} \qquad (1)$$

where,
n = record length in bits including check bits.
b = guaranteed single burst correction span in bits.
m = total number of check bits.

For many Fire codes, the miscorrection probability for double bursts where each burst is a single bit in error is given by the following equation, assuming all such errors are possible and equally probable.

$$P_{mcdb} \approx \frac{2*n*(b-1)}{c^2*(2^z-1)} \qquad (2)$$

where,
n and b are as defined above.
c = degree of the c(x) factor of the Fire code polynomial.
z = degree of the p(x) factor of the Fire code polynomial.

This equation is unique to the Fire Code. It is applicable only when the product of P_{mcdb} and the number of possible double-bit errors is much greater than one. When this is not true, a computer search should be used to determine P_{mcdb}.

The ratio of P_{mcdb} to P_{mc} provides a measure of pattern sensitivity for one particular double burst (each burst a single bit in error). Remember that the Fire code is sensitive to other short double bursts as well.

Properly selected computer-generated codes do not exhibit the pattern sensitivity of Fire codes. In fact, it is possible to select computer-generated codes that have a guaranteed double-burst detection span. The miscorrecting patterns of these codes are more random than those of Fire codes. They are selected by testing a large number of random polynomials of a particular degree. Provided the specifications are within certain bounds, some polynomials will satisfy them.

There are equations that predict the number of polynomials one must evaluate to meet a particular specification.

In some cases, thousands of computer-generated polynomials must be evaluated to find a polynomial with unique characteristics.

For a computer-generated code, correction and detection spans are determined by computer evaluation. Overall miscorrection probability is given by Equation #1.

To increase data accuracy, many disk controller manufacturers are switching from Fire codes to computer-generated codes.

ERROR RECOVERY STRATEGY

Error recovery strategies also have a significant influence on data accuracy. A strategy that requires data to be reread before attempting correction provides more accurate data than a strategy requiring the use of correction before rereading.

An equation for data inaccuracy is given below:

$$P_{ued} \approx P_e * P_c * P_{mc} \tag{3}$$

where,

P_{ued} = Probability of undetected erroneous data

Ratio of undetected erroneous data occurrences to total bits transferred. This is a measure of data inaccuracy.

P_e = Raw burst error rate

Ratio of raw burst error occurrences to total bits transferred.

P_c = Catastrophic probability

Probability that a given error occurrence exceeds the guaranteed capabilities of a code.

P_{mc} = Miscorrection probability

Probability that a given error occurrence, exceeding the guaranteed capabilities of a code, will result in miscorrection, assuming all errors are possible and equally probable.

It is desirable to keep the probability of undetected erroneous data (P_{ued}) as low as possible. The burst error rate, catastrophic probability or miscorrection probability must be reduced to reduce P_{ued}. (See Equation #3).

Miscorrection probability (P_{mc}) can be reduced by decreasing the record length and/or the correction span, or by increasing the number of check bits. Catastrophic probability (P_c) can be reduced by increasing the guaranteed capabilities of the code, or by reducing the percentage of error bursts that exceed the guaranteed code capabilities.

Burst error rate (P_e) can be reduced by using reread. Most disk products exhibit soft burst error rates several orders of magnitude higher than hard burst error rates. Rereading before attempting correction makes P_e (in Equation #3) the hard burst error rate instead of the soft burst error rate, reducing P_{ued} by orders of magnitude.

Rereading before attempting correction provides additional improvement in P_{ued} due to the different distributions of long error bursts and multiple error bursts in hard and soft errors.

Another strategy that reduces P_{ued} is to reread until an error disappears, or until there has been an identical syndrome for the last two reads. Correction is then attempted only after a consistent syndrome has been received.

DESIGN PARAMETERS

For data accuracy, a low miscorrection probability is desirable. Miscorrection probability can be reduced by decreasing the record length and/or correction span, or by increasing the number of check bits.

For most Winchester media, a five-bit correction span has been considered adequate. A longer correction span is needed if the drive uses a read/write modulation method that maps a single encoded bit in error into several decoded bits in error, such as group coded recording (GCR) and run-length limited (RLL) codes.

For several years, 32-bit codes were considered adequate for sectored Winchester disks provided that the polynomial was selected carefully, record lengths were short, correction span was low, correction was used only on hard errors, and the occurrence rate for hard errors exceeding the guaranteed capability of the code was low.

More recently, most disk controller developers have been using 48-, 56- and 64-bit codes in their new designs. Using more check bits increases data accuracy and provides flexibility for increasing the correction span when the product is enhanced. Using more check bits also allows other error-recovery strategies to be considered, such as on-the-fly correction.

Disk controller developers are also implementing redundant sector techniques and Reed-Solomon codes. Redundant sector techniques allow very long bursts to be corrected. Reed-Solomon codes allow multiple bursts to be corrected.

ECC CIRCUIT IMPLEMENTATION

Cyclic codes provide very poor protection when frame synchronization is lost, i.e., when synchronization occurs early or late by one or more bits.

One way to protect against this type of error is to initialize the shift register to a specially selected nonzero value. The same initialization constant must be used on read and write. Another method is to invert a specially selected set of check bits on write and read. Each method gives the ECC circuit another important feature - nonzero check bits are written for an all-zeros data record. This allows certain logic failures to be detected before inaccurate data is transferred. See Section 4.8.2 for further discussion of synchronization framing errors.

Still, some ECC circuit failures can result in transferring inaccurate data. If the probability of ECC logic failure contributes significantly to the probability of transferring inaccurate data, include some form of self-checking. See Section 6.5.

DEFECT MANAGEMENT STRATEGY

All defects should have alternate sectors assigned, either by the drive manufacturer or subsystem manufacturer, before the disk subsystem is shipped to the end user.

There are problems with a philosophy that leaves defects to be corrected by ECC on each read, instead of assigning alternate sectors. First, if correction before reread is used, a higher level of miscorrection results. This is because a soft error in a sector with a defect results in a double burst. Once a double burst occurs that exceeds the double-burst-detection span, miscorrection is possible. In the second case, if reread before correction is used, revolutions will be lost each time a defective sector is read.

ERROR RATES

Clearly, disk drive error rates also significantly influence data accuracy. If errors exceeding the guaranteed capability of the code never occurred, inaccurate data would never be transferred.

When a data separator is part of the controller, its design affects error rate and therefore data accuracy. While most drive manufacturers provide recommended data separator designs, there are also well-qualified consultants who specialize in this area.

SPECIFYING DATA ACCURACY

The probability of undetected erroneous data (P_{ued}) is a measure of data inaccuracy. Sophisticated developers of disk subsystems are now targeting 1.E-20 or less for P_{ued}.

Even when P_e and P_c are high, one can still achieve any arbitrarily low P_{ued} by carefully selecting the correction span, record length, and number of check bits. (See Equations #1 and #3).

ACHIEVING HIGHER DATA INTEGRITY

The following first appeared in slightly different form in the March 1988 issue of the ENDL Newsletter.

Horror stories about the consequences of a storage subsystem transferring undetected erroneous data have been circulating since the dawn of the computer age. As the computer industry matures, data integrity requirements for storage subsystems have increased along with capacity, throughput, and uptime requirements. To meet these higher demands, both the probability of uncorrectable error and the probability of transferring undetected erroneous data must decrease. As more and more powerful error detection and correction systems are implemented to protect data from higher media-related error rates, errors arising in other areas of the subsystem will come to dominate unless equivalent protection is provided. The most powerful media EDAC system is useless against errors occurring anywhere in the write path from the host interface to the input of the EDAC encoder or in the read path from the output of the EDAC decoder to the host interface.

One example of undetected erroneous data which the media EDAC system is powerless to detect is a single-bit soft error occurring in an unprotected data buffer after the EDAC system has corrected the data but before the data are transferred to the host. Another example is a subtle subsystem software error which causes a request for the wrong sector to be executed. The actual sector fetched may contain no media-related errors and so be accepted as correct by the media EDAC system, yet it is not the data which the host requested.

Data Systems Technology, Corp. (DST) has proposed a method to combat errors not covered by the media EDAC system. DST recommends that the host append a CRC redundancy field to each logical sector as it is sent to the storage subsystem and perform a CRC check on each logical sector as it is received from the storage subsystem. DST further recommends that a logical identification number containing at least the logical sector number, and perhaps the logical drive number as well, be placed within each logical sector written to a storage subsystem and that this number be required to match that requested when each logical sector is received from the storage subsystem.

It is possible to combine these two functions so that only four extra bytes per logical sector are needed to provide both thirty-two-bit CRC protection and positive sector/drive identification. Three methods are outlined below; whatever method is chosen for implementing the two functions, it must be selected with multiple-sector transfers in mind.

(1) Append to each logical sector within the host's memory a four-byte logical sector number field. Design the host adapter so that as each logical sector of a multiple-sector write is fetched from the host's memory, four bytes of CRC redundancy are computed across the data portion of the logical sector and then EXCLUSIVE-OR summed with the logical sector number field and transferred to the storage subsystem immediately behind the data. During a multiple-sector read, the host adapter would compute CRC redundancy over the data portion of each received logical sector and EXCLUSIVE-OR sum it with the received sum of the logical identification number and CRC redundancy generated on write, then store the result after the data portion of the logical sector in the host's memory. The host processor would then have to verify that the result for each logical sector of a multiple-sector transfer matches the identification number of the respective requested logical sector. If an otherwise undetected error occurs anywhere in a logical sector anywhere beyond the host interface which exceeds the guarantees of the host CRC code, including the fetching of the wrong sector, the logical identification number within the host's memory will be incorrect with probability 1-(2.33E-10).

(2) Keep data contiguous in the host's memory by instead recording the identification numbers of all of the logical sectors in a multiple sector transfer within the host adapter's memory, but process the data and identification numbers for the CRC code in the same manner as in (1). The host adapter would have the responsibility for checking that identification numbers match those requested. Equivalent error detection is achieved.

(3) Initialize the CRC shift register at the host interface with the identification number of each logical sector before writing or reading each logical sector of a multiple-sector transfer. The host adapter would require that on read the CRC residue for each logical sector be zero. Again equivalent error detection is achieved.

To implement the CRC/ID field approach toward achieving higher data integrity, computer builders will have to support the generation and checking of the extra four bytes of CRC redundancy. Storage subsystem suppliers accustomed to sector lengths which are powers of two will have to accommodate sector lengths which are greater by four bytes. If the storage subsystem architecture includes its own auxiliary CRC field of thirty-two or fewer bits, an option to disable it should be provided in order to minimize overhead when the storage subsystem is connected to a host which implements the CRC/ID field. The scope of coverage of the host CRC/ID field is much greater than that of an equivalent-length auxiliary CRC field which protects only against media errors, so data integrity can be greatly improved at no increase in overhead if the subsystem auxiliary CRC code is disabled and the host CRC/ID field is used instead.

Procedures like those outlined above can have a profound impact on data integrity in computer systems. They allow the computer builder to be in control of the integrity of data throughout the entire system without being concerned with the detailed designs of the storage subsystems connected to the system.

SUMMARY

When designing error correction for a disk controller, keep data accuracy high by using the techniques listed below:

- Use a computer-generated code to avoid pattern sensitivity.

- Reread before attempting error correction.

- Use the lowest possible correction span meeting the requirements of supported drives.

- Ensure that the ECC circuit provides adequate protection from sync framing errors.

- Design the ECC circuit to generate nonzero check bits for an all-zeros data record.

- Include self-checking, if it is required to meet the specification for probability of undetected erroneous data (P_{ued}).

- Use a manufacturer recommended data separator or get assistance from a consultant who specializes in this area.

- Assign alternate sectors for known defects.

- Establish a target for P_{ued}. Determine P_e and P_c by the manufacturer specification, measurement, and estimation. Select the number of check bits to meet the target for P_{ued}. In computing P_{ued}, derate P_e and P_c to account for error clustering and marginal drives.

4.5 PERFORMANCE REQUIREMENTS

Below are some of the parameters that should be specified for an error-control system.

DATA RECOVERABILITY

Specify permissible decoded hard error rate. For storage devices this specification is likely to be 1.E-13 or less.

DATA ACCURACY

Specify allowable undetected erroneous data rate. For storage devices this specification is likely to be 1.E-15 or less.

OPERATING SPEED

Specify data transfer rates that the error-control system must support.

DECODING SPEED

Specify allowable error-correction decoding times. These are times allowed for computing patterns and displacements when errors occur.

SELF-CHECKING

Specify the form of self-checking to be used, such as:

- Duplicated circuits
- Parity predict
- Periodic microcode or software testing.

This determination may have to be made after a code has been selected and the design is in progress. Use the reliability of circuit and packaging technologies along with parts counts to determine the reliability of the error-correction circuits. If the probability of error-correction circuit failure in a design contributes significantly to the probability of transferring undetected erroneous data, self-checking should be added to the design.

Once error rates and the nature of errors have been characterized and the performance requirements established, code selection can begin.

4.6 PATTERN SENSITIVITY

When selecting a code for a particular application it is important to consider 0pattern sensitivity.

Some error detecting and error correcting codes are more likely to misdetect or miscorrect on certain classes of error patterns than others. This is called pattern sensitivity. If these classes of errors are also the most likely to occur, then protection provided by these codes may not be as good as expected. In this section several examples of pattern sensitivity are discussed.

PATTERN SENSITIVITY OF ERROR DETECTION CODES

Some error detection codes have pattern sensitivity. Consider for example, the error detection code defined by the circuit below.

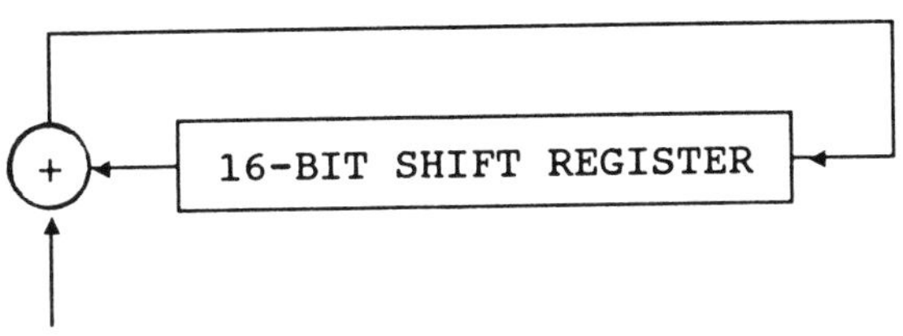

The polynomial for this circuit is $(x^{16} + 1)$. Of all possible error bursts, this circuit will fail to detect one out of 65,536. Any degree 16 polynomial would have the same misdetection probability for all possible error bursts. However, this circuit has a pattern sensitivity. It will fail to detect one out of every sixteen possible error patterns, consisting of two bits in error, separated by more than sixteen bits.

To understand the pattern sensitivity, consider reading a data record that is zeros except for two bits in error, sixteen bits apart. The shift register will be all zeros until the first error bit arrives. After arrival of the first error bit, the shift register will contain '0.....01'. After receiving the fifteen zeros separating the error bits, the shift register will contain '10.....0'. After receiving the second error bit, the shift register will again contain all zeros, due to the cancellation of the high-order bit by the second error bit.

This circuit is 4000 times more likely to fail to detect an error pattern consisting of two bits in error, separated by more than sixteen bits, than it is to fail to detect a pattern consisting of many random bits in error.

The pattern sensitivity of this circuit is obvious. Nevertheless, it was implemented by a large computer manufacturer on the 2314 magnetic disk device in the mid 1960's. After the product was in the field, additional checking was installed to correct the problem.

PATTERN SENSITIVITY OF ERROR CORRECTION CODES

The Fire code is used for single burst correction. Many Fire codes have a high pattern sensitivity for short double bursts. See Section 4.4 for a discussion of the Fire code's pattern sensitivity.

Many interleaved error correcting codes have a pattern sensitivity for multiple short bursts. The 3370 code (see Section 5.2) is such a code. It uses a single symbol error correcting, double symbol error detecting Reed-Solomon code interleaved to depth three. Symbols are one byte wide. Its miscorrection probability is 2.2E-16 for all possible error bursts. However, the miscorrection probability is 2.6E-3 for all possible errors exceeding code guarantees and affecting a single interleave.

OTHER FORMS OF PATTERN SENSITIVITY

Many codes are sensitive to the error patterns caused by circuit or power supply failures. For example, if the line supplying data bits to a magnetic-disk error correction circuit fails, the failure may not be detected by these circuits. One way to protect against this form of pattern sensitivity is to make sure nonzero check bytes are guaranteed for an all zeros data record. See also Section 4.4 and Chapter 6.

A semiconductor memory error correction circuit may not detect the error when a word of all zeros (data and check bits) is erroneously read from memory, due to a circuit or power supply failure. Again, a solution is to cause nonzero check bits to be generated for an all zeros data word.

4.7 K-BIT-SERIAL TECHNIQUES

Clocking error-correction circuits once per data bit limits operating speed. To operate at higher speeds, it is necessary to clock these circuits once per symbol. A symbol is some convenient cluster of bits, for example a byte or word.

There are at least two ways to do this. A code such as the Reed-Solomon code can be selected that inherently operates on symbols; or the shift-register for a code such as the Fire code can be transformed from bit-serial to k-bit-serial. The k-bit-serial shift register operates on k input bits and accomplishes k bit shifts per clock. A special case of k-bit-serial is byte-serial (k=8).

The higher operating speed of k-bit-serial shift registers is attained at the expense of added complexity.

There are two methods for implementing k-bit-serial shift register divide circuits. The first method adds the necessary XOR gates to shift k bits per clock. The second method uses 2^k:k bit tables to accomplish k bit shifts per clock.

For both k-bit serial methods discussed in this section, circuitry is shown for computing the remainder only. If the quotient is required, additional circuitry must be added.

XOR GATE METHOD

The transformation discussed here is for the internal-XOR form of shift register. Extension to the external-XOR form of shift register is straightforward.

The procedure for this transformation was developed intuitively as follows. Assume the shift register below is to be transformed.

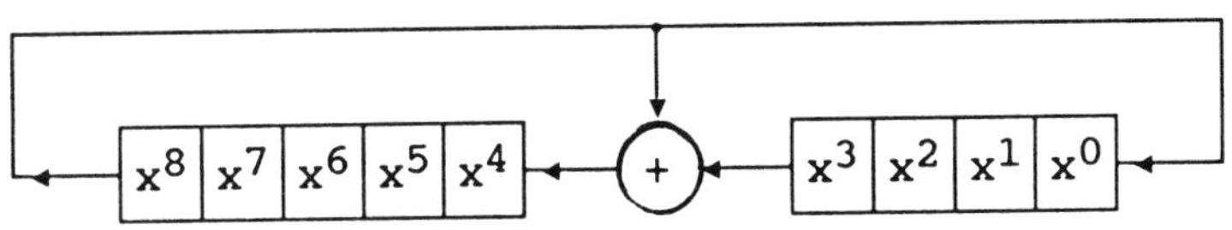

Recognize that in k shifts of the bit-serial shift register, the bits influencing the new shift register contents via the feedback network, are the high order k bits. To determine the contribution of any one of these bits, bit j for example, do the following. Clear the shift register, set bit j to 1, and shift k times. The resulting 1 bits in the shift register is the contribution of bit j.

The other contributor to the new state of each bit, when the shift register is shifted k times, is the bit itself shifted k bits to the right.

The result of this intuitive development is the basis for the following procedure.

PROCEDURE

Let i represent the polynomial degree and k the desired number of shifts per clock. The following steps transform a bit-serial shift register into a k-bit-serial shift register.

Simulate the bit-serial shift register. Initialize the high-order bit of the simulated shift register to 1 and clear the remaining bits. Shift k times. After each shift, record the new state of the shift register.

The first state in the sequence recorded is the contribution of shift register stage x^{i-k} to the feedback network. The second state is the contribution of stage x^{i-k+1}, and so on. The last state in the sequence is the contribution of shift register stage x^{i-1}.

The next step is to add circuitry for the k information bits. It will be clear from the examples how this is done.

The last step is to minimize logic.

EXAMPLE #1

BIT-SERIAL SHIFT REGISTER

Premultiply by x^9 and divide by $x^9 + x^4 + 1$

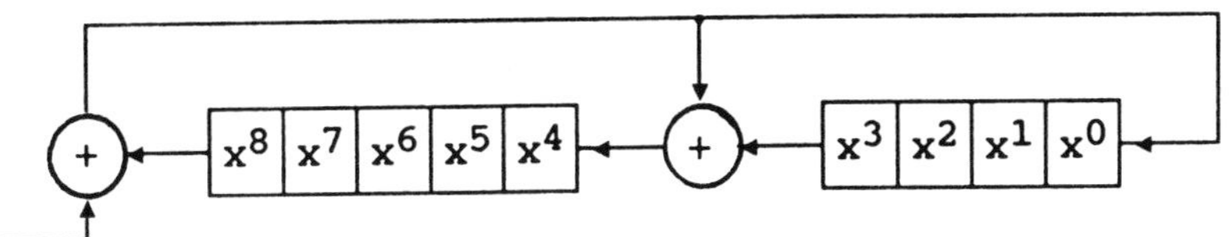

PROCEDURE FOR k=3

```
1
-
000010001  contribution of x^6
000100010  contribution of x^7
001000100  contribution of x^8
```

k-BIT-SERIAL SHIFT REGISTER, k=3

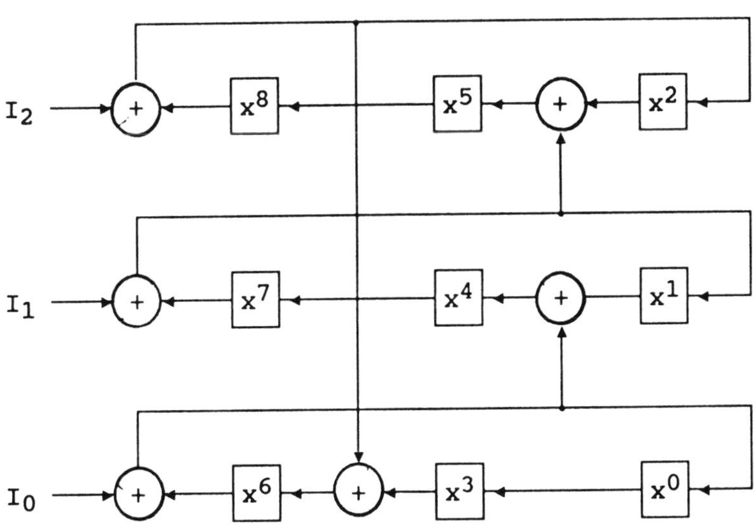

EXAMPLE #2

BIT-SERIAL SHIFT REGISTER

Divide by $x^8 + x^6 + x^5 + x + 1$

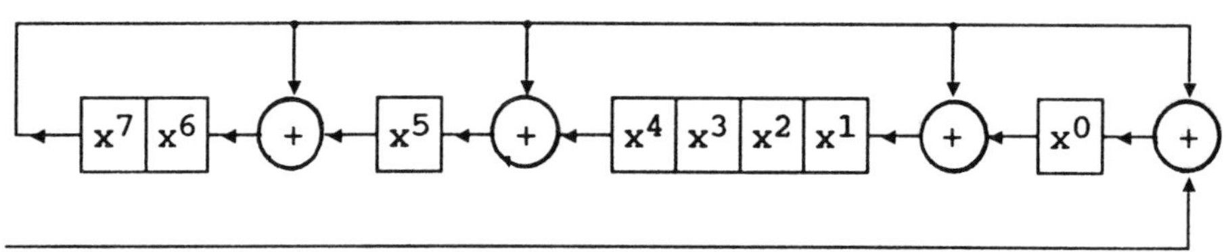

PROCEDURE FOR k=3

```
1
-
01100011  contribution of x^5
11000110  contribution of x^6
11101111  contribution of x^7
```

k-BIT-SERIAL SHIFT REGISTER, k=3

(Logic has not been minimized)

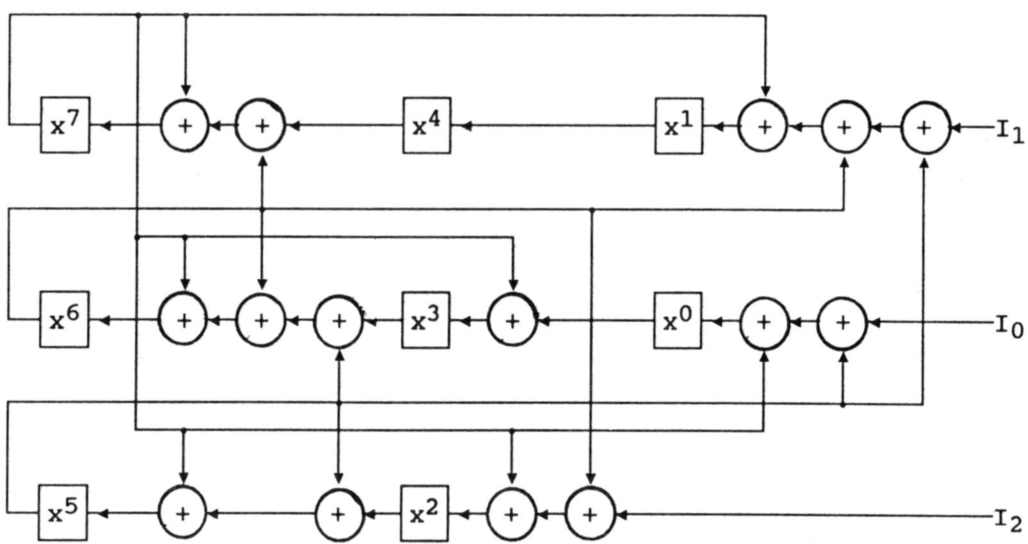

TABLE METHOD OF k-BIT-SERIAL IMPLEMENTATION

This method will be illustrated by example. The circuit of this example premultiplies by x^{32} and divides by:

$$g(x)=x^{32}+x^{28}+x^{26}+x^{19}+x^{17}+x^{10}+x^{6}+x^{2}+1$$

k-BIT SERIAL SHIFT REGISTER, k=8

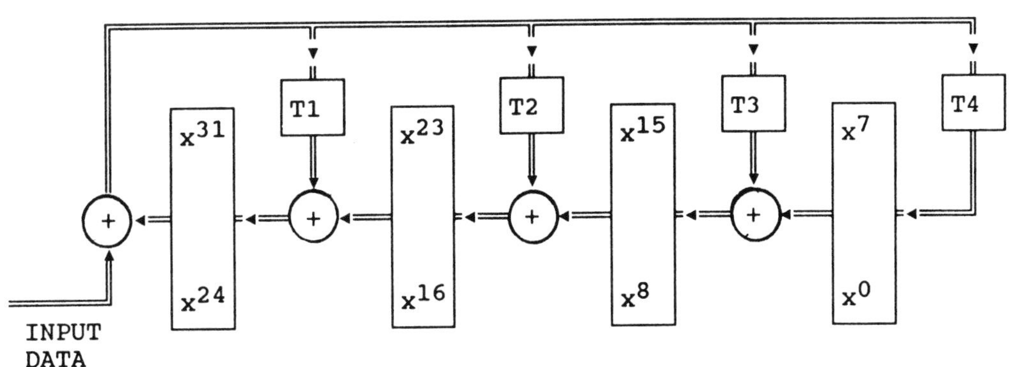

PROCEDURE FOR GENERATING THE TABLES

To generate the tables for the circuit above, simulate a left shifting bit-serial shift register (internal-XOR form) in software using the polynomial above. For each address of the tables (0-255), place the address in the eight most significant (left-most) bits of the shift register and clear the remaining bits. Shift eight times, then store the four bytes of shift register contents in tables T1 through T4 at the location indexed by current address. The coefficient of x^{31} is stored as the high-order bit of T1; the coefficient of x^{0} is stored as the low-order bit of T4. Tables for the above shift register and polynomial are shown on the following pages. Each table is 256:8 bits.

The circuit above can be modified to divide by g(x) without the premultiply: move the input XOR circuit to the input of the low order k-bits.

A similar procedure could be used to implement a right shifting k-bit serial shift register. Extension to external XOR k-bit serial shift registers is straightforward.

Implementation examples for the above polynomial are given in Section 5.3, where byte-serial tables are used for decoding by reverse clocking using its reciprocal polynomial.

BYTE-SERIAL TABLE T1

	0	1	2	3	4	5	6	7	8	9	A	B	C	D	E	F
00	00	14	28	3C	50	44	78	6C	A0	B4	88	9C	F0	E4	D8	CC
10	54	40	7C	68	04	10	2C	38	F4	E0	DC	C8	A4	B0	8C	98
20	A9	BD	81	95	F9	ED	D1	C5	09	1D	21	35	59	4D	71	65
30	FD	E9	D5	C1	AD	B9	85	91	5D	49	75	61	0D	19	25	31
40	46	52	6E	7A	16	02	3E	2A	E6	F2	CE	DA	B6	A2	9E	8A
50	12	06	3A	2E	42	56	6A	7E	B2	A6	9A	8E	E2	F6	CA	DE
60	EF	FB	C7	D3	BF	AB	97	83	4F	5B	67	73	1F	0B	37	23
70	BB	AF	93	87	EB	FF	C3	D7	1B	0F	33	27	4B	5F	63	77
80	8D	99	A5	B1	DD	C9	F5	E1	2D	39	05	11	7D	69	55	41
90	D9	CD	F1	E5	89	9D	A1	B5	79	6D	51	45	29	3D	01	15
A0	24	30	0C	18	74	60	5C	48	84	90	AC	B8	D4	C0	FC	E8
B0	70	64	58	4C	20	34	08	1C	D0	C4	F8	EC	80	94	A8	BC
C0	CB	DF	E3	F7	9B	8F	B3	A7	6B	7F	43	57	3B	2F	13	07
D0	9F	8B	B7	A3	CF	DB	E7	F3	3F	2B	17	03	6F	7B	47	53
E0	62	76	4A	5E	32	26	1A	0E	C2	D6	EA	FE	92	86	BA	AE
F0	36	22	1E	0A	66	72	4E	5A	96	82	BE	AA	C6	D2	EE	FA

BYTE-SERIAL TABLE T2

	0	1	2	3	4	5	6	7	8	9	A	B	C	D	E	F
00	00	0A	14	1E	28	22	3C	36	50	5A	44	4E	78	72	6C	66
10	AA	A0	BE	B4	82	88	96	9C	FA	F0	EE	E4	D2	D8	C6	CC
20	54	5E	40	4A	7C	76	68	62	04	0E	10	1A	2C	26	38	32
30	FE	F4	EA	E0	D6	DC	C2	C8	AE	A4	BA	B0	86	8C	92	98
40	A3	A9	B7	BD	8B	81	9F	95	F3	F9	E7	ED	DB	D1	CF	C5
50	09	03	1D	17	21	2B	35	3F	59	53	4D	47	71	7B	65	6F
60	F7	FD	E3	E9	DF	D5	CB	C1	A7	AD	B3	B9	8F	85	9B	91
70	5D	57	49	43	75	7F	61	6B	0D	07	19	13	25	2F	31	3B
80	46	4C	52	58	6E	64	7A	70	16	1C	02	08	3E	34	2A	20
90	EC	E6	F8	F2	C4	CE	D0	DA	BC	B6	A8	A2	94	9E	80	8A
A0	12	18	06	0C	3A	30	2E	24	42	48	56	5C	6A	60	7E	74
B0	B8	B2	AC	A6	90	9A	84	8E	E8	E2	FC	F6	C0	CA	D4	DE
C0	E5	EF	F1	FB	CD	C7	D9	D3	B5	BF	A1	AB	9D	97	89	83
D0	4F	45	5B	51	67	6D	73	79	1F	15	0B	01	37	3D	23	29
E0	B1	BB	A5	AF	99	93	8D	87	E1	EB	F5	FF	C9	C3	DD	D7
F0	1B	11	0F	05	33	39	27	2D	4B	41	5F	55	63	69	77	7D

BYTE-SERIAL TABLE T3

	0	1	2	3	4	5	6	7	8	9	A	B	C	D	E	F
00	00	04	08	0C	11	15	19	1D	22	26	2A	2E	33	37	3B	3F
10	40	44	48	4C	51	55	59	5D	62	66	6A	6E	73	77	7B	7F
20	80	84	88	8C	91	95	99	9D	A2	A6	AA	AE	B3	B7	BB	BF
30	C0	C4	C8	CC	D1	D5	D9	DD	E2	E6	EA	EE	F3	F7	FB	FF
40	04	00	0C	08	15	11	1D	19	26	22	2E	2A	37	33	3F	3B
50	44	40	4C	48	55	51	5D	59	66	62	6E	6A	77	73	7F	7B
60	84	80	8C	88	95	91	9D	99	A6	A2	AE	AA	B7	B3	BF	BB
70	C4	C0	CC	C8	D5	D1	DD	D9	E6	E2	EE	EA	F7	F3	FF	FB
80	08	0C	00	04	19	1D	11	15	2A	2E	22	26	3B	3F	33	37
90	48	4C	40	44	59	5D	51	55	6A	6E	62	66	7B	7F	73	77
A0	88	8C	80	84	99	9D	91	95	AA	AE	A2	A6	BB	BF	B3	B7
B0	C8	CC	C0	C4	D9	DD	D1	D5	EA	EE	E2	E6	FB	FF	F3	F7
C0	0C	08	04	00	1D	19	15	11	2E	2A	26	22	3F	3B	37	33
D0	4C	48	44	40	5D	59	55	51	6E	6A	66	62	7F	7B	77	73
E0	8C	88	84	80	9D	99	95	91	AE	AA	A6	A2	BF	BB	B7	B3
F0	CC	C8	C4	C0	DD	D9	D5	D1	EE	EA	E6	E2	FF	FB	F7	F3

BYTE-SERIAL TABLE T4

	0	1	2	3	4	5	6	7	8	9	A	B	C	D	E	F
00	00	45	8A	CF	14	51	9E	DB	28	6D	A2	E7	3C	79	B6	F3
10	15	50	9F	DA	01	44	8B	CE	3D	78	B7	F2	29	6C	A3	E6
20	2A	6F	A0	E5	3E	7B	B4	F1	02	47	88	CD	16	53	9C	D9
30	3F	7A	B5	F0	2B	6E	A1	E4	17	52	9D	D8	03	46	89	CC
40	11	54	9B	DE	05	40	8F	CA	39	7C	B3	F6	2D	68	A7	E2
50	04	41	8E	CB	10	55	9A	DF	2C	69	A6	E3	38	7D	B2	F7
60	3B	7E	B1	F4	2F	6A	A5	E0	13	56	99	DC	07	42	8D	C8
70	2E	6B	A4	E1	3A	7F	B0	F5	06	43	8C	C9	12	57	98	DD
80	22	67	A8	ED	36	73	BC	F9	0A	4F	80	C5	1E	5B	94	D1
90	37	72	BD	F8	23	66	A9	EC	1F	5A	95	D0	0B	4E	81	C4
A0	08	4D	82	C7	1C	59	96	D3	20	65	AA	EF	34	71	BE	FB
B0	1D	58	97	D2	09	4C	83	C6	35	70	BF	FA	21	64	AB	EE
C0	33	76	B9	FC	27	62	AD	E8	1B	5E	91	D4	0F	4A	85	C0
D0	26	63	AC	E9	32	77	B8	FD	0E	4B	84	C1	1A	5F	90	D5
E0	19	5C	93	D6	0D	48	87	C2	31	74	BB	FE	25	60	AF	EA
F0	0C	49	86	C3	18	5D	92	D7	24	61	AE	EB	30	75	BA	FF

4.8 SYNCHRONIZATION

4.8.1 SYNCHRONIZATION CODES

In order to recover data, we must be able to determine where the data are recorded on the medium, or equivalently, when data begin and end in the read bit stream; this is called data framing or frame synchronization. This is normally accomplished by detecting a special pattern called a sync mark. This process is called byte synchronization and it is preceded by frequency and phase lock. Several types of synchronization errors arise. A synchronization failure occurs when it is known that we have been unable to establish initial synchronization; this is a serious error situation but one which is detected. A synchronization framing error occurs when we erroneously believe we have established correct synchronization; this is worse than synchronization failure in that undetected erroneous data could be transferred, as many error detection and correction codes have a weakness for this type of error. A loss of synchronization occurs when synchronization has been achieved and is later lost; the ease and speed of re-synchronization are heavily implementation-dependent.

It is common for data storage device track formats to include a sector mark and one or more sync marks in front of each sector for achieving initial synchronization.

A sector mark is used to establish coarse synchronization to a sector. The sector mark is unique and very different from data. It may be chosen so that it is impossible for data to emulate it and very difficult for a defect to emulate it. Sector marks are generally detected before data acquisition and therefore must be detected asynchronously. After coarse synchronization has been established, the general location of the sync mark is known and the search for the sync mark can be restricted to a window spanning the time around which it is expected to occur.

Ideally, the sync mark is unique and we are assured that no combination of valid channel bits can emulate it. To achieve this, the sync mark might include a run-length violation or an invalid decode. An invalid decode is a sequence which satisfies the run-length constraints but which cannot be emulated by any valid combination of channel words. When the sync mark is unique, the misdetection probability in the absence of error is zero. In some cases, the sync mark is not unique and there is a valid data bit sequence which can emulate it, but with sufficiently low misdetection probability. In such a case there would generally be additional sync mark detection qualification criteria.

In selecting a sync mark strategy, it is desirable to minimize overhead yet maximize the probability of successful decoding and minimize the probability of false decoding. These conflicting goals require that trade-offs be made in selecting sync mark parameters. Typical parameters include:

- Detection window width
- Error tolerance of the mark
- Mark length
- Mark pattern

Detection window width and error tolerance of the mark may be changed for retry reads. A narrow detection window is desirable in order to minimize the probability of false detection. However, if the detection window is established by a counter running off a reference clock then spindle speed variations, eccentricity, and mechanical oscillations will influence timing accuracy and will therefore influence window width as well.

Increasing the error tolerance of the sync mark while keeping its length constant increases the probability of successful decoding but also increases the possibility of false decoding.

Increasing the sync mark length decreases the probability of false decoding but increases overhead.

The sync mark pattern is selected to minimize the probability of false decoding when defects exist within and/or preceding the sync mark. To accomplish this, the pattern is selected to maximize the number of error bits and/or the error burst length that are required to cause a sync mark to be falsely detected in front of the true mark. This selection can be accomplished with a computer.

If we assume that the bit stream preceding and following a mark is random, we are motivated to use a sequence which does not resemble itself when shifted one or more bits, so that it is impossible for a small number of errors to cause false detection. As an illustration, consider a sequence of all '1's. If the bit immediately preceding is random, there is a 50% chance of falsely detecting this sequence one bit early.

The autocorrelation function of a sequence is used to measure the degree to which a sequence resembles itself. Conceptually, one copy of the sequence is "slid past" another. At each offset i, the autocorrelation R(i) is the number of corresponding bits which are identical minus the number which differ. R(0) is of course equal to the number of bits in the sequence, n; the maximum value of R(i) is $n-|i|$, with lower values being preferred. The class of sequences called Barker codes has the so-called "perfect" property $|R(i)| \leq 1$ for $i \neq 0$. Only eight Barker codes are known to exist, with lengths 2, 3, 4, 5, 7, 11, and 13.

BARKER CODES

n	Sequence*	Autocorrelation (i=0 to n-1)
2	10	2,-1
3	110	3, 0,-1
4a	1101	4,-1, 0, 1
4b	1110	4, 1, 0,-1
5	11101	5, 0, 1, 0, 1
7	1110010	7, 0,-1, 0,-1, 0,-1
11	11100010010	11, 0,-1, 0,-1, 0,-1, 0,-1, 0,-1
13	1111100110101	13, 0, 1, 0, 1, 0, 1, 0, 1, 0, 1, 0, 1

* Including reversals and complements

Barker codes can be combined to form longer codes which have good, though not "perfect" autocorrelations. To construct such a combined Barker code, each bit of a Barker code is replaced with the entire sequence of another (possibly the same) Barker code, the sequence being inverted if the bit being replaced is zero. Longer sequences with autocorrelations which are nearly as desirable (Barker-like codes) also exist.

In practice, a sync mark is detected by counting the number of matching bits, without subtracting the number of mismatched bits. The sync mark is considered detected when the count of matching bits meets or exceeds a threshold, which may be variable so that it can be changed for read retries. In discrete designs, an efficient implementation may include PROM circuits; in integrated designs, logic gates may be preferable.

Window width can be increased and misdetection can be reduced by writing a known bit pattern (preamble) preceding the mark. A mark pattern is then selected for minimum correlation with the preamble and with itself. This preamble-sync mark combination is equivalent to a sync mark which is detected by searching only for its last half. An example is 16 zero-bits followed by the 16-bit mark '0001111100110101' (3 zero bits followed by the 13-bit Barker code) and followed by random data. When detected in a window from 16 bits before the position of the mark up to 5 bits after and requiring 13 bits (out of 16) to match, this pattern is guaranteed to be detected and guaranteed not to be falsely detected when not more than 3 random bits (out of 16) are in error or when a single error burst of length 3 bits or less exists. There are other patterns besides this one which have the same error tolerance using the same detection method.

Note that a preamble of all one-bits could be used as well, in which case each bit of the mark would be inverted. The preamble need not be all zero-bits or all one-bits; satisfactory codes can be selected for any given preamble pattern.

An extension of the above technique would be to write known patterns both preceding and following the sync mark. Selecting the pattern following the mark for minimum correlation with the sync mark would increase the acceptable window width after the position of the mark.

Sync marks can be decoded in either the data-bit domain or the channel-bit domain; the error propagation of the RLL decoding process motivates us to decode in the channel bit domain when possible, particularly if the detection criteria have been relaxed to achieve error tolerance. The desire to have error tolerant clock phasing also motivates us to decode in the channel-bit domain. In this case clock phasing and byte synchronization are established simultaneously with the detection of the sync mark.

Some implementations do not detect sync marks using a bit-by-bit comparison, but by comparing groups of bits. This reduces the circuitry required to implement majority-vote detection. Such a code has been proposed for use in optical disk. The 48 channel-bit mark is made up of 12 groups of 4 bits, each group containing a single one-bit. The whole mark obeys (2,7) run-length constraints and is preceded by the highest-frequency (2,7) pattern. The mark is detected in the channel bit domain using 4-bit groups. The correlation function for the sync mark sequence against the preamble-sync mark-random data sequence on a 4-bit basis, counted as the number of matches (plus the number of possible matches when correlating with random data at positive offsets) is

```
Offset: -15 . . . . . . . . . . . . . .-1  0 1 . . . . . . . . . .11
4-bit:    2 3 3 4 4 0 4 4 3 2 2 3 4 0 0 12 0 0 5 4 3 2 5 5 4 4 5
```

If a detection threshold of 9 is used, 4 groups-in-error are required before failure to detect is possible, while 5 groups-in-error are required before false detection is possible. If a detection threshold of 8 is used, 5 groups-in-errors are required before failure to detect is possible, while 4 groups-in-error are required before false detection is possible. This suggests the following strategy for sync mark detection: on the first try, using a threshold of 9 will insure that the mark will not be falsely detected, while on read re-try, using a threshold of 8 will insure that those cases of 4 groups-in-error that fail on the first try will be detected, subject to only a low probability of early false detection.

In the general case, 4 groups-in-error could be caused by an error burst of 10 channel bits, but analysis of the specific correlation bit patterns for this code reveals that this detection scheme handles any error burst of not more than 12 channel bits in length on the first pass, and will very likely handle any error burst of 16 channel bits or less on retry. This is the same performance which would be obtained using a bit-by-bit majority vote criterion of 42 channel bits (out of 48; there is a peak correlation of 35 bits at an offset of -3) but with much lower implementation cost.

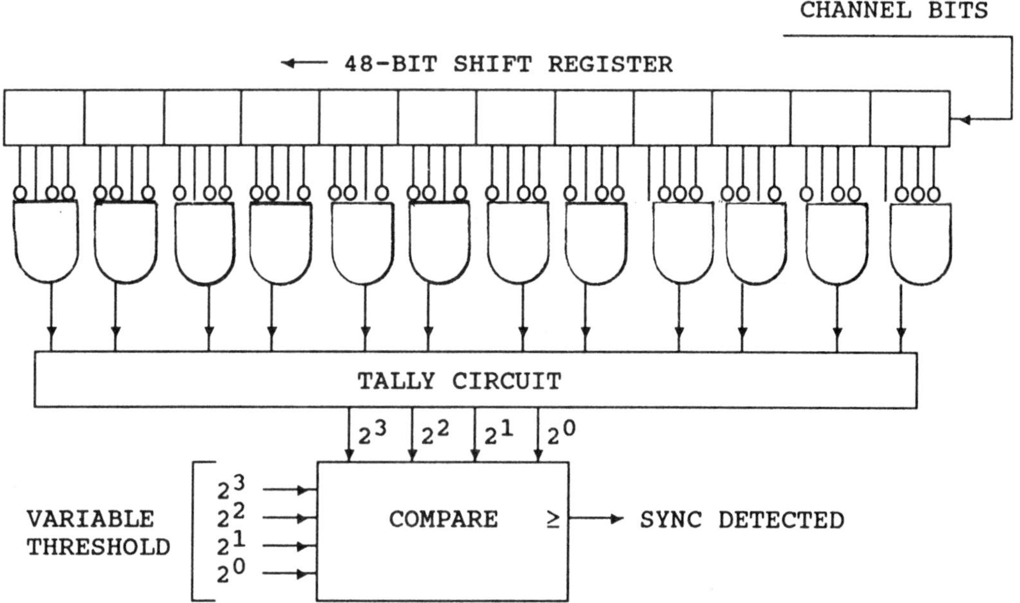

One possible decoding alternative for the X3B11 data field sync code.

As another example consider a 32 data-bit sync mark that is composed of four 8-bit groups A, B, C, and D, preceded by all zeros and followed by random data, to be detected in the data-bit domain when any one of the pairs A-B, C-D, or A-D is detected. It is possible to construct a mark which will be detected in the presence of a burst of not more than 9 data bits (out of 32 data bits) and will not be falsely detected in the presence of a burst of not more than 10 data bits (out of 32 data bits) in length when detected in a window from 16 bits before to 16 bits after the mark.

Using the same pair-wise detection method in the channel bit domain, it is possible to construct a 32 channel-bit mark subject to a (1,7) run-length constraint and preceded by 32 bits of the maximum-frequency (1,7) pattern which will be detected and will not be falsely detected in the presence of a burst of not more than 9 channel bits (out of 32 channel bits) when detected in the channel-bit domain in this pair-wise fashion. Similarly, 32-bit marks have been constructed using a (2,7) run-length constraint which will be detected in the presence of a burst of not more than 9 channel bits (out of 32 channel bits) and will not be falsely detected in the presence of a burst of not more than 8 channel bits (out of 32 channel bits) in length.

For a given detection method, it is possible to use a computer to select mark patterns which satisfy the desired error tolerance requirements, if such patterns exist. The most straightforward method is to successively generate random patterns (using run-length constraints, if the mark is to be detected in the channel-bit domain), analyze them, and record the best performers.

RESYNC MARKS

When the probability of loss of synchronization is high, due for example to long defects, some applications require one or more sync marks preceding each sector and resync marks interspersed at regular intervals within each sector. The sync marks are used for achieving initial clock phasing and byte synchronization and the resync marks are used for restoring clock phasing and byte synchronization after a loss of sync (when the PLL has slipped cycles).

Many resync marks may be required per sector, so it is very important to minimize resync mark length to minimize overhead. In clever implementations it is not necessary for each resync mark to be detected, so the resync mark itself need not be error tolerant. To minimize the false detection of resync marks, their detection window is made very narrow. In addition they are typically assigned a channel bit pattern that cannot be emulated by a channel-bit sequence encoded from data. This guarantees that correct data will never emulate a resync mark.

4.8.2 *SYNCHRONIZATION FRAMING ERRORS*

In order to properly frame data, a read system must know where data begins. This is normally accomplished by detecting a sync mark, a process called byte synchronization. A defect can emulate a sync mark at an incorrect position on the media. It is possible (depending on windowing, etc.) for this to result in an incorrect assumption about the starting position of data. This is called a synchronization framing error. The probability of a sync framing error increases as sync mark length decreases, as sync mark error tolerance increases, and as the length of the sync mark detection window increases. A sync framing error may be detected as an uncorrectable error or it may incorrectly cause data to appear correctable or error-free. If the data appears correctable or error free, the transfer of undetected erroneous data may result which could have disastrous consequences.

In order to keep the probability of transferring undetected erroneous data low it is very important to detect sync framing errors with high probability. In some systems the responsibility for detecting such errors is placed on the error detection and correction circuitry.

Most codes used for error detection and correction in data storage systems for computers are shortened cyclic codes. Cyclic codes are linear codes with the property that each cyclic (i.e. wraparound) shift of each codeword is also a codeword. Shortened cyclic codes are not truly cyclic. However, the codewords of a shortened cyclic code when shifted (left or right) a few symbol positions will either form another shortened codeword or form a sequence that differs from another shortened codeword in only a few symbol positions. This property of shortened cyclic codes causes them to have poor detection capability for sync framing errors.

Shortened cyclic codes are often modified by some method in order to increase their capability to detect sync framing errors. The degree to which capability of the modified code to detect sync framing errors is increased depends highly on the specific method of modification selected.

Ideally, a code modification method will assure that all sync framing errors result is an error pattern that exceeds correction guarantees but not detection guarantees of the code. If this cannot be achieved, then as a very minimum it is desirable that the probability of transferring undetected erroneous data be no greater for sync framing errors than for all other types of errors that exceed detection guarantees.

Some frequently used codes for performing error detection and/or correction are listed below. In some cases these codes are cyclic codes but most often they are shortened cyclic codes. Problem analysis and the selection of a method for code modification is similar between the different types of codes.

- Error detection codes using a polynomial with binary coefficients.
- Single-burst error correcting codes using a polynomial with binary coefficients.
- Single- and multiple-burst error correcting Reed-Solomon codes.
- Interleaved Reed-Solomon codes.

Binary error detection/correction codes operate on single-bit symbols while Reed-Solomon codes operate on multiple-bit symbols, typically byte-wide (eight-bit) symbols. Reed-Solomon codes are cyclic but only on a symbol basis: cyclic rotation of a Reed-Solomon codeword by a number of bits which is not a multiple of the symbol width does not generally produce another codeword; an obvious counter-example is the all-zeros Reed-Solomon codeword. This property allows us to discuss binary codes and non-interleaved Reed-Solomon codes together. We shall then apply similar methods to interleaved Reed-Solomon codes.

Let us use the following notation to represent a non-interleaved codeword of a binary code or a Reed-Solomon code:

•••pppddd•••dddrrr•••rrrggg•••

where 'p' is a preamble/sync symbol, 'd' is a data symbol, 'r' is a redundancy symbol, and 'g' is a gap symbol. '0' will represent a symbol whose bits are all zeros, '1' will represent a symbol whose bits are all ones, and 'X' will represent a symbol whose bits are neither. In the case of a Reed-Solomon code, each symbol is a group of w bits. In the case of a binary error correction code, each symbol is one bit (w=1).

LATE SYNCHRONIZATION IN UNMODIFIED SHORTENED CYCLIC CODES

Consider the case of late synchronization by one symbol. There are four combinations for the values of the data symbol skipped and the gap symbol read.

```
                Codeword read
           ┌──────────┴──────────┐
1)  •••ppp0dd•••dddrrr•••rrr0gg•••
```

The pattern read is a multiple of the codeword written. This is also a codeword, so the pattern read appears to be error free and the sync framing error is not detected.

```
                Codeword read
           ┌──────────┴──────────┐
2)  •••pppXdd•••dddrrr•••rrr0gg•••
```

The pattern read is a multiple of the codeword written with a symbol in error at symbol -1 of the codeword (i.e. the symbol before the first data symbol of the codeword). When shortened codewords are used, the error appears to be outside the bounds of the codeword and the correction algorithm will post it as uncorrectable.

In random data the probability that the first symbol of a codeword is zero is 2^{-w}, so from the 1) and 2) above analyses we conclude that this is also the probability that the read pattern will appear to be error free when synchronization occurs late by one symbol and a zero symbol is read following the codeword. By similar reasoning, if synchronization occurs late by k symbols and the first k gap symbols are all zeros then the codeword read will appear to be error free if the first k symbols of the codeword written were all zeros. This should occur in random data with probability $2^{-(k*w)}$.

If synchronization occurs late by k symbols, the first k gap symbols are all zeros, and the first k symbols of the codeword written were not all zeros, then there will appear to be an error burst of length k or fewer symbols preceding the codeword read. If the guaranteed detection capability of the code is equal to or greater than the apparent error created by the pattern of non-zero symbols missed, then there will appear to be an error burst of length k or fewer symbols preceding the codeword read and the correction algorithm will post the error as uncorrectable, since the error burst appears to be beyond the bounds of the shortened codeword. If in the same situation the apparent error created by the pattern of non-zero symbols missed exceeds the guaranteed detection and correction capabilities of the code, then the error will appear to be correctable with probability P_{mc}, where P_{mc} is the miscorrection probability of the code. Equivalently, the error will appear to be uncorrectable with probability $1\text{-}P_{mc}$.

```
              Codeword read
         ┌──────────┴──────────┐
3)  •••ppp0dd•••dddrrr•••rrrXgg•••
```

The pattern read is that of a multiple of the codeword written with a symbol in error in the last symbol position. A code which performs only error detection will therefore detect the sync framing error, but an error correction code will not.

```
              Codeword read
         ┌──────────┴──────────┐
4)  •••pppXdd•••dddrrr•••rrrXgg•••
```

The read remainder will be that of two symbols in error, one at the symbol before the first symbol of the codeword and one at the last symbol of the codeword. If this double-burst error is within the detection guarantees of the code, then an uncorrectable error will be posted by the error correction algorithm. If this error pattern exceeds the detection guarantees of the code, then the error will appear to be correctable with probability P_{mc} and will appear to be uncorrectable with probability $1\text{-}P_{mc}$.

By similar reasoning, if synchronization occurs late by k symbols, the first k gap symbols are not all zeros, and the guaranteed correction capability of the code is equal to or greater than the number of non-zero gap symbols, then the codeword read will appear to be correctable if the first k symbols of the codeword written were all zeros. This should occur in random data with probability $2^{-(k*w)}$.

If synchronization occurs late by k symbols, the first k gap symbols are not all zeros, and the first k symbols of the codeword written were not all zeros, then there will appear to be an error burst preceding the codeword read and an error burst at the end of the codeword. If this double-burst error is within the detection guarantees of the code, then an uncorrectable error will be posted by the error correction algorithm. If this error pattern exceeds the detection guarantees of the code, then the error will appear to be correctable with probability P_{mc} and will appear to be uncorrectable with probability $1-P_{mc}$.

EARLY SYNCHRONIZATION IN UNMODIFIED SHORTENED CYCLIC CODES

Consider the case of early synchronization by one symbol. Again there are four combinations for the values of the preamble/sync symbol read and the redundancy symbol missed.

```
              Codeword read
         ┌──────────┴──────────┐
1)  •••pp0ddd•••dddrrr•••rr0ggg•••
```

The pattern read is a multiple of the codeword written. The sync framing error will not be detected.

```
              Codeword read
         ┌──────────┴──────────┐
2)  •••pp0ddd•••dddrrr•••rrXggg•••
```

The read remainder will be that of a single symbol in error at a location corresponding to the first symbol of the full-length codeword. Since this is beyond the bounds of the shortened codeword, the error correction algorithm will post an uncorrectable error. Given random data, the probability that the last redundancy symbol is zero is 2^{-w}, so this is also the probability that the read pattern will appear to be error free when synchronization occurs early by one symbol and a zero symbol is read preceding the codeword read. By similar reasoning, if synchronization occurs early by k symbols and the last k preamble/sync symbols are all zeros then the codeword read will appear to be error free if the last k symbols of the codeword written were all zeros. This should occur in random data with probability $2^{-(k*w)}$.

If synchronization occurs early by k symbols, the last k preamble/sync symbols are all zeros, and the last k symbols of the codeword written were not all zeros, then on read there will appear to be an error burst of length k or fewer symbols near the beginning of the full-length codeword. If the guaranteed detection capability of the code is equal to or greater than the apparent error created by the pattern of non-zero symbols missed, then the correction algorithm will post an uncorrectable error since the errors appear to be outside the bounds of the shortened codeword. If the apparent error created by the pattern of non-zero symbols missed exceeds the guaranteed detection capability of the code, then the error will appear to be correctable with probability P_{mc} and will appear to be uncorrectable with probability $1\text{-}P_{mc}$.

If synchronization occurs early by k symbols, the last k preamble/sync symbols are not all zeros and the correction capability of the code is equal to or greater than the number of non-zero preamble/sync symbols read, then the codeword read will appear to be correctable if the last k symbols of the codeword written were all zeros. This should occur in random data with probability $2^{-(k*w)}$.

If synchronization occurs early by k symbols, the last k preamble/sync symbols are not all zeros, and the last k symbols of the written codeword were not all zeros, then on read there will appear to be two error bursts, one near the beginning of the full-length codeword and one near the beginning of the shortened codeword. If this double-burst error is within the detection guarantees of the code, then an uncorrectable error will be posted by the error correction algorithm. If this error pattern exceeds the detection guarantees of the code, then the error will appear to be correctable with probability P_{mc} and will appear to be uncorrectable with probability $1\text{-}P_{mc}$.

```
           Codeword read
     ┌───────────┴───────────┐
3)  •••ppXddd•••dddrrr•••rr0ggg•••
```

The pattern read is that of a multiple of the codeword written with a symbol in error at the first symbol of the codeword read. A code used only for error detection would therefore detect the sync framing error while an error correction code would not.

```
           Codeword read
     ┌───────────┴───────────┐
4)  •••ppXddd•••dddrrr•••rrXggg•••
```

The read remainder will be that for two symbols in error, one at the first symbol of the full-length codeword and one at the first symbol of the shortened codeword. If this double-burst error is within the detection guarantees of the code, then an uncorrectable error will be posted by the error correction algorithm. If this error pattern exceeds the detection guarantees of the code, then the error will appear to be correctable with probability P_{mc} and will appear to be uncorrectable with probability $1\text{-}P_{mc}$.

By similar reasoning, if synchronization occurs early by k symbols, the last k preamble/sync symbols are not all zeros, and the guaranteed correction capability of the code equals or exceeds the number of non-zero preamble/sync symbols read, then the codeword read will appear to be correctable if the last k symbols of the codeword written were all zeros. This should occur in random data with probability $2^{-(k*w)}$. If the last k symbols of the codeword written were not all zeros, then on read there will appear to be an error burst at the beginning of the full-length codeword and an error burst at the beginning of the shortened codeword. If this double-burst error is within the detection guarantees of the code, then an uncorrectable error will be posted by the error correction algorithm. If this error pattern exceeds the detection guarantees of the code, then the error will appear to be correctable with probability P_{mc} and will appear to be uncorrectable with probability $1-P_{mc}$.

INITIALIZING THE ECC SHIFT REGISTER TO ALL ONES

One method of code modification that is used to improve the detectability of sync framing errors is to initialize the ECC shift register that implements the error correcting code to all ones prior to any write or read. This is equivalent to inverting the first m symbols, where m is the degree of the code, before they are processed by the ECC shift register. Let us modify our representation by showing symbols which appear to be inverted to the ECC shift register in uppercase:

```
                  Codeword written
          ┌─────────────────┴─────────────────┐
•••pppDDD•••DDDddd•••dddrrr•••rrrggg•••
          └────┬────┘
m inverted symbols
```

If there is no sync framing error on read then the inversions cancel:

```
                  Codeword read
          ┌─────────────────┴─────────────────┐
•••pppddd•••dddddd•••dddrrr•••rrrggg•••
          └────┬────┘
m re-inverted symbols
```

If a sync framing error occurs then read inversions will cancel write inversions except at the end points of the inversion.

In the case of late synchronization by one symbol, after read inversion the pattern read is:

```
                 Codeword read
      ┌─────────────────┴────────────────┐
•••pppDdd•••dddDdd•••dddrrr•••rrrggg•••
      └────┬────┘
 m re-inverted symbols
```

The read remainder will reflect one error at symbol m-1 of the codeword read, and may reflect errors at the symbol before the first data symbol and at the last symbol of the codeword read, depending on the value of the first symbol of the codeword written and the value of the gap symbol read as part of the codeword read, respectively.

Let us examine the four combinations for late synchronization by one symbol when the ECC shift register is initialized to all ones.

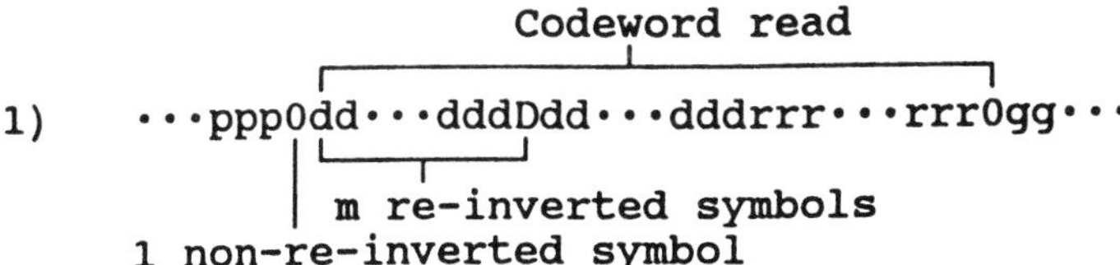

When the bits of the skipped data symbol were all ones, the inversion caused it to appear as a zero on write, so the read remainder reflects only the symbol in error at symbol m-1 of the codeword read. The sync framing error will go undetected by an error correction code.

```
                     Codeword read
          ┌─────────────────┴─────────────────┐
2)    •••pppXdd•••dddDdd•••dddrrr•••rrr0gg•••
          └────┬────┘
   m re-inverted symbols
```

When the bits of the skipped data symbol were not all ones then the inversion causes the read remainder to appear to be that of two symbols in error, one at symbol -1 and one at symbol m-1 of the codeword read. If this double-burst error pattern is within the detection guarantees of the code, then an uncorrectable error will be posted by the error correction algorithm. If this error pattern exceeds the detection guarantees of the code, then the error will appear to be correctable with probability P_{mc} and will appear to be uncorrectable with probability $1\text{-}P_{mc}$. For random data all bits of the first data symbol will be ones with probability 2^{-w} and therefore this is the probability that late synchronization by one symbol will be undetected by a code guaranteed to detect a double-burst error when a zero gap symbol is read. Under similar assumptions the probability is $2^{-(k^*w)}$ that late synchronization by k symbols will be undetected.

```
                     Codeword read
          ┌─────────────────┴─────────────────┐
3)    •••ppp0dd•••dddDdd•••dddrrr•••rrrXgg•••
          │└────┬────┘
          │ m re-inverted symbols
   1 non-re-inverted symbol
```

When the bits of the skipped data symbol were all ones and the first gap symbol is non-zero, the inversion causes the read remainder to appear to be that of two symbols in error, one at symbol m-1 and one at the last symbol of the codeword. If this double-burst error pattern is within the detection guarantees of the code, then an uncorrectable error will be posted by the error correction algorithm. If this error pattern exceeds the detection guarantees of the code, then the error will appear to be correctable with probability P_{mc} and will appear to be uncorrectable with probability $1\text{-}P_{mc}$.

```
                     Codeword read
          ┌─────────────────┴─────────────────┐
3)    •••pppXdd•••dddDdd•••dddrrr•••rrrXgg•••
          └────┬────┘
   m re-inverted symbols
```

When the bits of the first data symbol were not all ones and the first gap symbol is non-zero, the inversion causes the read remainder to appear to be that of three symbols in error, one at symbol -1, one at symbol m-1, and one at the last symbol of the codeword. If this triple-burst error pattern is within the detection guarantees of the code, then an uncorrectable error will be posted by the error correction algorithm. If this error pattern exceeds the detection guarantees of the code, then the error will appear to be correctable with probability P_{mc} and will appear to be uncorrectable with probability $1-P_{mc}$. Late synchronization by k symbols can be analyzed in a similar manner.

In the case of early synchronization by one symbol, after read inversion the pattern read is:

```
                Codeword read
        ┌─────────────┴─────────────┐
•••ppPddd•••ddDddd•••dddrrr•••rrrggg•••
        └───┬───┘
m re-inverted symbols
```

The read remainder will reflect one error at symbol m of the codeword read, and may reflect errors at symbol 0 and at the first symbol of the full-length codeword, depending on the value of the preamble/sync symbol read as part of the codeword read and the value of the written redundancy symbol, respectively.

Analysis of early synchronization when the ECC shift register is initialized to all ones is affected in much the same way as that of late synchronization. It is left as an exercise for the reader.

INVERTING REDUNDANCY SYMBOLS

Inverting the redundancy symbols is another method of code modification that is used to improve the detectability of sync framing errors. This method is essentially the mirror image of initializing the ECC shift register to all ones. Its effect on reducing the probability of undetected sync framing errors is the same as that of initializing the ECC shift register to all ones.

INITIALIZING THE ECC SHIFT REGISTER WITH A SPECIALLY SELECTED PATTERN

An approach which is capable of providing much better detectability for sync framing errors than those discussed above is initializing the ECC shift register with a pattern including both ones and zeros. If the pattern is carefully selected, an early or late sync slippage of one or more symbols will produce a remainder on read which differs significantly from that of any correctable codeword, enhancing the probability of detecting sync framing errors even in the presence of data errors within the record.

We are motivated to use for the initialization pattern a sequence which does not resemble itself when shifted one or more symbols, so that many errors result when read inversions are not perfectly aligned with write inversions as a result of a sync framing error. Pattern selection is influenced by the symbol patterns written immediately before (preamble/sync symbols) and after (gap symbols) the codeword symbols. Assuming no errors other than those causing the sync framing error it is possible to use simulation to determine for each candidate initialization pattern and given conditions an integer k such that all sync framing errors caused by synchronizing up to k symbols early or late will be detected. k will be a function of the candidate initialization pattern, the polynomial, the symbol patterns written immediately before a codeword (preamble/sync symbols) and after a codeword (gap symbols), and the record lengths. It is also possible to find integers k and b for each candidate initialization pattern and given conditions such that all sync framing errors caused by synchronizing up to k symbols early or late will be detected even if there is a burst of length b or fewer symbols anywhere within the codeword, or to find integers k and e such that all sync framing errors caused by synchronizing up to k symbols early or late will be detected even if there are e random symbols in error.

INVERTING A SPECIALLY SELECTED SET OF REDUNDANCY SYMBOLS

Another good approach for providing better detectability of sync framing errors is inverting a specially selected set of redundancy symbols. Again, this approach is essentially the mirror image of initializing the ECC shift register to a specially selected pattern and it provides equivalent protection against sync framing errors. Pattern selection would be accomplished in about the manner.

INTERLEAVED REED-SOLOMON CODES

We shall illustrate with three-way interleaving. Let us use the following notation to represent the set of interleaved Reed-Solomon codewords:

• • • pppdef• • • defrst• • • rstggg• • •

where 'p' is a preamble/sync symbol, 'd', 'e', and 'f' are data symbols of the three codewords, 'r', 's', and 't' are redundancy symbols of the three codewords, and 'g' is a gap symbol.

Consider the case of late synchronization by one symbol:

```
                    Codewords read
              ┌───────────┴───────────┐
• • •pppdef• • •defrst• • •rstggg• • •
```

The second and third codewords written are read as the first and second codewords and contain no errors caused by the sync framing error. The first codeword written is read as the third codeword. The same analysis performed above for late synchronization by one symbol of a single codeword applies to the apparent third codeword.

Consider the case of early synchronization by one symbol.

```
                  Codewords read
            ┌───────────┴───────────┐
• • •pppdef• • •defrst• • •rstggg• • •
```

The first and second codewords written are read as the second and third codewords and contain no errors caused by the sync framing error. The third codeword written is read as the first codeword. The same analysis performed above for early synchronization by one symbol of a single codeword applies to the apparent first codeword.

Analysis of late or early synchronization by any number of bits can be performed in a similar fashion; there is no qualitative difference in the effect on individual codewords between the interleaved and non-interleaved cases given the same amount of sync slippage per codeword.

INITIALIZING THE ECC SHIFT REGISTER TO ALL ONES

The effect of initializing the ECC shift register to all ones can be extrapolated from the non-interleaved to the interleaved case in the same way:

```
                  Codewords written
       ┌──────────────────┴────────────────┐
•••pppDEF•••DEFdef•••defrst•••rstggg•••
       └─────┬──────┘
3*m inverted symbols
```

If there is no sync framing error on read then the inversions cancel:

```
                  Codewords read
       ┌──────────────────┴────────────────┐
•••pppdef•••defdef•••defrst•••rstggg•••
       └─────┬──────┘
3*m re-inverted symbols
```

If a sync framing error occurs then read inversions will cancel write inversions except at the end points of the inversion.

In the case of late synchronization by one symbol, after read inversion the pattern read is:

```
                  Codewords read
        ┌─────────────────┴─────────────────┐
•••pppDef•••defDef•••defrst•••rstggg•••
        └─────┬──────┘
 3*m re-inverted symbols
```

Aside from misidentification, two of the codewords are not affected by the sync framing error. The read remainder for the other will reflect one error at symbol m-1 of the codeword read, and may reflect errors at the symbol before the first data symbol and at the last symbol of the codeword read, depending on the value of the first symbol of the codeword written and the value of the gap symbol read as part of the codeword read, respectively. The rest of the analysis is identical.

In the case of early synchronization by one symbol, after read inversion the pattern read is:

```
                 Codewords read
      ┌──────────────────┴────────────────┐
•••ppPdef•••deFdef•••defrst•••rstggg•••
      └─────┬──────┘
3*m re-inverted symbols
```

Again aside from misidentification, two of the codewords are not affected by the sync framing error. The read remainder for the other will reflect one error at symbol m of the codeword read, and may reflect errors at symbol 0 and at the first symbol of the full-length codeword, depending on the value of the first symbol of the preamble-/sync symbol read as part of the codeword read and the value of the written redundancy symbol, respectively. The rest of the analysis is identical.

Analysis of late or early synchronization by any number of bits can be performed in a similar fashion; there is no qualitative difference in the effect on individual codewords between the interleaved and non-interleaved cases given the same amount of sync slippage per codeword.

Initializing the ECC shift register to all ones (or inverting all redundancy symbols) has no qualitative difference between the interleaved and non-interleaved cases, and use of a specially selected pattern is called for. When a high degree of interleaving or a code of high degree is used, it might be permissible to initialize a selected set of symbol-wide registers to all ones (or to invert a selected set of redundancy symbols). However, best results would be achieved if each bit of the ECC shift register could be independently initialized to one (or a selected set of redundancy bits could be inverted).

RANDOMIZING DATA

More complete protection against sync framing errors can be achieved by implementing a shift register which generates a pseudo-random sequence, which is initialized to a known state before writing or reading each data record. The EXCLUSIVE-OR sum of the data-bit stream and the pseudo-random-bit sequence is fed to the ECC shift register instead of the data bit stream itself. Again an all-zeros data record produces non-zero redundancy, and if no sync framing error occurs the effects of the pseudo-random bit sequence on write and read cancel out. A sync framing error of any number of bits except the period of the pseudo-random sequence can be guaranteed to produce errors throughout the data record in excess of the correction capability of the EDAC code, so a sync framing error is no more subject to misdetection than any other uncorrectable error.

PROTECTING THE SYNC MARK WITH THE ERROR DETECTION/CORRECTION CODE

A different method for enhancing sync framing error protection is to include the sync mark in the symbols protected by the error detection/correction code. The effectiveness of this approach decreases as the length of the sync mark decreases.

Consider the case where the sync mark is protected by the error detection/correction code and synchronization occurs late by one or more symbols. If the gap symbols read due to the slippage are all zeros, the pattern read will appear to be that of a multiple of the codeword written plus some error pattern of about the same length as the sync mark at a location which includes the symbols of the assumed sync mark plus one or more symbols before the assumed sync mark. If the gap symbols read due to the slippage are not all zeros, the read remainder will reflect the same error burst as above plus an error burst in the redundancy symbols. In the former case if the error pattern does not exceed the correction guarantees of the code, the correction algorithm will detect the presence of error in the assumed sync mark or outside the bounds of the shortened codeword and raise an uncorrectable error flag. In either case, if the error pattern exceeds the correction guarantees but not the detection guarantees of the code, the correction algorithm will still raise an uncorrectable error flag. If the error pattern exceeds all correction and detection guarantees of the code, the sync framing error will appear to be correctable with probability P_{mc} and will appear to be uncorrectable with probability 1-P_{mc}. As the amount of synchronization slippage increases, the length of the apparent error burst(s) also increases.

Consider the case where the sync mark is protected by the error detection/correction code and synchronization occurs early by one or more symbols. The pattern read will appear to be that of a multiple of the codeword written plus some error pattern of about the same length as the sync mark at a location which includes the symbols of the assumed sync mark plus one or more symbols following the assumed sync mark. If the error pattern does not exceed the correction guarantees of the code, the correction algorithm will detect the presence of error in the assumed sync mark and raise an uncorrectable error flag since an error at the location of the sync mark implies that the original detection of the sync mark was mistaken and a sync framing error must have occurred. If the error pattern exceeds the correction guarantees but not the detection guarantees of the code, the correction algorithm will still raise an uncorrectable error flag. If the error pattern exceeds all correction and detection guarantees of the code, the sync framing error will appear to be correctable with probability P_{mc} and will appear to be uncorrectable with probability 1-P_{mc}. As the amount of synchronization slippage increases, the length of the apparent error burst also increases.

CONCLUSIONS

Based on the material presented above DST recommends that all cyclic and shortened cyclic error detection and correction codes be modified by either:

(1) Initializing the ECC shift register to a specially selected pattern prior to each write and read, or

(2) EXCLUSIVE-OR-ing a specially selected pattern against the redundancy bits on each write and read.

(3) Feeding the ECC shift register with the EXCLUSIVE-OR sum of data and a pseudo-random sequence on each write and read.

Including the sync mark in the bits covered by the error detection/correction code and insuring that codewords are preceded and followed by non-zero symbols could provide additional protection.

The measures (a), (b), and (c) below have been used in the past to provide increased sync framing error protection. If economic reasons dictate the use in new designs of existing IC's or other hardware for which it is not feasible to implement (1), (2) or (3) above, DST recommends the use of all of provisions for sync framing error protection (a)-(c) below whose implementation is possible:

(a) Initializing ECC shift register to all ones prior to each write or read.

(b) Inverting redundancy on each write and read.

(c) Including the sync mark within the ECC check on each write and read.

(d) Insuring that codewords are preceded and followed by non-zero symbols.

4.9 INTERLEAVED, PRODUCT, AND REDUNDANT-SECTOR CODES

4.9.1 INTERLEAVED CODES

Interleaving is a technique used to geographically disperse data for each codeword over a larger area of media in order to spread error bursts over multiple codewords. In this way, the error contribution to any one codeword from a long defect is minimized.

As an example, consider a two-dimensional array with C bytes per row and N bytes per column, in which each column is a codeword of a Reed-Solomon code. As bytes are written to the media, they are also processed by the redundancy-generating circuitry. Bytes 0, C, 2C, etc. are processed by the circuitry for interleave 0. Bytes 1, C+1, 2C+1, etc. are processed by the circuitry for interleave 1, and so on. As bytes are read, operation is identical except that syndromes are generated rather than redundancy. If necessary, the correction algorithm is performed, after which the data is released to the host. In this example, any error burst must span more than C bytes before affecting more than one byte from any one codeword (interleave).

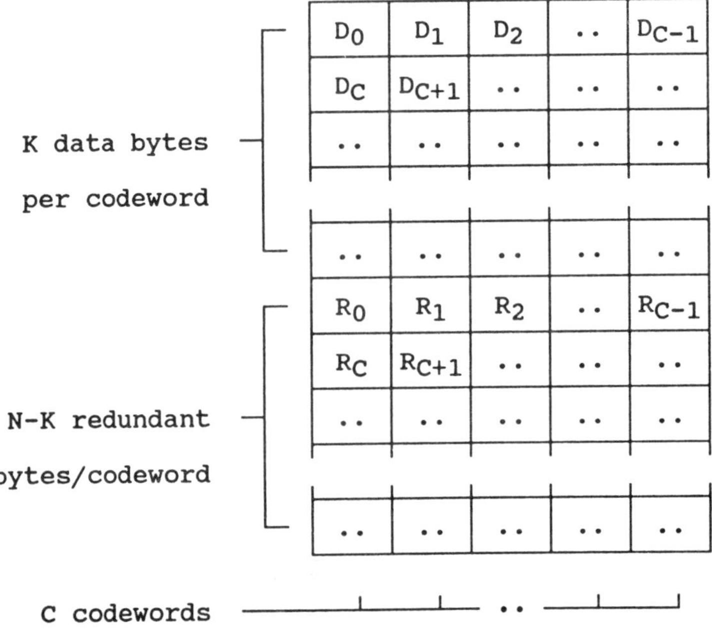

There are many interleaving techniques. Selection of a technique for a particular application involves tradeoffs between cost, code performance, transfer rate, block size, and correction time.

4.9.2 PRODUCT CODES

Product codes perform error correction on a block of data in more than one dimension. Consider an array of symbols organized into rows and columns, with each row treated as a codeword of some code C_1 and each column as a codeword of another (possibly the same) code C_2. The resulting overall code is called a product code. It is common to see Reed-Solomon codes used as the component codes of product codes.

There are many techniques for loading and unloading the array of product code. As an example, consider an array which on write is loaded one row at a time from the source. After all redundancy in both dimensions has been calculated, the array is unloaded diagonally to the device. On read, the data from the device is loaded diagonally, then after correction, the array is unloaded one row at a time to the destination. The diagonal unloading and loading accomplishes geographical dispersion of data in a manner that minimizes the number of error bytes that a long burst can contribute to any codeword in either dimension.

D_0 D_1 D_2 . . .

ROW REDUNDANCY

COLUMN REDUNDANCY

CHECKS ON REDUNDANCY

There are many decoding techniques for product codes, one of which is to correct rows first, then correct columns. Another technique is to iterate row and column correction; errors in an uncorrectable codeword from one dimension may belong to correctable codewords in the other dimension, and after they are corrected, the uncorrectable codeword may become correctable. Another technique is to combine row/column iteration with erasure correction; the row [column] numbers of codewords in error are used as erasure pointers for column [row] correction. There are other decoding techniques for product codes as well. The correction capability of product codes is very dependent on the precise decoding techniques used.

Product codes have been popular with the Japanese companies and have been implemented on a number of digital audio products, including both optical disk and magnetic tape products for consumer and commercial use.

4.9.3 REDUNDANT-SECTOR CODES

Redundant-sector codes can handle very long error bursts. As an example, consider an implementation with one redundant sector for each K data sectors. Each sector has its own sync field, and uses CRC for error detection. Each byte of the redundant sector is generated by EXCLUSIVE-OR-ing together the corresponding bytes of the K data sectors i.e. computing a parity sector. If on reading a data sector, a CRC error is detected, its contents can be regenerated by EXCLUSIVE-OR-ing the remaining data sectors with the redundant sector. This technique can correct even a long burst which wipes out a sync mark.

SYNC	DATA SECTOR #1	CRC
•	• • •	•
SYNC	DATA SECTOR #K	CRC
SYNC	PARITY SECTOR #1	CRC

An extension of this technique is to use interleaving e.g. one redundant sector for even sectors and another for odd sectors. This will allow correction of a long burst spanning any two adjacent sectors, or correction of any two random sectors in error provided that one is even and the other is odd.

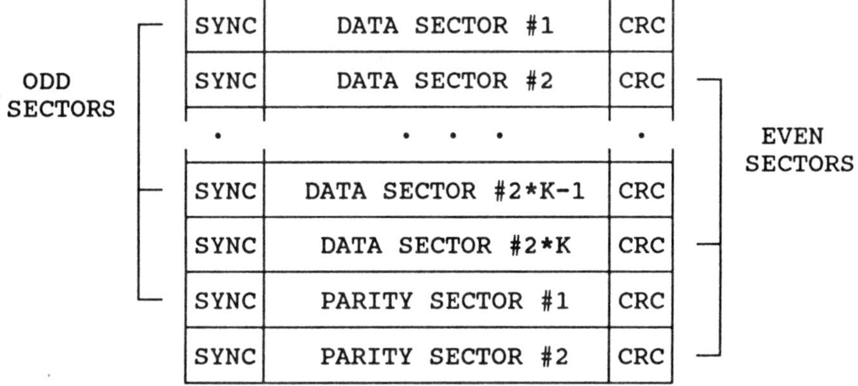

A more powerful technique is to implement a Reed-Solomon code across the corresponding bytes of the K data sectors, allowing the correction of multiple sectors in error within a codeword-long set of sectors. CRC error information can be used as erasure pointers by the Reed-Solomon correction algorithm, so the number of redundant sectors needed is just the number of errors to be corrected, not twice this number.

SYNC	DATA SECTOR #1	CRC
•	• • •	•
SYNC	DATA SECTOR #K	CRC
SYNC	REDUNDANT SECTOR #1	CRC
•	• • •	•
SYNC	REDUNDANT SECTOR #N-K	CRC

Redundant-sector techniques can be combined with other ECC techniques to form a more powerful EDAC scheme. For example, the CRC shown for each sector can be replaced with an ECC code which can correct single (or multiple) small bursts at the sector level, and redundant-sector techniques can be used to correct the much lower rate of long bursts. This is in effect a product code, with the individual sectors comprising the row codewords and corresponding bytes from the individual sectors comprising the column codewords.

Redundant-sector techniques have been used on Bernoulli disks, read-only optical disks, and numerous tape devices.

CHAPTER 5. SPECIFIC APPLICATIONS

5.1 EVOLUTION OF EDAC SCHEMES

5.1.1 *EVOLUTION OF OEM MAGNETIC DISK EDAC SCHEMES*

In the early 1970's, 32-bit Fire codes were widely used for error correction in OEM magnetic disk devices. These codes were easy to define and required a moderate amount of hardware to implement. However, their sensitivity to multiple short bursts posed a serious data accuracy problem. Some error recovery procedures in use at the time performed correction on soft as well as hard errors; this worsened the problem. By the late 1970's, many companies had dropped 32-bit Fire codes in favor of 32-bit computer-generated codes that were selected to be insensitive to multiple short bursts. They also changed their error recovery procedures to correct hard errors only, and had taken other steps to achieve better data accuracy. As the 5¼ inch hard disk industry developed, form factor pressure on controller builders pushed implementation efficiency to the point where 32-bit computer-generated codes were implemented using just five and one-half standard TTL IC's.

Over the last four or five years, many hard disk developers have implemented the (2,7) RLL modulation code. The error-propagation properties of this code necessitates a larger correction span, which has prompted many companies to switch to more powerful 48-bit, 56-bit and 64-bit computer-generated ECC codes to maintain good data accuracy. Several hard disk controller IC's developed during this period implement programmable polynomial generators that support 48-bit codes, and at least one supports codes up to 64 bits in length.

Hard disk controller IC developers (including Cirrus Logic) are now incorporating two symbol error correcting Reed-Solomon codes in their new designs in order to handle higher raw-error-rate media by correcting two independent error bursts within a sector.

5.1.2 *EVOLUTION OF IBM MAGNETIC DISK EDAC SCHEMES*

In 1970, IBM introduced the 3330 magnetic disk drive, which uses a 56-bit Fire code to correct single 11-bit bursts in variable length records of up to approximately 13,000 bytes. This code's generator polynomial was selected to allow fast computation of error location using the Chinese Remainder Theorem. However, the structure that permitted fast correction also introduced a pattern sensitivity to multiple short bursts.

The IBM 3340 (the first drive using Winchester technology) and 3350 magnetic disk drives were introduced in 1973 and 1975, respectively. They use the same 48-bit Fire code to correct single bursts (up to 4 bits for the 3340, 5 bits for the 3350) in variable record lengths (up to 19,000 bytes for the 3350).

In 1979, IBM introduced the 3370 magnetic disk drive, which employs a three-way interleaved, Reed-Solomon code on fixed-length sectors of 512 bytes. Three redundancy bytes are used in each of the three interleaves, giving single symbol (byte) error correction and double symbol error detection in each interleave. IBM uses this code to guarantee the correction of any single burst of 9 bits or less, the detection of any single burst of 41 bits or less, and the detection of any two bursts, each of 17 bits or less.

This code has a high miscorrection probability for cases in which multiple short bursts cause a single interleave to have more than two symbols in error. The existence of this pattern sensitivity is clear when one considers that for all possible errors, a sector has nine bytes of redundancy protecting it from miscorrection, versus only three bytes of redundancy for single-interleave errors.

In 1980, IBM introduced the 3380 magnetic disk drive, which employs a Reed-Solomonlike, two-way interleaved code to correct single bursts in variable length records of up to approximately 48,000 bytes. Operating on 16-bit symbols, this code will correct any single burst contained within two contiguous symbols and detect any single burst contained within three contiguous symbols. Twelve bytes of redundancy are used; four bytes are associated with each interleave, and an additional four bytes are shared between the two interleaves. This sharing of redundancy between interleaves reduces the miscorrection probability for single-interleave errors.

In 1987, IBM announced the 3380K magnetic disk drive, which employs a novel multiple-burst error correcting code that dedicates more than six percent redundancy to error detection and correction and accommodates a raw error rate much higher than for earlier versions of the 3380. Other features of the code include minimum data delay and a unique supplementary error detection method. The higher track densities achieved by the 3380K may have motivated IBM to use multiple-burst correction. DST expects to see even more powerful codes of the same class implemented on future high-end magnetic (and possibly optical) devices.

5.1.3 *EVOLUTION OF IBM MAGNETIC TAPE EDAC SCHEMES*

In 1973, IBM introduced the 3420 (models 4, 6, and 8) Group Code Recording (GCR) magnetic tape drives. These employ a Reed-Solomon-like code over nine tracks to correct errors in one track without erasure pointers or two tracks with erasure pointers.

In 1984, IBM introduced the 3480 eighteen track magnetic tape cartridge drive. It employs a code which uses parity on bit-vectors in three dimensions (vertical, left diagonal, and right diagonal). The eighteen tracks are divided into two nine-track sets. Each set contains seven data tracks, its own vertical parity track, and a diagonal parity track which is shared with the other set. The code is capable of correcting the following error situations:

- Up to two tracks in error with a pointer, or one track in error without a pointer, in each of the two sets

- Up to two tracks, one of which has a pointer, in one set, and up to one track without a pointer in the other set

- Up to three tracks with pointers in one set, and up to one track with a pointer in the other set

Since the 3480 ECC employs redundancy sharing between the two sets of nine tracks, it is more powerful than that of the 3420 models 4, 6, and 8, even though both employ the same percentage of redundant tracks. The 3480 ECC is also the simpler and less expensive of the two methods. We expect IBM to continue to use this code as new versions of the 3480 are offered. Other companies developing eighteen-track magnetic cartridge tape products are likely to use it as well.

5.1.4 *HIGH-PERFORMANCE EDAC SCHEMES FOR MAGNETIC DISK*

As mentioned above, several hard-disk controller IC's developed recently, as well as others currently under development, employ Reed-Solomon codes for correction of random single and double symbol errors. There is also a segment of the industry interested in employing more powerful Reed-Solomon codes to allow the use of so-called "horrible" media containing hundreds of defects per platter.

Digital Equipment Corporation implemented a very powerful Reed-Solomon code in its UDA-50 magnetic disk controller. This code corrects up to eight 10-bit symbols in error within a 512-byte record. The UDA-50 also employs error-tolerant headers and sync marks. Other companies in the same market have implemented similar codes.

We at Data Systems Technology feel that magnetic disk drive manufacturers which develop products in the future will be making trade-offs between media costs and the cost of high-performance ECC IC's. The optical storage industry has proven the technical feasibility of using high error-rate media; the magnetic storage industry is likely to follow suit, using ECC parts developed for optical products.

5.1.5 *HIGH-PERFORMANCE EDAC SCHEMES FOR OPTICAL DISK*

There are three major types of optical media: read-only, write-once, and erasable. Each type has different ECC requirements, but all require high performance ECC.

For stamped media, there is no possibility of sector retirement. Therefore all initial defects as well as end-of-life defects must be handled by the ECC. This requires higher performance ECC and more geographic dispersion of data than would be necessary if retirement were possible.

Since the mastering of stamped optical media is performed only once for each unique set of data, the ECC redundancy generation process can be more complex than for the other two types of media. Since the full content of each track is known at the time of mastering, greater geographic dispersion is possible than for the other two types of media, where each data sector may be written at different times.

Product codes have been popular for stamped media; the CD digital audio players and CD-ROM digital data storage devices employ product codes. One stamped-media device employs a three-dimensional product code over each track, geographically dispersing each sector over the entire track.

A number of companies in the U.S. and Japan support the use of single-dimension interleaved Reed-Solomon codes (also referred to as long distance Reed-Solomon codes (LDC)) for ECC on 90 mm and 130 mm, WORM and rewritable optical media. Such a code has been approved by the U.S. Accredited Standards Committee X3B11. Several companies have developed or are developing LSI parts using DST's ECC technology to support this code.

DST expects to continue to be at the forefront of EDAC technology for optical storage. We support the use of long distance Reed-Solomon codes for 90 mm and 130 mm, WORM and rewritable optical and developed our NG-8510, NG-8520, and CL-SH8530 IC's especially for this application. The NG-8510/8520 approach splits the error correction task between logic within the IC and logic within support software. The CL-SH8530 performs correction real-time in hardware.

5.2 APPLICATION TO LARGE-SYSTEMS MAGNETIC DISK

5.2.1 *CAPABILITY OF DISK CODES*

(See glossary for definitions)

	3330	3340	3350	OEM	3370
ECC Bits	56	48	48	32	72
Rec Length (Bits)	104240	70320	152552	4644	4168
Correction Span	11	3	4	11	9
Detection Span Before Correction	56	48	48	32	65
Published Det. Span After Corr.	22	11	10	32	16
Actual Det. Span After Correction	28	30	26	13	41
P_{mc}*	1.5E-9	1.E-9	4.3E-9	1.1E-3	2.2E-16
P_{md}*	1.4E-17	3.6E-15	3.6E-15	2.3E-10	2.1E-22

* Assuming all errors are possible and equally probable.

POLYNOMIALS

3330

$$(x^{22} + 1)\cdot(x^{11} + x^7 + x^6 + x^1 + 1)\cdot(x^{11} + x^9 + x^7 + x^6 + x^5 + x^1 + 1)$$
$$\cdot(x^{12} + x^{11} + x^{10} + x^9 + x^8 + x^7 + x^6 + x^5 + x^4 + x^3 + x^2 + x^1 + 1)$$

3340/3350

$$(x^{13} + 1)\cdot(x^{35} + x^{23} + x^8 + x^2 + 1)$$

OEM

$$(x^{21} + 1)\cdot(x^{11} + x^2 + 1)$$

3370

Field Generator: $x^8 + x^6 + x^5 + x^4 + 1$

Code Generator: $(x + \alpha^0)\cdot(x + \alpha^1)\cdot(x + \alpha 3^{-1})$

5.2.2 *THE 3330 MAGNETIC DISK CODE*

CODE DEFINITION

The 3330 code is a generalized Fire code. It has a single-burst correction span of 11 bits and a single-burst detection span of 22 bits. Decoding uses the Chinese Remainder Method for displacement calculation. This method requires only a fraction of the shifts required by clocking around a sequence. The code is defined by the following polynomial:

$$\begin{aligned} g(x) = {} & x^{56} + x^{55} + x^{49} + x^{45} + x^{41} + x^{39} + x^{38} + x^{37} \\ & + x^{36} + x^{31} + x^{22} + x^{19} + x^{17} + x^{16} + x^{15} \\ & + x^{14} + x^{12} + x^{11} + x^{9} + x^{5} + x + 1 \end{aligned}$$

g(x) has the following relatively prime factors:

$$\begin{aligned} P_0(x) &= x^{22} + 1 \\ P_1(x) &= x^{11} + x^{7} + x^{6} + x + 1 \\ P_2(x) &= x^{12} + x^{11} + x^{10} + x^{9} + x^{8} + x^{7} + x^{6} \\ & \quad + x^{5} + x^{4} + x^{3} + x^{2} + x + 1 \\ P_3(x) &= x^{11} + x^{9} + x^{7} + x^{6} + x^{5} + x + 1 \end{aligned}$$

Each of the factors has a different period as shown below.

FACTOR	PERIOD
$P_0(x)$	22
$P_1(x)$	89
$P_2(x)$	13
$P_3(x)$	23

IMPLEMENTATION

Encoding is performed by a shift register implementing:

$$g(x) = P_0(x) \cdot P_1(x) \cdot P_2(x) \cdot P_3(x)$$

The shift register premultiplies by x^{56} and simultaneously divides by g(x). The remainder is inverted and appended to the data as check bytes.

For decoding, the hardware is modified to divide the received data polynomials by each of the factors of g(x) independently. There are four independent shift registers generating four independent syndromes. The shift registers are named P_0, P_1, P_2 and P_3, according to the factor of the generator polynomial implemented. There is no premultiplying during decode.

On read, if all four syndromes are zero, the data is considered to be correct. If there are both zero and non-zero syndromes, the error exceeds the guaranteed correction capability of the code. If all four syndromes are non-zero, correction is attempted using the Chinese Remainder Theorem.

CORRECTION PROCEDURE

(1) Shift P_0 until the 11 high order bits of the shift register are zeros and the lowest order bits of the shift register are nonzero. Save the shift count (n_0). If the above alignment is not achieved in less than 22 shifts, the error is uncorrectable.

(2) Shift P_1 (with feedback) until a match with P_0 is achieved. Save the shift count (n_1). If a match is not found in less than 89 shifts, the error is uncorrectable.

(3) Shift P_2 (with feedback) until a match with P_0 is achieved. Save the shift count (n_2). If a match is not found in less than 13 shifts, the error is uncorrectable.

(4) Shift P_3 (with feedback) until a match with P_0 is achieved. Save the shift count (n_3). If a match is not found in less than 23 shifts, the error is uncorrectable.

The error pattern is in P_0. The error displacement, measured from the last check bit to the last error bit (low order of P_0) is given by:

$$d = [-(k_0*n_0+k_1*n_1+k_2*n_2+k_3*n_3)] \text{ MOD } e$$

where,

k_0 = 452,387
k_1 = 72,358
k_2 = 315,238
k_3 = 330,902
e = 585,442 = LCM(22,89,13,23)

HARDWARE SELF-CHECKING

Self-checking of the shift registers is performed with parity predict circuits. See Section 6.5 for information on parity predict.

SYNC FRAMING ERROR PROTECTION

The inversion of check bytes provides protection against sync framing errors. This also provides protection for some types of hardware failures by making the check bytes nonzero for an all-zeros record.

5.2.3 *THE 3350 MAGNETIC DISK CODE*

CODE DEFINITION

The 3350 code is a shortened Fire code. It is defined by the following generator polynomial:

$$g(x) = x^{48} + x^{36} + x^{35} + x^{23} + x^{21} + x^{15} + x^{13} + x^{8} + x^{2} + 1$$

The polynomials below are factors of g(x):

$$c(x) = x^{13} + 1$$

$$p(x) = x^{35} + x^{23} + x^{8} + x^{2} + 1$$

The c(x) factor is composite and has a period of 13. The p(x) factor is irreducible and has a period of 34,359,738,367. The period of g(x) is the least common multiple of the periods of c(x) and p(x), which is 446,676,598,771. Fire codes are discussed in Section 3.1. Decoding of shortened codes is discussed in Section 2.4.

CODE CAPABILITY

When the 3350 code is used for detection only, any single burst not exceeding 48 bits in length is guaranteed to be detected. In addition, any combination of double bursts is guaranteed to be detected provided the sum of the burst lengths is no greater than 14. This number comes from the Fire code theory and is a lower bound only. It is very conservative since record length is very short compared to the period of g(x) (see Section 3.1). Misdetection probability for bursts exceeding the code guarantees is 3.55E-15.

In the 3350 implementation, the code is used to correct bursts through four bits in length on records up to 19,069 bytes in length. With this correction span and record length, the code is guaranteed to detect any single burst not exceeding 26 bits in length. This number was determined by a computer search. The Fire code theory gives the detection span as only ten bits. For 19,069 byte records, the miscorrection probability is 4.3E-9 for error bursts exceeding code guarantees, assuming all errors are possible and equally probable.

CODE DESCRIPTION

The 3350 code is shortened by the premultiplication of the data polynomial. This requires a shift-register circuit that multiplies and divides simultaneously. These circuits are discussed in Section 1.3.2.

The multiplier polynomial is determined by computing the reciprocal of the polynomial that is the residue of:

$$(x^{156352-1+48}) \text{ MOD } g'(x)$$

where $g'(x)$ is the reciprocal polynomial of g(x).

The multiplier polynomial is:

$$x^{47} + x^{39} + x^{35} + x^{32} + x^{30} + x^{25} + x^{21} + x^{20} + x^{17} + x^{15} + x^{13} + x^{9} + x^{7} + x^{6} + x^{2}$$

The multiplier polynomial is used only during read, since shortening of the code applies only to the read case.

The logical shift-register configurations used for write and read are shown in Figures 5.2.3.1 and 5.2.3.2 respectively. Although the write and read configurations are shown separately, the physical implementation is a single 48-bit shift register. As seen in Figure 5.2.3.2, there are three separate groups of bits feeding the XOR gates of the shift register in the read configuration:

B (BOTH): Feedback terms that are common to both the multiplier polynomial and the generator polynomial.

I (INPUT): Feedback terms unique to the multiplier polynomial.

F (FEEDBACK): Feedback terms unique to the generator polynomial.

Figure 5.2.3.3 shows a circuit equivalent to that shown in Figure 5.2.3.2. This circuit is easier to understand. It is shown in the same form as circuits performing simultaneous multiplication and division in Section 1.3.2. A close comparison of the circuits of Figures 5.2.3.2 and 5.2.3.3 reveals that splitting the read configuration feedback logic into three parts is a way to save logic. The feedback logic for the write configuration can be obtained from the feedback logic for the read configuration by OR'ing the BOTH and FEEDBACK lines and adding gating functions.

WRITE OPERATION

There are two write modes:

a. WRITE DATA BITS
b. WRITE CHECK BITS

During the WRITE DATA BITS mode, serial data bits are written to the disk. Simultaneously, the ECC shift register, with write feedbacks enabled, receives the serial data bits and calculates write check bits. During the WRITE CHECK BITS mode, the feedbacks are disabled and check bits are shifted out of the register, complemented, and written to the disk.

READ OPERATION

There are three read modes:

a. READ DATA BITS
b. READ CHECK BITS
c. CORRECTION

During the READ DATA BITS mode, the ECC shift register, with read feedbacks enabled, receives serial read data bits. A syndrome is partially computed. During the

READ CHECK BITS mode, read feedbacks remain enabled. The complements of check bits are received by the ECC shift register and the computation of the syndrome is completed. After processing the read check bits, the ECC shift register should be all zeros if no error occurred and nonzero if a detectable error occurred.

The CORRECTION mode is entered at the end of a read when the ECC shift register contents (syndrome) is found to be nonzero. The shift register is shifted with read feedbacks enabled, until bits 4-47 are zero. When this occurs, the error pattern is in bits 0-3. Shifting continues to the next byte boundary to place the error pattern in byte alignment. The shift count is used to calculate an error displacement. The error is uncorrectable if all zeros are not found in bits 4-47 of the shift register within 156,352 (19,544*8) shifts.

MICROCODE CORRECTION ALGORITHM

Part of the 3350 correction algorithm is implemented in microcode at the storage control unit. When correction is required, the microcode initializes a counter to 19,544. As the ECC shift register is shifted during the CORRECTION mode, a flag is raised to the storage control unit microcode once every eight shifts. The microcode decrements its counter by one. When ECC hardware finds bits 4-47 of the shift register zero, the microcode is alerted at the next byte boundary. The microcode counter then contains the error displacement in bytes from the last check byte to the first byte in error. The error pattern is obtained from bits 0-15 of the ECC shift register.

HARDWARE SELF-CHECKING

The 3350 employs parity predict for self-checking of error correcting circuits. These techniques are discussed in Section 6.5.

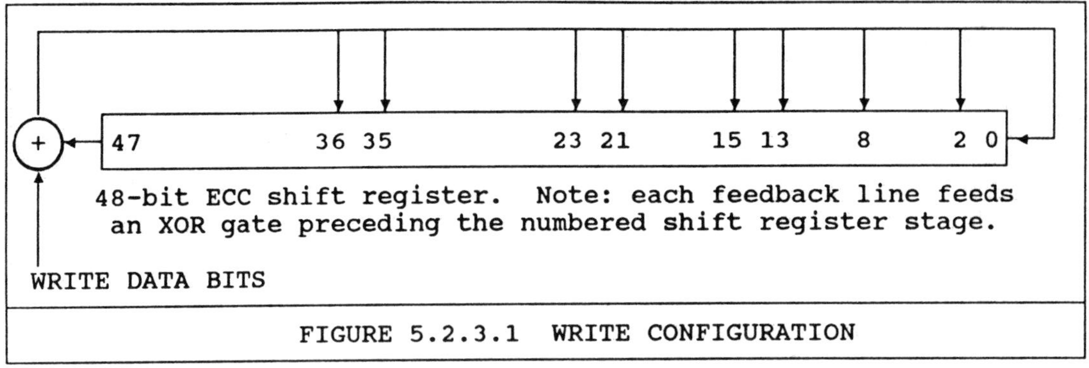

FIGURE 5.2.3.1 WRITE CONFIGURATION

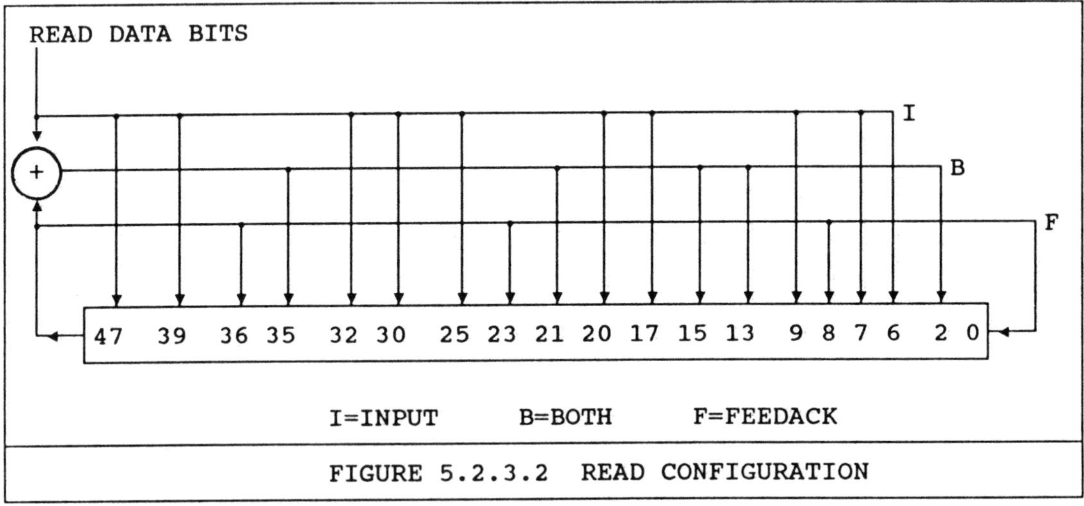

FIGURE 5.2.3.2 READ CONFIGURATION

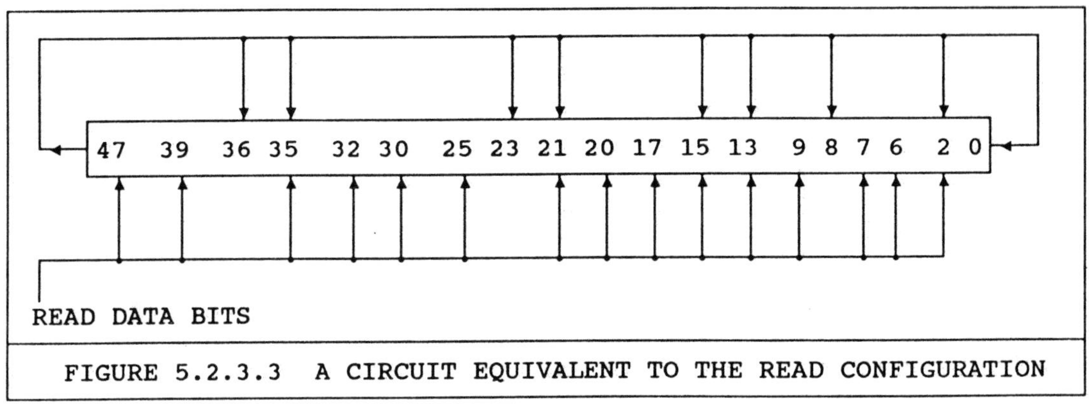

FIGURE 5.2.3.3 A CIRCUIT EQUIVALENT TO THE READ CONFIGURATION

5.2.4 *THE 3370 MAGNETIC DISK CODE*

INTRODUCTION

The 3370 magnetic-disk code is a single-error correcting, double-error detecting extended Reed-Solomon code interleaved to depth three. There are three logical circuits (interleaves), each sharing the same physical hardware. Each logical circuit protects one third of the data. The interleave protecting a particular byte is determined by its byte count modulo three.

Byte Count	Byte count modulo y	
0	0	Interleave 0
1	1	Interleave 1
2	2	Interleave 2
3	0	
4	1	
5	2	
6	0	
7	1	
8	2	
.	.	
.	.	

For the 512-byte data fields, there are nine check bytes, three for each interleave. Two interleaves contain 171 data bytes each and the remaining interleave contains 170 data bytes.

ID fields are protected by a three-byte detection-only code that is a subset of the data field code.

The check bytes are stored in a memory organized logically into three areas referred to as RAM1, RAM2 and RAM3.

	Interleave 0	Interleave 1	Interleave 2
RAM1	G1(x) MEM LOC 0	G1(x) MEM LOC 1	G1(x) MEM LOC 2
RAM2	G2(x) MEM LOC 3	G2(x) MEM LOC 4	G2(x) MEM LOC 5
RAM3	G3(x) MEM LOC 6	G3(x) MEM LOC 7	G3(x) MEM LOC 8

OVERALL CODE CAPABILITY

The three syndromes of the 3370 implementation give the code a minimum distance of four, which is sufficient to correct single errors and detect double errors.

The single-error correction and double-error detection interleaved to depth three results in the following overall capability for the code.

CAPABILITIES (See glossary for definitions)

1. Guaranteed correction span: 9 bits

   NOTE: The structure of the 3370 code provides the capability to correct any single burst up to 17 bits in length. However, the correction span as implemented is limited to nine bits.

2. Guaranteed detection span without correction: 65 bits

3. Guaranteed detection span with correction:

   Single-burst: 41 bits
   Double-burst: 17 bits

4. Misdetection probability (Pmd) = 2.1E-22

5. Miscorrection probability (Pmc) = 2.2E-16

ENCODING

Normally, the encoding for a Reed-Solomon code with the capability of the 3370 code would be accomplished with circuits implementing the following encode equation:

$$G(x) = (x + \alpha^{-1})\cdot(x + \alpha^{0})\cdot(x + \alpha^{+1})$$

However, in the 3370 implementation, each write check byte is generated separately by dividing the data polynomial by each factor of G(x).

In the 3370 implementation, α is defined by the primitive polynomial

$$x^8 + x^6 + x^5 + x^4 + 1.$$

SYNDROME GENERATION

On read, the encoding process is repeated. Syndromes are the XOR sum of check bytes generated on write and check bytes generated on read. The single-error syndrome equations are shown below:

$$S_0 = E1$$
$$S_1 = E1 \cdot \alpha^{L1}$$
$$S_{-1} = E1 \cdot \alpha^{-L1}$$

where,

E1 = Error value

L1 = Error location (displacement from the end of an interleave; the last data byte of an interleave is location zero).

HARDWARE IMPLEMENTATION

The figures on the following page show the 3370 hardware implementation. The operation is as follows:

WRITE DATA:

SR1,RAM1 = $D(x)$ MOD $(x + \alpha^0)$

SR2,RAM2 = $D(x)$ MOD $(x + \alpha^1)$

SR3,RAM3 = $D(x)$ MOD $(x + \alpha^{-1})$

WRITE CHECK BYTES:

Nine check bytes are written via the MUX in the following order:

- 3 bytes from RAM1
- 3 bytes from RAM2
- 3 bytes from RAM3

READ DATA:

Same as write data above.

READ CHECK BYTES:

The nine check bytes read are added modulo-2 (XOR) to the nine check bytes generated (the contents of the RAM's). The resulting syndromes are stored in the RAM's.

3370 ECC HARDWARE

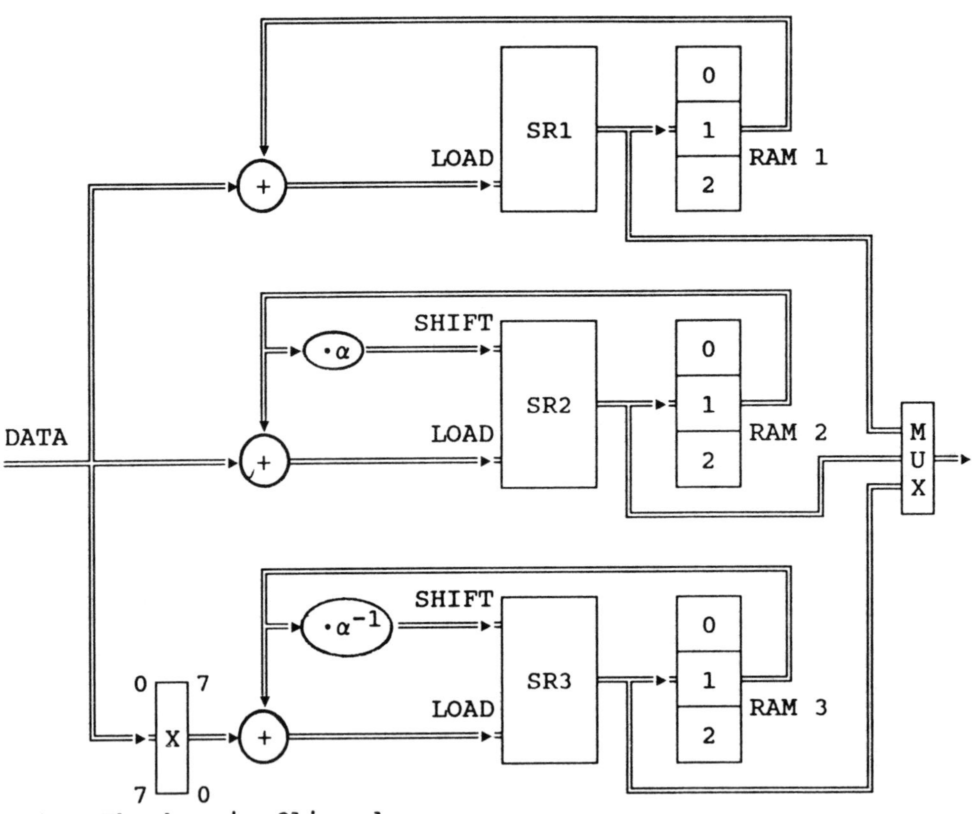

Note: The bus is flipped end-for-end at this point

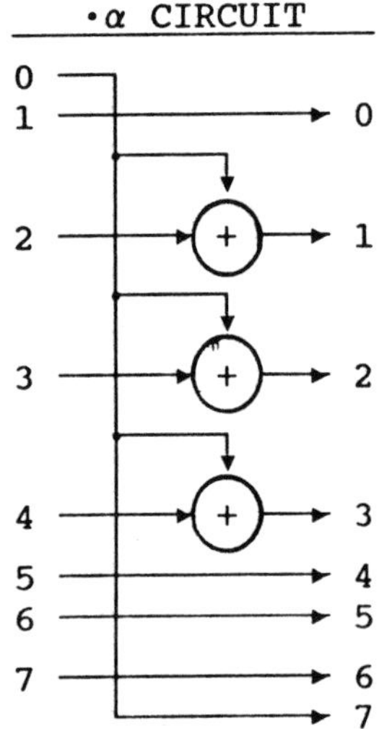

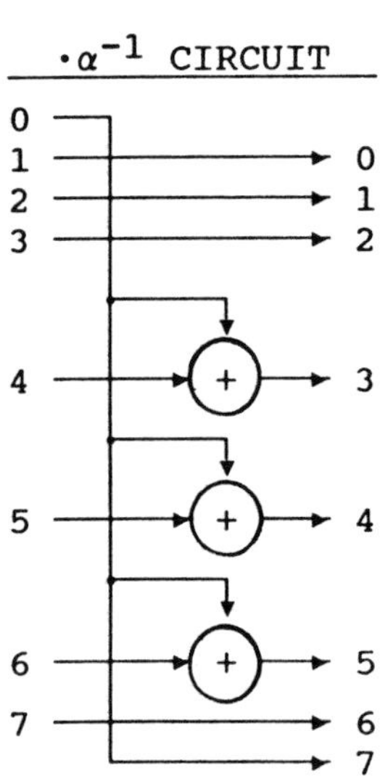

CORRECTION ALGORITHM

In the 3370 implementation, each interleave is decoded separately. The patterns and displacements within each interleave are analyzed to compute a single pattern and displacement within the 512-byte data block.

Interleave Correction Algorithm

Single-Error Correction: The correction algorithm for each interleave is simple and fast. When there is a single-symbol error, the syndrome equations form a system of equations in two unknowns, error value E1 and error location L1.

$$S_0 = E1$$
$$S_1 = E1 \cdot \alpha^{L1}$$
$$S_{-1} = E1 \cdot \alpha^{-L1}$$

The error value is S_0. The error location can be determined by substituting S_0 for E1 in the equation for S_1.

$$E1 = S_0$$
$$L1 = LOG_\alpha(S_1) - LOG_\alpha(S_0)$$

In the 3370 implementation the LOG_α function is accomplished with a ROM table.

Double-Error Detection: Double errors can be detected by verifying that:

$$\frac{S_1}{S_0} + \frac{S_0}{S_{-1}} = 0$$

This relationship will not be zero if a double error occurs.

To do the test above requires finite field division capability. The test below is equivalent and requires only the use of a log table. This version of the test is implemented on the 3370.

$$LOG_\alpha(S_1)\text{-}LOG_\alpha(S_0) = LOG_\alpha(S_0)\text{-}LOG_\alpha(S_{-1})$$

If the relationship above is true, a single error is assumed. If it is not true, a double error is assumed.

OVERALL CORRECTION ALGORITHM

Functions of the overall correction routine include:

1. Comparing displacements for the three interleaves to determine if a single error burst occurred.

2. Comparing displacements to determine which interleave contains the first byte of error pattern.

3. Converting the patterns and displacements within each interleave to a single pattern and displacement within the 512-byte data block.

3370 LOG TABLE (INPUT IS α^n, OUTPUT IS n)

	0	1	2	3	4	5	6	7	8	9	A	B	C	D	E	F
00	--	00	01	E7	02	CF	E8	3B	03	23	D0	9A	E9	14	3C	B7
10	04	9F	24	42	D1	76	9B	FB	EA	F5	15	0B	3D	82	B8	92
20	05	7A	A0	4F	25	71	43	6A	D2	E0	77	DD	9C	F2	FC	20
30	EB	D5	F6	87	16	2A	0C	8C	3E	E3	83	4B	B9	BF	93	5E
40	06	46	7B	C3	A1	35	50	A7	26	6D	72	CB	44	33	6B	31
50	D3	28	E1	BD	78	6F	DE	F0	9D	74	F3	80	FD	CD	21	12
60	EC	A3	D6	62	F7	37	88	66	17	52	2B	B1	0D	A9	8D	59
70	3F	08	E4	97	84	48	4C	DA	BA	7D	C0	C8	94	C5	5F	AE
80	07	96	47	D9	7C	C7	C4	AD	A2	61	36	65	51	B0	A8	58
90	27	BC	6E	EF	73	7F	CC	11	45	C2	34	A6	6C	CA	32	30
A0	D4	86	29	8B	E2	4A	BE	5D	79	4E	70	69	DF	DC	F1	1F
B0	9E	41	75	FA	F4	0A	81	91	FE	E6	CE	3A	22	99	13	B6
C0	ED	0F	A4	2E	D7	AB	63	56	F8	8F	38	B4	89	5B	67	1D
D0	18	19	53	1A	2C	54	B2	1B	0E	2D	AA	55	8E	B3	5A	1C
E0	40	F9	09	90	E5	39	98	B5	85	8A	49	5C	4D	68	DB	1E
F0	BB	EE	7E	10	C1	A5	C9	2F	95	D8	C6	AC	60	64	AF	57

SIMULATION OF 3370 ECC IMPLEMENTATION

ERROR CASE

BEGIN READ DATA PART OF SIMULATION
(DATA PART OF RECORD IS ALL ZEROS EXCEPT FOR ERROR)
(CHECK BYTES ARE '00 FF FF 00 08 08 00 A3 A3')

BYTE CNT	MOD 3	ERROR	---RAM 1---			---RAM 2---			---RAM 3---		
			FF	FF	00	FF	FF	00	FF	FF	00
0											
/			MOD($x+\alpha^0$)			MOD($x+\alpha^1$)			MOD($x+\alpha^{-1}$)		
/											
501	0		FF	FF	00	01	B8	00	FE	7F	00
502	1		FF	FF	00	01	01	00	FE	FE	00
503	2		FF	FF	00	01	01	00	FE	FE	00
504	0		FF	FF	00	02	01	00	E1	FE	00
505	1		FF	FF	00	02	02	00	E1	E1	00
506	2	OF	FF	FF	OF	02	02	1E	E1	E1	FD
507	0		FF	FF	OF	04	02	1E	DF	E1	FD
508	1		FF	FF	OF	04	04	1E	DF	DF	FD
509	2		FF	FF	OF	04	04	3C	DF	DF	E7
510	0		FF	FF	OF	08	04	3C	A3	DF	E7
511	1		FF	FF	OF	08	08	3C	A3	A3	E7

FINISHED DATA, NOW READ CHECK BYTES

INTERLEAVE 2,1, 0

BYTE CNT	MOD 3	ERROR	---RAM 1---			---RAM 2---			---RAM 3---		
512	2		FF	FF	OF	08	08	3C	A3	A3	E7
513	0		00	FF	OF	08	08	3C	A3	A3	E7
514	1		00	00	OF	08	08	3C	A3	A3	E7
515	2		00	00	OF	08	08	3C	A3	A3	E7
516	0		00	00	OF	00	08	3C	A3	A3	E7
517	1		00	00	OF	00	00	3C	A3	A3	E7
518	2		00	00	OF	00	00	3C	A3	A3	E7
519	0		00	00	OF	00	00	3C	00	A3	E7
520	1		00	00	OF	00	00	3C	00	00	E7

END OF CHECK BYTES, BEGIN CORRECTION ALGORITHM

CORRECTABLE ERROR AT DISPLACEMENT 5 FROM END OF RECORD,
COUNTING LAST DATA BYTE AS 0.
CORRECTABLE PATTERN IS '0F 00 00'

END OF SIMULATION

SIMULATION OF 3370 ECC IMPLEMENTATION

NO ERROR CASE

BEGIN READ DATA PART OF SIMULATION
(DATA PART OF RECORD IS ALL ZEROS)
(CHECK BYTES ARE '00 FF FF 00 08 08 00 A3 A3')

BYTE CNT	MOD 3	ERROR	---RAM 1---			---RAM 2---			---RAM 3---		
			FF	FF	00	FF	FF	00	FF	FF	00
0											
/			MOD$(x+\alpha^0)$			MOD$(x+\alpha^1)$			MOD$(x+\alpha^{-1})$		
/											
501	0		FF	FF	00	01	B8	00	FE	7F	00
502	1		FF	FF	00	01	01	00	FE	FE	00
503	2		FF	FF	00	01	01	00	FE	FE	00
504	0		FF	FF	00	02	01	00	E1	FE	00
505	1		FF	FF	00	02	02	00	E1	E1	00
506	2		FF	FF	00	02	02	00	E1	E1	00
507	0		FF	FF	00	04	02	00	DF	E1	00
508	1		FF	FF	00	04	04	00	DF	DF	00
509	2		FF	FF	00	04	04	00	DF	DF	00
510	0		FF	FF	00	08	04	00	A3	DF	00
511	1		FF	FF	00	08	08	00	A3	A3	00

FINISHED DATA, NOW READ CHECK BYTES

BYTE CNT	MOD 3	ERROR	---RAM 1---			---RAM 2---			---RAM 3---		
512	2		FF	FF	00	08	08	00	A3	A3	00
513	0		00	FF	00	08	08	00	A3	A3	00
514	1		00	00	00	08	08	00	A3	A3	00
515	2		00	00	00	08	08	00	A3	A3	00
516	0		00	00	00	00	08	00	A3	A3	00
517	1		00	00	00	00	00	00	A3	A3	00
518	2		00	00	00	00	00	00	A3	A3	00
519	0		00	00	00	00	00	00	00	A3	00
520	1		00	00	00	00	00	00	00	00	00

END OF CHECK BYTES, BEGIN CORRECTION ALGORITHM

SYNDROME ALL ZEROS, NO ERROR DETECTED

END OF SIMULATION

5.3 APPLICATION TO SMALL-SYSTEMS MAGNETIC DISK

This section describes a discrete error-correction implementation that for several years was ideal for small-systems magnetic-disk controllers. It is part of many existing controller designs in both discrete and LSI form. The code discussed in this section employs a 32-bit polynomial. However, new designs are likely to employ polynomials of degree 48, 56 or 64.

Methods for hardware and software implementation are described. Included are unique methods for

(a) realizing the divide circuit
(b) detecting the error and
(c) passing the syndrome to software.

All real time operations are performed by the error-correction hardware. Computation of the error pattern and the displacement is performed by software. Two software algorithms, bit-serial and byte-serial, are described. Approximately 120 instructions are required to implement the software algorithms on the Z80 and similar 8-bit processors. In addition, 1K bytes of table space are required if the byte-serial software algorithm is selected.

For the Z80 and similar microprocessors, typical correction time is four milliseconds, (eight maximum), if the byte-serial software algorithm is used. If the bit-serial algorithm is used, typical correction time for the Z80 is 30 milliseconds, 60 maximum. These correction times are for a 256 byte record. For longer records, divide the record length by 256 and multiply the result by the times given. If, as recommended, correction is used only on hard errors, the bit-serial software algorithm will be fast enough for most applications. Bit slice implementations are typically four to five times faster than the Z80.

The polynomial used in this implementation is a computer-generated polynomial, selected for its insensitivity to short double bursts, good detection span and eight feedback terms. It was optimized for correction spans of five and eight bits on record lengths of 256 and 512 bytes, although its capabilities exceed this.

The forward polynomial is:

$$x^{32} + x^{28} + x^{26} + x^{19} + x^{17} + x^{10} + x^{6} + x^{2} + x^{0}$$

The reciprocal polynomial is:

$$x^{32} + x^{30} + x^{26} + x^{22} + x^{15} + x^{13} + x^{6} + x^{4} + x^{0}$$

5.3.1 *POLYNOMIAL CAPABILITIES*

The capabilities specified below represent the extremes for which the polynomial has been tested. Further testing is required if the polynomial is to be used beyond these extremes.

If you plan to use this polynomial, read Section 4.4 DATA ACCURACY to understand miscorrection probability (number 8 below) before selecting the correction span.

1. Maximum record length (r) = 8*1038 bits (including check bits)

2. Maximum correction span (b) = 11 bits.

3. Degree of polynomial (m) = 32.

4. Single-burst detection span when the code is used for error detection only = 32 bits.

5. Single-burst detection span (d) when the code is used for error correction:

```
= 20 bits for b=5  and r=8*270
= 14 bits for b=8  and r=8*270
= 13 bits for b=11 and r=8*270
= 19 bits for b=5  and r=8*526
= 14 bits for b=8  and r=8*526
= 12 bits for b=11 and r=8*526
= 11 bits for b=11 and r=8*1038
```

6. Double-burst detection span when the code is used for error correction:

```
= 4 bits for b=5 and r=8*270
= 2 bits for b=8 and r=8*270
= 3 bits for b=5 and r=8*526
= 2 bits for b=8 and r=8*526
```

7. Nondetection probability = 2.3E-10

8. Miscorrection probability:

```
= 8.00E-6 for b=5  and r=8*270
= 6.40E-5 for b=8  and r=8*270
= 5.12E-4 for b=11 and r=8*270
= 1.57E-5 for b=5  and r=8*526
= 1.25E-4 for b=8  and r=8*526
= 1.00E-3 for b=11 and r=8*526
= 2.01E-3 for b=11 and r=8*1038
```

5.3.2 *HARDWARE IMPLEMENTATION*

Several examples of encoder and decoder circuits are described below. Although they are shown separately, circuitry can obviously be shared between encoder and decoder.

BIT-SERIAL ENCODER USING THE INTERNAL-XOR FORM OF SHIFT REGISTER

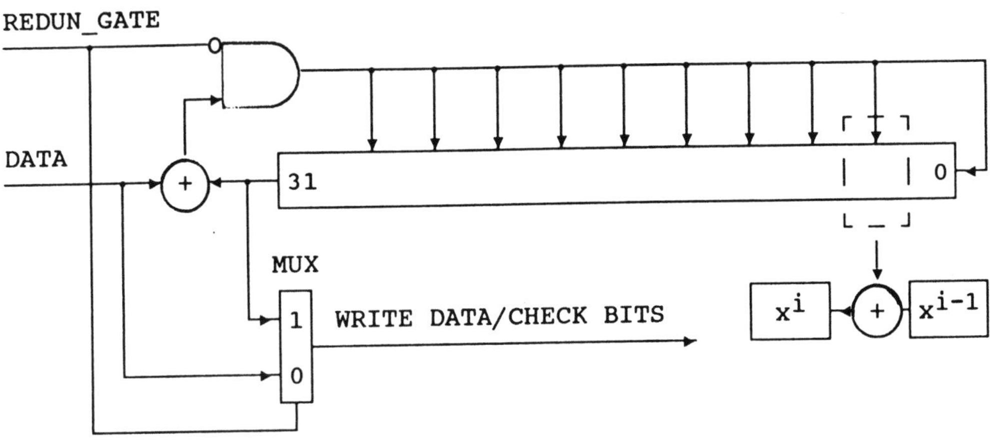

The shift register has an XOR gate feeding the input of each stage which has a non-zero coefficient in the generator polynomial (except stage 0). For initialization considerations, see Section 4.8.2.

After all DATA bits have been clocked into the shift register, REDUN_GATE is asserted. The AND gate then disables feedback, allowing the check bits to be shifted out of the shift register, and the MUX passes the check bits to the device.

BIT-SERIAL ENCODER USING THE EXTERNAL-XOR FORM OF SHIFT REGISTER

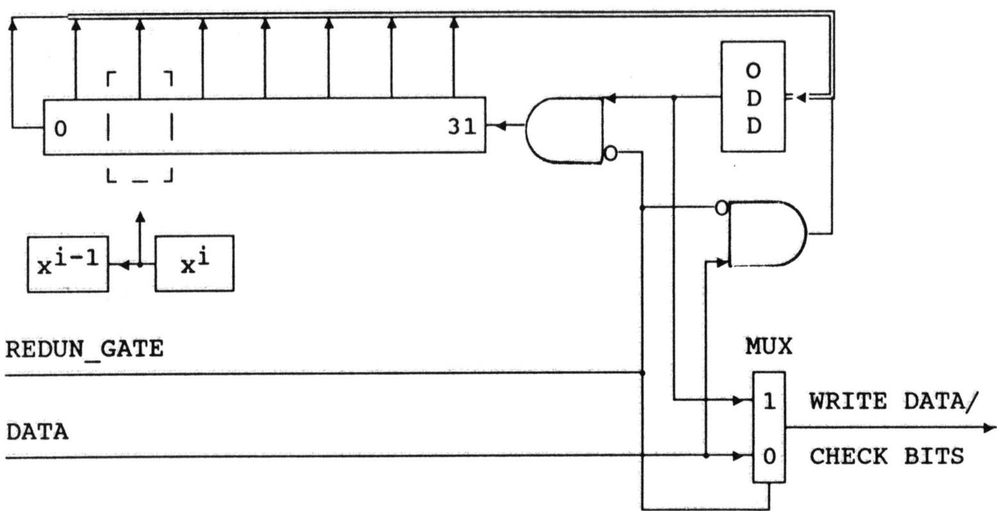

The shift register is tapped at the output of each stage which has a non-zero coefficient in the generator polynomial. For initialization considerations, see Section 4.8.2.

After all DATA bits have been clocked into the shift register, REDUN_GATE is asserted. The upper AND gate then disables feedback and the lower AND gate blocks extraneous DATA input to the ODD parity tree, whose output the MUX passes as check bits to the device.

CIRCUITS TO GENERATE SYNDROMES TO BE USED IN SOFTWARE CORRECTION

CASE 1: SYNDROME IS OUTPUT BEHIND DATA

BIT-SERIAL DECODER USING THE INTERNAL-XOR FORM OF SHIFT REGISTER

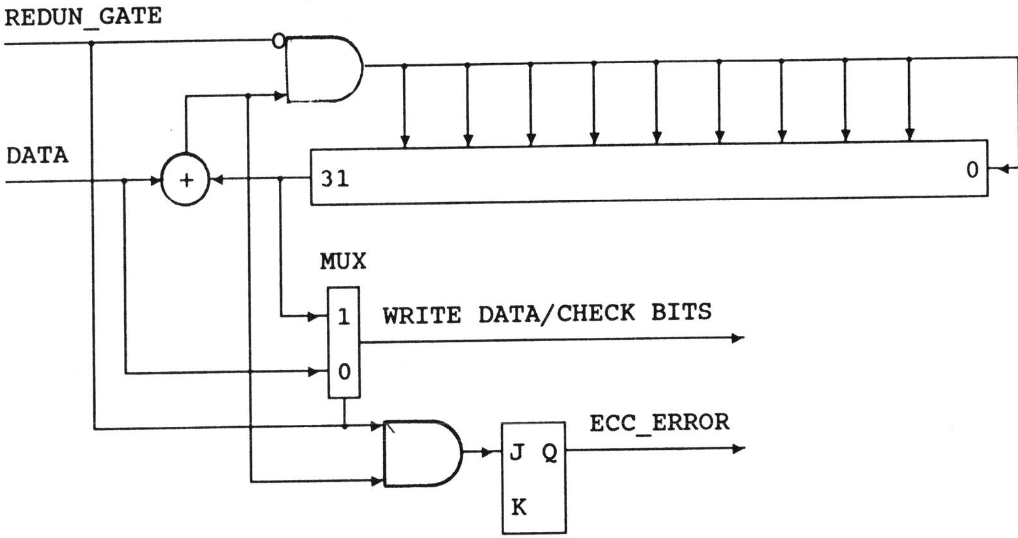

The shift register has an XOR gate feeding the input of each stage which has a non-zero coefficient in the generator polynomial (except stage 0). The shift register must be initialized to the same state used before write.

After all DATA bits have been clocked into the shift register, REDUN_GATE is asserted. The upper AND gate then disables feedback, allowing the check bits to be shifted out of the shift register, and the MUX passes the syndrome bits to the buffer. The lower AND gate allows any non-zero syndrome bit to latch the JK flip-flop, asserting the ECC_ERROR signal.

BIT-SERIAL DECODER USING THE EXTERNAL-XOR FORM OF SHIFT REGISTER

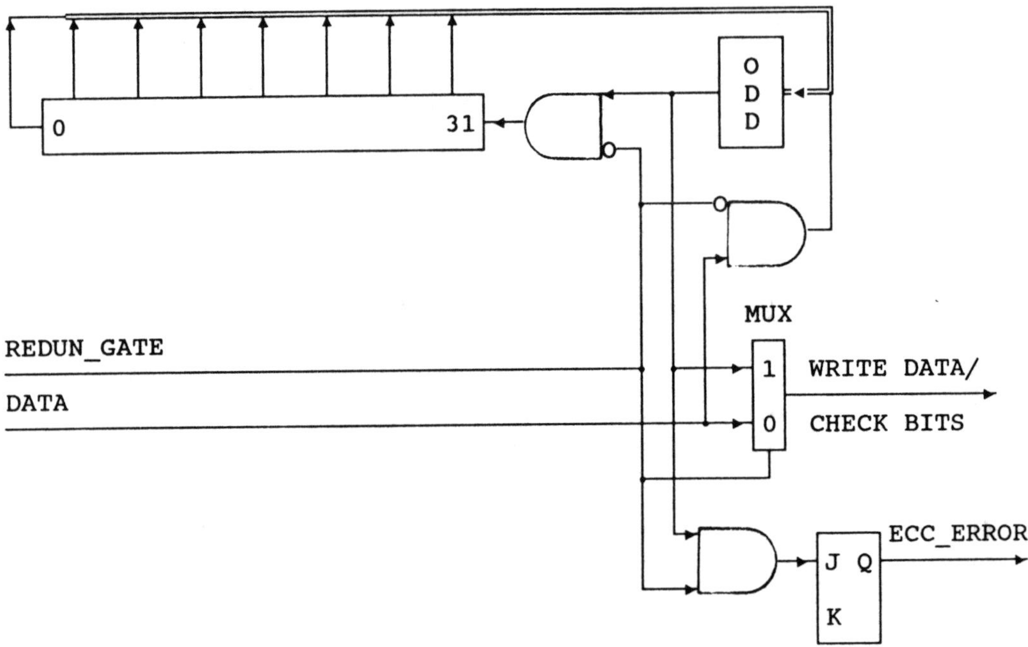

The shift register is tapped at the output of each stage which has a non-zero coefficient in the generator polynomial. The shift register must be initialized to the same state used before write.

After all DATA bits have been clocked into the shift register, REDUN_GATE is asserted. The upper AND gate then disables feedback and the lower AND gate blocks extraneous DATA input to the ODD parity tree, whose output the MUX passes as syndrome bits to the buffer. The bottom AND gate allows any non-zero syndrome bit to latch the JK flip-flop, asserting the ECC_ERROR signal.

CASE 2: SYNDROME IS FETCHED FROM SHIFT REGISTER

BIT-SERIAL DECODER USING THE INTERNAL-XOR FORM OF SHIFT REGISTER

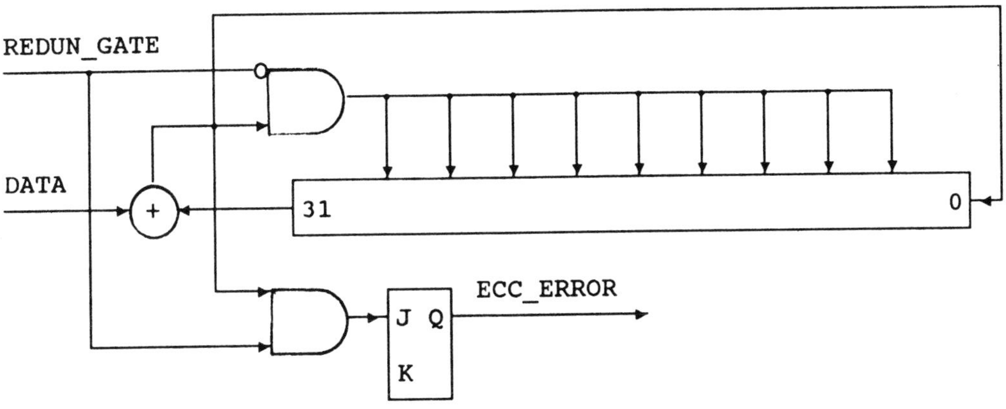

The shift register has an XOR gate feeding the input of each stage which has a non-zero coefficient in the generator polynomial (except stage 0). The shift register must be initialized to the same state used before write.

After all DATA bits have been clocked into the shift register, REDUN_GATE is asserted. The upper AND gate then disables feedback. The upper-most path, leading from the XOR gate to stage 0, allows the shift register to collect the syndrome bits for later retrieval. The lower AND gate allows any non-zero syndrome bit to latch the JK flip-flop, asserting the ECC_ERROR signal.

DETAILED IMPLEMENTATION EXAMPLE #1

The hardware of Figure 5.3.2.1 is used on write to generate check bits and on read to generate an error syndrome. The error syndrome is stored in memory via the deserializer during check bit time. It has the following format, where x^0 is the high order bit of the first byte stored:

Lowest memory address	$x^0 - x^7$
	$x^8 - x^{15}$
	$x^{16} - x^{23}$
Highest memory address	$x^{24} - x^{31}$

This format assumes that the high-order bit of a byte is serialized and deserialized first. The bits are numbered here for the software flow chart. Bits numbered 0-7 above are bits 31-24 of the syndrome from hardware.

As the data is written, data bits are directed to pin 10 via the 2:1 circuit. At the same time, check bits are generated in the shift register in a transformed format. During write checkbit time, the transformed check bits in the shift register are converted to true check bits (some inverted) by the odd circuit and are directed to pin 10 via the 2:1 circuit.

As the data is read, data bits are directed to pin 9 via the 2:1 circuit. At the same time, syndrome bits are generated in the shift register in a transformed format. During read check-bit time, the transformed syndrome bits in the shift register are converted to true syndrome bits by the odd circuit and are directed to pin 9 via the 2:1 circuit.

During read check-bit time, the flip-flop (LS74) will be clocked to its error state if any of the syndrome bits are nonzero. At the end of any read, pin 11 will indicate if an error occurred.

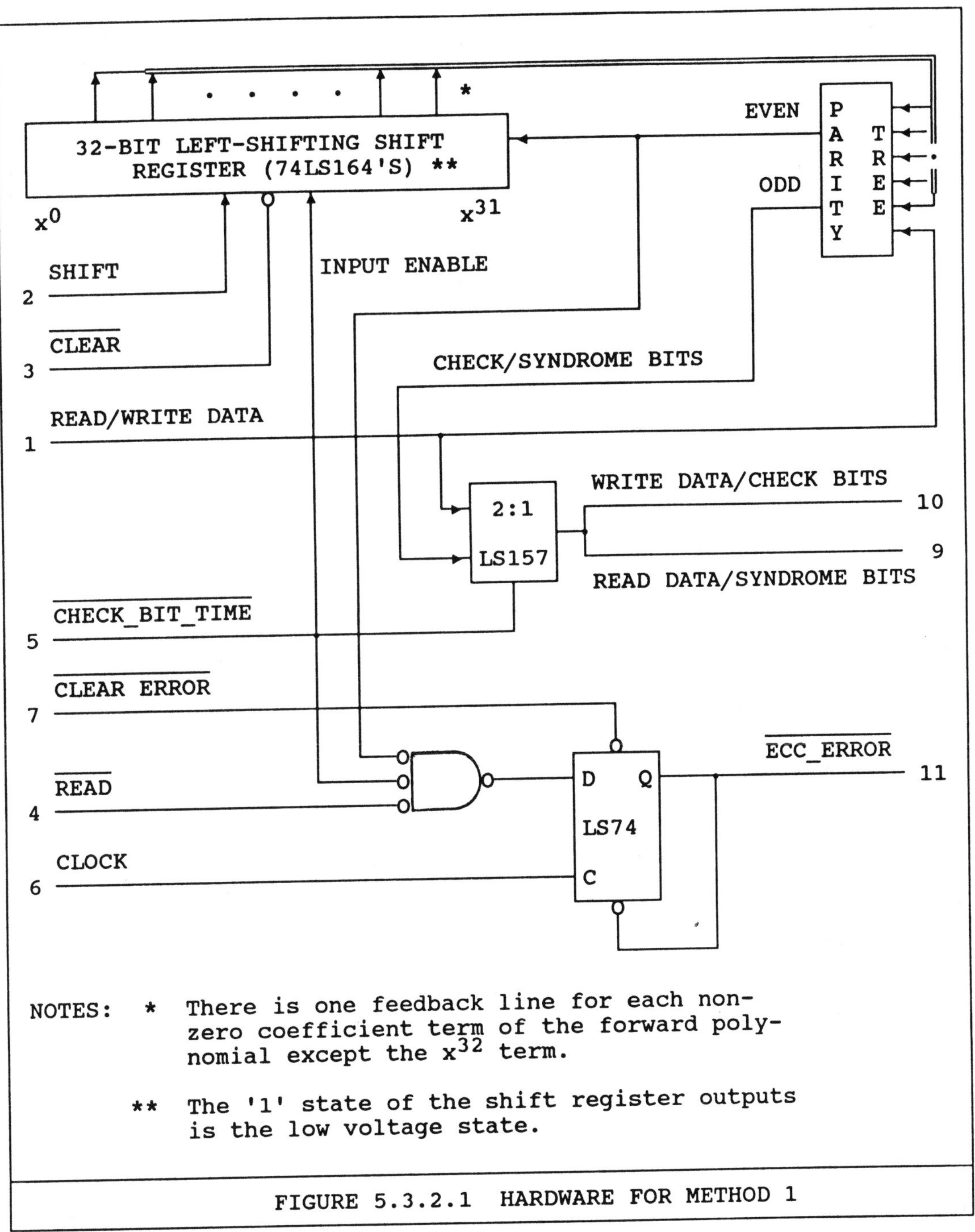

NOTES: * There is one feedback line for each non-zero coefficient term of the forward polynomial except the x^{32} term.

** The '1' state of the shift register outputs is the low voltage state.

FIGURE 5.3.2.1 HARDWARE FOR METHOD 1

DETAILED IMPLEMENTATION EXAMPLE# 2

The hardware of Figure 5.3.2.2 is similar to that for method 1 except that ECC error is detected by software. After a record and syndrome have been read and stored, software fetches the four byte syndrome and checks for zero. If the syndrome is nonzero, an error has occurred and the software ECC algorithm must be performed.

In summary, if method 2 is used, software performs the function that the flip-flop (LS74) performed in method 1.

IMPLEMENTATION SUBTLETIES

Listed below are some points that have been misunderstood by engineers implementing the hardware.

1. There is a shift register stage for x^0 through x^{31}. There is no shift register stage for x^{32}.
2. The input end of the shift register is labeled x^{31} and the direction of shift is towards x^0. This is not an arbitrary assignment. It is required for this particular form of the polynomial shift register.
3. There is a feedback path from the shift register to the parity tree for each nonzero coefficient term of the forward polynomial except for the x^{32} term.
4. The forward polynomial is implemented in hardware and the reverse polynomial is implemented in software.
5. The read/write data line (pin 1 of Figures 5.3.2.1 and 5.3.2.2) must be inactive while writing check bytes.
6. After activating the clear line of Figures 5.3.2.1 and 5.3.2.2, the circuit should not be clocked until the first bit is ready to be processed.

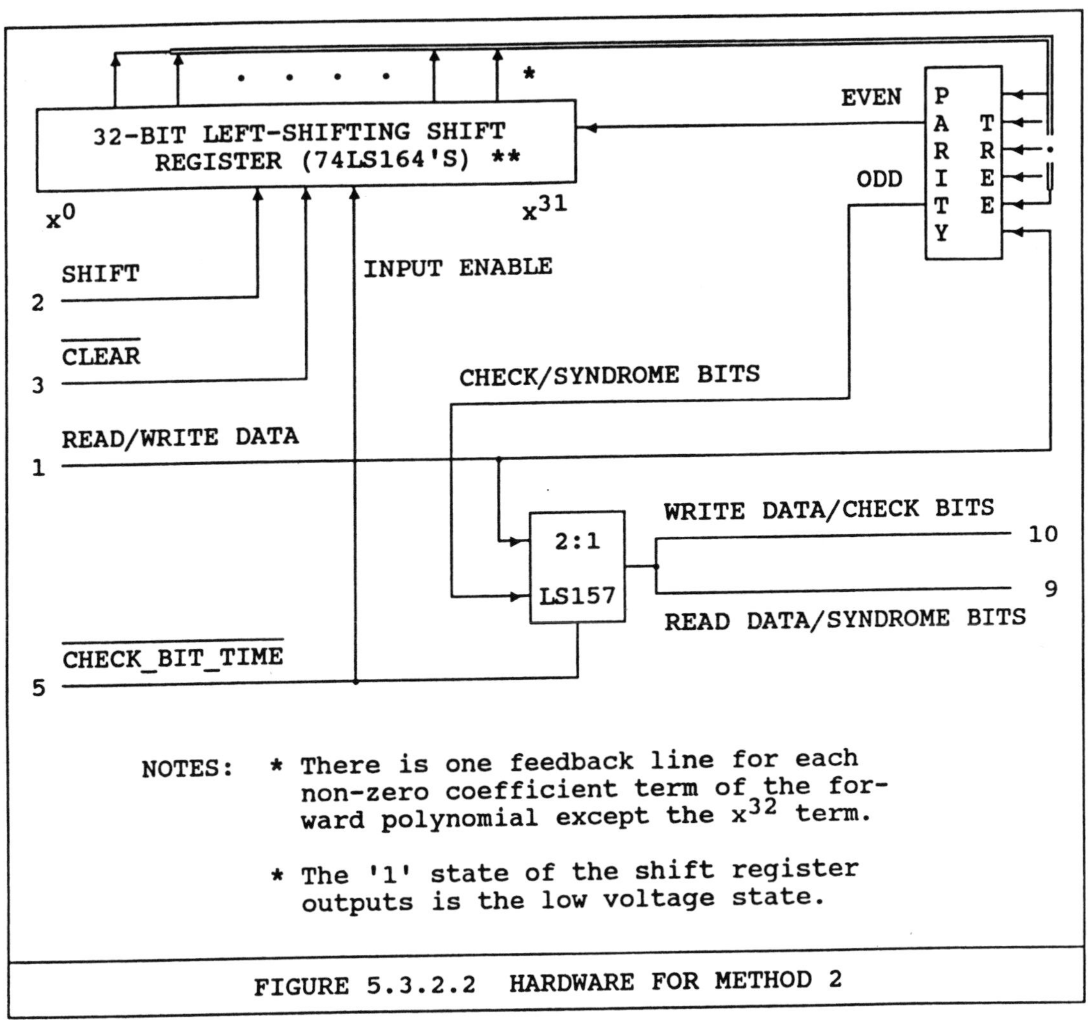

FIGURE 5.3.2.2 HARDWARE FOR METHOD 2

5.3.3 *SOFTWARE IMPLEMENTATION*

<u>*ERROR DETECTION*</u>

At the completion of a read, the existence of any non-zero bit in the syndrome indicates the existence of an error or errors. A non-zero syndrome is typically, and most efficiently, detected with sequential logic as shown above. However, if a shift register of the internal-XOR form with a separate path to stage 0 (see the circuit for Case 2 above) is used, it is possible (and has been done in the past) to use combinatorial logic (*e.g.*, a 48-input OR gate) rather than the AND gate/JK latch shown.

<u>*ERROR CORRECTION*</u>

When a non-zero syndrome indicates the presence of an error, correction is accomplished by shifting the syndrome until the error pattern is found. This shifting may be done bit-serially or byte-serially, in hardware or by software. Bit-serial and byte-serial software algorithms are given below. For discussion of byte-serial hardware implementations Section 4.7.

The required shifting may be performed either forward along the code's shift-register sequence using the code's generator polynomial, or reverse along the code's shift register sequence using the reciprocal of the code's generator polynomial.

When forward shifting is implemented, pre-multiply must be used to shorten the code. For a discussion of code-shortening, see Section 2.4. Use the following expression for the pre-multiply polynomial:

$$P_{mult}(x) = x^{m-1} \cdot F(1/x)$$

where

$$F(x) = x^{n-1} \text{ MOD } g'(x)$$

Forward shifting requires either the use of a different Pmult(x) for each sector length, or that Pmult(x) must be selected for the largest sector-length to be used. In the latter case extra shifts are required for the shorter sector-lengths.

When reverse shifting using the reciprocal polynomial is implemented, then if the shift register shifts left [right] during read, then either

a) The shift register must shift right [left] during correction or

b) The syndrome must be flipped end-for-end before correction, and the shift register must continue to shift left [right] during correction.

DETERMINING ERROR PATTERN AND LOCATION

The error pattern is found by shifting until a given number of consecutive zeros appears in one end of the shift register. When this occurs, the error pattern is aligned with the other end of the shift register. Which end of the shift register is aligned with the error pattern is a matter of implementation choice. See Sections 2.3 and 2.4 for examples of pattern alignment.

Error displacement is calculated by counting the number of shifts executed while locating the error pattern. The details of displacement calculation depend on which end of the shift register is used to align the error pattern.

The detection of consecutive zeros to indicate that a valid error pattern has been found can be accomplished using either combinatorial or sequential logic. Combinatorial logic would consist of a many-input OR gate.

Sequential logic circuitry for an internal-XOR shift register implementation would include a counter that is incremented by each '0' that appears at the output of the high-order stage and is reset by any '1' that appears. When the counter reaches the given threshold, the error pattern has been found.

It is also possible to simulate such a counter in software; the software would control an output line to initiate each shift of the hardware shift register and receive the output of the high order stage. The software can simultaneously simulate the displacement counter.

OTHER CONSIDERATIONS

1) The detection of consecutive zeros that surround the error pattern is more complex when error correction is performed using byte-serial hardware or software. See Figure 3.1.1 for an example of byte-serial hardware. A byte-serial software algorithm is given below.

2) When error correction is performed in hardware, the internal-XOR form os shift register is typically used. However, it is also possible to perform error correction in hardware when the external-XOR form is shift register is used.

3) Feedback could be left enabled during redundancy time. If the reverse-shifting correction method is used, the error location process would then require 48 additional shifts. If the forward-shifting correction process is used, a different Pmult(x) would be used.

SOFTWARE ALGORITHMS

The software algorithms use the syndrome to generate the correction pattern and displacement for correctable errors, or to detect uncorrectable errors.

In the correction algorithms, a simulated shift register implements the reciprocal polynomial. The simulated shift register is loaded with the syndrome and shifted until a correctable pattern is found or the error is determined to be uncorrectable. Flow charts of the correction algorithms are shown at the end of this section.

The maximum record length for this polynomial is 1038 bytes (including check bytes). The flow charts and software listings have been designed so that the record length can be varied by changing a single constant (K1).

The flow charts cover the algorithms through determination of pattern and displacement. Both forward (FWD) and reverse (REV) displacements are computed. FWD displacement starts at the beginning of the record counting the first byte as zero. REV displacement begins with the end of the record counting the last byte as zero.

The pattern is in R2, R3, and R4. R2 is XOR'd with the record byte indicated by byte displacement. R3 is XOR'd with the byte one address higher than the byte displacement. R4 is XOR'd with the byte two addresses higher than the byte displacement.

If the correction span selected is nine bits or less, the pattern is in R2 and R3. No action is required for R4.

Once an error pattern and displacement have been computed, there are several special displacement cases that must be handled. For example, the error may be in check bytes or it may span data and check bytes. The error may be a header field or a data field. Some formats combine header information with the data field. The data field in this case, has several overhead bytes, containing header information, preceding data. This adds additional special displacement cases.

The software routines defined in this section contain logic for separate and combined header and data fields.

The procedures below handle the special displacement cases of four overhead bytes. In a particular implementation there may be more, less, or even no overhead bytes.

FORWARD DISPLACEMENT LESS THAN 4

0-1 Error burst in overhead bytes. Correct overhead bytes in RAM. If overhead bytes are not contiguous in RAM, handle the boundaries.

2 Error burst spans overhead bytes and data. XOR R2 with next to last overhead byte. XOR R3 with last overhead byte. XOR R4 with first data byte.

3 Error burst spans overhead bytes and data. XOR R2 with last overhead byte, XOR R3 with first data byte. XOR R4 with second data byte.

REVERSE DISPLACEMENT LESS THAN 6

0-3 Error burst in check bytes. No action required.

4 Error burst spans data and check bytes. XOR R2 with last data byte. No action required for R3 or R4.

5 Error burst spans data and check bytes. XOR R2 with next to last data byte. XOR R3 with last data byte. No action required for R4.

FORWARD DISPLACEMENT EQUAL OR GREATER THAN 4 AND REVERSE DISPLACEMENT EQUAL OR GREATER THAN 6

Error burst in data bytes. Correct data bytes in RAM. If the data is not contiguous in RAM, handle the boundaries. Generate displacement from the first data byte to the first byte in error by subtracting the number of overhead bytes from the forward displacement. XOR R2 with the data byte indicated by the displacement just computed. XOR R3 with the data byte one address higher than the displacement. XOR R4 with the data byte two addresses higher than the displacement.

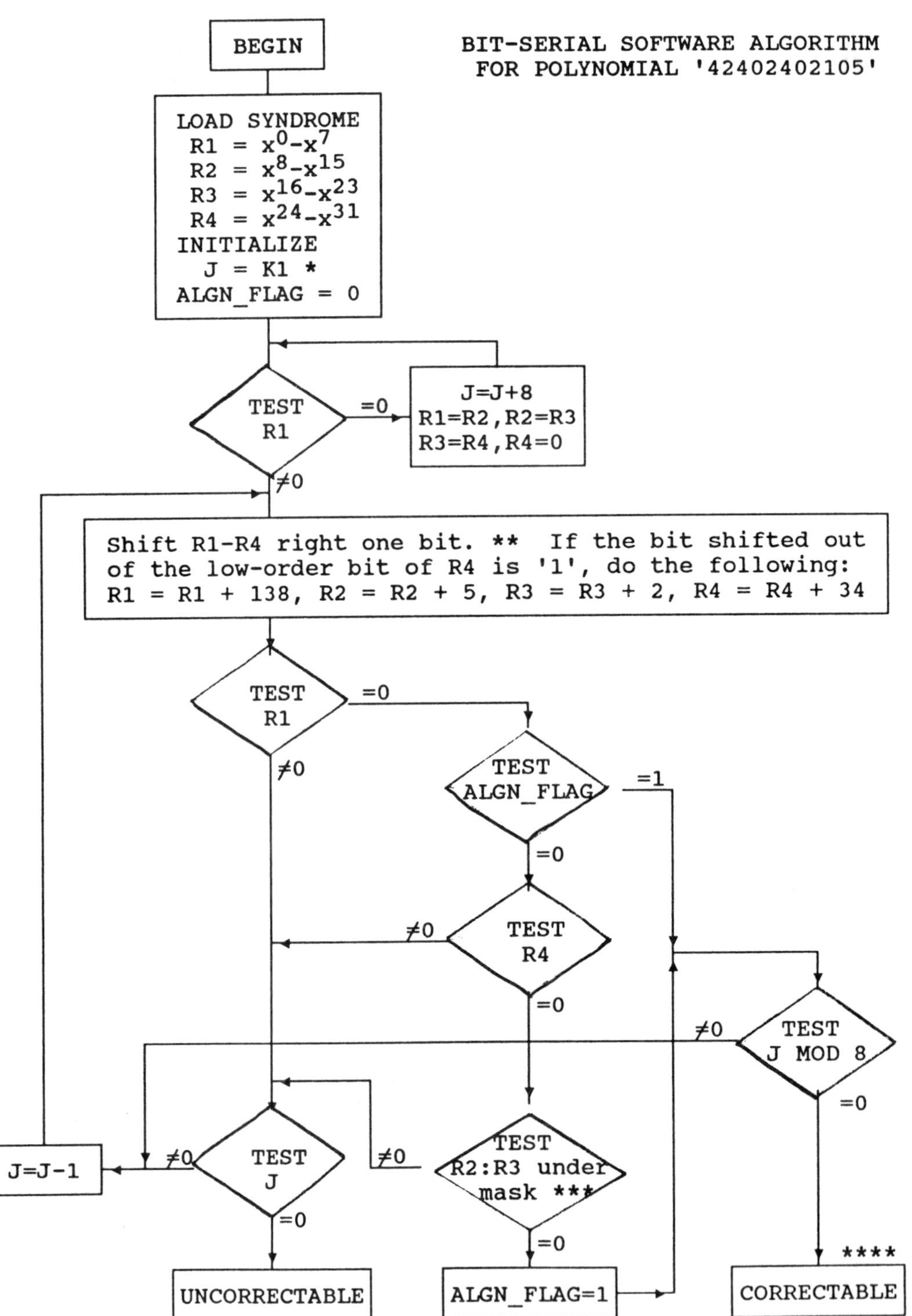

BIT-SERIAL SOFTWARE ALGORITHM
FOR POLYNOMIAL '42402402105'
BEGIN
LOAD SYNDROME
R1 = x0-x7
R2 = x8-x15
R3 = x16-x23
R4 = x24-x31
INITIALIZE
J = K1 *
ALGN_FLAG = 0
TEST
R1
=0
J=J+8
R1=R2,R2=R3
R3=R4,R4=0
≠0
Shift R1-R4 right one bit. ** If the bit shifted out of the low-order bit of R4 is '1', do the following:
R1 = R1 + 138, R2 = R2 + 5, R3 = R3 + 2, R4 = R4 + 34
TEST
R1
=0
≠0
TEST
ALGN_FLAG
=1
=0
≠0
TEST
R4
=0
≠0
TEST
J MOD 8
=0
J=J-1
≠0
TEST
J
≠0
TEST
R2:R3 under
mask ***
=0
=0
UNCORRECTABLE
ALGN_FLAG=1
****
CORRECTABLE

NOTES FOR BIT-SERIAL CORRECTION ALGORITHM

* K1 = Record length in bits minus 25. Record length includes all bits covered by ECC including the check bits.

** When shifting, the low-order bit of a register is shifted into the high-order bit of the next higher-numbered register. '+' here means EXCLUSIVE-OR; the constants are a form of the reciprocal polynomial in decimal.

*** Mask for given correction span:

Span	Mask R2:R3
1	01111111:11111111
2	00111111:11111111
3	00011111:11111111
4	00001111:11111111
5	00000111:11111111
6	00000011:11111111
7	00000001:11111111
8	00000000:11111111
9	00000000:01111111
10	00000000:00111111
11	00000000:00011111

**** On correctable exit, J is the forward bit displacement and J/8 is the forward byte displacement. The reverse byte displacement is (K1+25-J)/8-1. The error pattern is in R2:R3:R4.

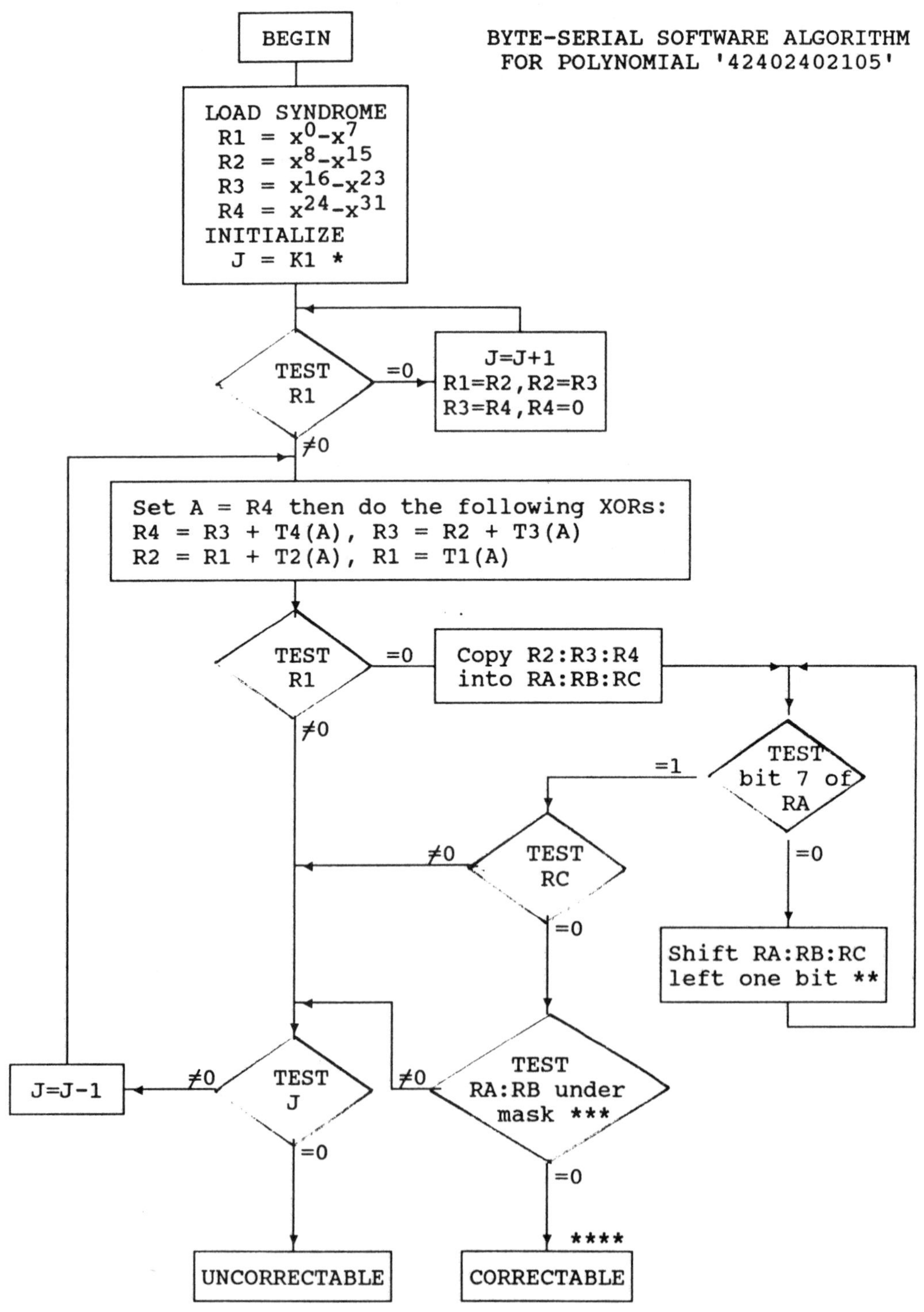
BYTE-SERIAL SOFTWARE ALGORITHM
FOR POLYNOMIAL '42402402105'
BEGIN
LOAD SYNDROME
R1 = x0-x7
R2 = x8-x15
R3 = x16-x23
R4 = x24-x31
INITIALIZE
J = K1 *
TEST
R1
=0
J=J+1
R1=R2,R2=R3
R3=R4,R4=0
≠0
Set A = R4 then do the following XORs:
R4 = R3 + T4(A), R3 = R2 + T3(A)
R2 = R1 + T2(A), R1 = T1(A)
TEST
R1
=0
Copy R2:R3:R4
into RA:RB:RC
≠0
TEST
bit 7 of
RA
=1
=0
TEST
RC
≠0
=0
Shift RA:RB:RC
left one bit **
J=J-1
≠0
TEST
J
=0
≠0
TEST
RA:RB under
mask ***
=0
****
UNCORRECTABLE
CORRECTABLE

NOTES FOR BYTE-SERIAL CORRECTION ALGORITHM

* K1 = Record length in bytes minus 4. Record length includes all bytes covered by ECC including the check bytes.

** When shifting, the high-order bit of a register is shifted into the low-order bit of the next higher-lettered register.

*** Mask for given correction span:

Span	Mask RA:RB
1	01111111:11111111
2	00111111:11111111
3	00011111:11111111
4	00001111:11111111
5	00000111:11111111
6	00000011:11111111
7	00000001:11111111
8	00000000:11111111
9	00000000:01111111
10	00000000:00111111
11	00000000:00011111

**** On correctable exit, J is the forward byte displacement. The reverse byte displacement is (K1+3-J). The error pattern is in R2:R3:R4.

```
;       POLYNOMIAL - '42402402105' (OCTAL)
;       Z80 CODE FOR BIT-SERIAL ALGORITHM
;       --------------------------------
;
;       THIS ROUTINE PERFORMS ALL THE FUNCTIONS OF THE
;       BIT-SERIAL ALGORITHM (SEE FLOWCHART)
;
;       TIMING IN THE 'SHIFT AND XOR' AREA OF THE CODE IS CRITICAL.
;
;       EXECUTE THIS ROUTINE AFTER ATTEMPTING
;       REREADS AND FINDING THE SAME SYNDROME ON 2
;       CONSECUTIVE READS.
;
;-----------------------------------------------------------------
;       IMPLEMENTATION CONSTANTS
;
;       DEFINE POLYNOMIAL - DECIMAL CONSTANTS, SEE FLOW CHART
P1         EQU      138
P2         EQU      5
P3         EQU      2
P4         EQU      34
;       DEFINE CONSTANTS K1 AND K2 (SEE FLOW CHART)
K1         EQU      ----           ;INSERT DATA FIELD CONSTANT K1
K2         EQU      ----           ;INSERT HEADER FIELD CONSTANT K2
;       DEFINE NUMBER OF OVERHEAD BYTES
OV         EQU      ----           ;INSERT # OF OVERHEAD BYTES
;       DEFINE CORRECTION SPAN MASK
CSM1       EQU      ----           ;INSERT APPROPRIATE MASK BELOW
                                   ; CORR SPAN  1  -  MASK '01111111'
                                   ;            2  -       '00111111'
                                   ;            3  -       '00011111'
                                   ;            4  -       '00001111'
                                   ;            5  -       '00000111'
                                   ;            6  -       '00000011'
                                   ;            7  -       '00000001'
                                   ;            8  -       '00000000'
                                   ;            9  -       '00000000'
                                   ;           10  -       '00000000'
                                   ;           11  -       '00000000'
CSM2      EQU      ----            ;INSERT APPROPRIATE MASK BELOW
                                   ; CORR SPAN  1  -  MASK '11111111'
                                   ;            2  -       '11111111'
                                   ;            3  -       '11111111'
                                   ;            4  -       '11111111'
                                   ;            5  -       '11111111'
                                   ;            6  -       '11111111'
                                   ;            7  -       '11111111'
                                   ;            8  -       '11111111'
                                   ;            9  -       '01111111'
                                   ;           10  -       '00111111'
                                   ;           11  -       '00011111'
;-----------------------------------------------------------------
```

```
;       INITIALIZE PSEUDO SHIFT REGS AND SHIFT COUNT (J)
;
INIT     LD      HL,(nn)          ;FETCH 1ST 2 SYNDROME BYTES
         LD      B,L              ;SYNDROME BITS x0-x7
         LD      C,H              ;SYNDROME BITS x8-x15
         LD      HL,(nn)          ;FETCH 2ND 2 SYNDROME BYTES
         LD      D,L              ;SYNDROME BITS x16-x23
         LD      E,H              ;SYNDROME BITS x24-x31
INIT05   LD      A,(FLDFLG)       ;LOAD FIELD FLAG
         OR      A
         JP      NZ,INIT20        ;JP TO INIT20 IF CORRECTING HEADER
;       INITIALIZE FOR DATA FIELD
INIT10   LD      HL,nn-OV         ;SAVE DATA BUFFER ADDRESS
         LD      (BUFFADR),HL     ;  MINUS NUMBER OF OVERHEAD BYTES
         LD      HL,(K1+25)/8-1;SAVE
         LD      (RLBMO),HL       ;  DATA FIELD LENGTH IN BYTES-1
         LD      HL,K1            ;LOAD J WITH K1 (CONST FOR DATA)
         JP      CALGN
;       INITIALIZE FOR HEADER FIELD
INIT20   LD      HL,nn            ;SAVE
         LD      (BUFFADR),HL     ;  HEADER BUFFER ADDRESS
         LD      HL,(K2+25)/8-1;SAVE
         LD      (RLBMO),HL       ;  HEADER LENGTH IN BYTES MINUS 1
         LD      HL,K2            ;LOAD J WITH K2 (CONST FOR HEADER)
;
;       CLEAR ALGN-FLAG
;
CALGN    XOR     A                ;CLEAR A
         LD      (ALGNFLG),A      ;CLEAR ALGN-FLAG
;
;       LEFT JUSTIFY FIRST NON-ZERO SYNDROME BYTE IN 'B'
;
JUST     XOR     A
         OR      B                ;TEST 'B' FOR ZERO
         JP      NZ,SHIFT         ;BRANCH ON NONZERO
         LD      A,L
         ADD     8                ;J=J+8
         LD      L,A
         JP      NC,JUST10
         INC     H
JUST10   LD      B,C
         LD      C,D
         LD      D,E
         LD      E,0
         JP      JUST
;
;
;       SHIFT PSEUDO SHIFT REG UNTIL CORRECTABLE PATTERN FOUND
;
SHIFT    SRL     B                ;
         RR      C                ;SHIFT RIGHT
         RR      D                ;
```

```
         RR      E               ;
         JP      NC,SHIFT10      ;BRANCH IF NO BIT SHIFTED OUT
         LD      A,E
         XOR     P4              ;
         LD      E,A             ;
         LD      A,D             ;
         XOR     P3              ;XOR DECIMAL CONSTANTS
         LD      D,A             ;  (SHIFT REG FEED-BACK)
SHIFT05  LD      A,C             ;
         XOR     P2              ;
         LD      C,A             ;
         LD      A,B             ;
         XOR     P1              ;
         LD      B,A             ;
SHIFT10  LD      A,B
         OR      A
         JP      Z,PTRNTST
SHIFT20  XOR     A
         OR      L
         JP      NZ,SHIFT30
         OR      H
         JP      NZ,SHIFT30
         JP      UNCORR          ;UNCORRECTABLE
SHIFT30  DEC     HL              ;DECREMENT SHIFT COUNT ('J')
         JP      SHIFT           ;
;
;        TEST FOR CORRECTABLE PATTERN
;
PTRNTST  LD      A,(ALGNFLG)     ;LOAD ALGN-FLAG
         OR      A
         JP      NZ,PTRNTST5     ;BRANCH IF ALGN-FLAG NONZERO
         OR      E
         JP      NZ,SHIFT20      ;BRANCH IF CORR PTRN NOT YET FOUND
         LD      A,C
         AND     CSM1            ;SEE DEFINITION OF CSM1 ABOVE
         JP      NZ,SHIFT20      ;BRANCH IF CORR PTRN NOT YET FOUND
         LD      A,D
         AND     CSM2            ;SEE DEFINITION OF CSM2 ABOVE
         JP      NZ,SHIFT20      ;BRANCH IF CORR PTRN NOT YET FOUND
;        GET HERE TO START BYTE ALIGNMENT
         LD      A,1
         LD      (ALGNFLG),A     ;SET ALGN-FLAG TO NONZERO
PTRNTST5 LD      A,L
         AND     7               ;TEST 'J' MODULO 8
         JP      NZ,SHIFT30      ;JP IF BYTE ALIGN NOT COMPLETE
```

```
;
;       CORRECT BYTES IN ERROR
;
CORRECT LD      B,C             ;MOVE
        LD      C,D             ;  PATTERN
        LD      D,E             ;
        LD      A,3
CORR10  SRL     H               ;
        RR      L               ;DIVIDE BIT DISPLACEMENT BY 8
        DEC     A               ;  TO GET FWD BYTE DISPLACEMENT
        JP      NZ,CORR10
;       COMPUTE REV BYTE DISPLACEMENT
        SCF
        CCF
        PUSH    DE
        EX      DE,HL
        LD      HL,(RLBMO)
        SBC     HL,DE
        POP     DE
;       TEST REVERSE DISPLACEMENT CASES
        LD      A,H
        OR      A
        JP      NZ,CORR40       ;BR IF HI BYTE OF REV DISP NONZERO
        LD      A,L
        CP      6
        JP      NC,CORR40       ;BR IF REV DISP EQ OR GTH THAN 6
        CP      5
        JP      Z,CORR25        ;BR IF REV DISP EQ 5
        CP      4
        JP      Z,CORR30        ;BR IF REV DISP EQ 4
;       GET HERE IF ERROR IN CHECK BYTES
CORR20  JP      EXIT            ;IGNORE CORR ERR IN CHECK BYTES
;       GET HERE IF ERROR STARTS IN NEXT TO LAST DATA BYTE
CURR25  LD      A,(nn)
        XOR     B               ;CORRECT NEXT TO LAST DATA BYTE
        LD      (nn),A
        LD      B,C
;       GET HERE IF ERROR STARTS IN LAST DATA BYTE
CORR30  LD      A,(nn)
        XOR     B               ;CORRECT LAST DATA BYTE
        LD      (nn),A
        JP      EXIT            ;DONE
;       RECOMPUTE FWD BYTE DISPLACEMENT
CORR40  SCF
        CCF
        PUSH    DE
        EX      DE,HL
        LD      HL,(RLBMO)
        SBC     HL,DE
        POP     DE
```

```
;       TEST FWD DISPLACEMENT CASES
        LD      A,H
        OR      A
        JP      NZ,CORR45       ;BR IF HI BYTE OF FWD DISP NONZERO
        LD      A,L
        CP      4
        JP      NC,CORR45       ;BR IF FWD DISP EQ OR GTH THAN 4
        CP      3
        JP      Z,CORR60        ;BR IF FWD DISP EQ 3
        CP      2
        JP      Z,CORR55        ;BR IF FWD DISP EQ 2
        JP      CORR70
;
;       GET HERE IF ERROR IN DATA BYTES
CORR45  PUSH    DE
        LD      DE,(BUFFADR)    ;LOAD BUFFER ADDRESS
CORR50  ADD     HL,DE           ;ADD DATA BUFFER
                                ;  ADDR TO DISPLACEMENT
        POP     DE
        LD      A,(HL)
        XOR     B               ;CORRECT 1ST DATA BYTE IN ERROR
        LD      (HL),A
        INC     HL
        LD      A,(HL)
        XOR     C               ;CORRECT 2ND DATA BYTE IN ERROR
        LD      (HL),A
        INC     HL
        LD      A,(HL)
        XOR     D               ;CORRECT 3RD DATA BYTE IN ERROR
        LD      (HL),A
        JP      EXIT            ;DONE
;
;       ERROR STARTS IN NEXT TO LAST OVHD BYTE
CORR55  LD      A,(nn)
        XOR     B               ;CORRECT NEXT TO LAST OVHD BYTE
        LD      (nn),A
        LD      B,C
        LD      C,D
        LD      D,0
;
;       ERROR STARTS IN LAST OVHD BYTE
CORR60  LD      A,(nn)
        XOR     B               ;CORRECT LAST OVERHEAD BYTE
        LD      (nn),A
        LD      A,(nn)
        XOR     C               ;CORRECT FIRST DATA BYTE
        LD      (nn),A
        LD      A,(nn)
        XOR     D               ;CORRECT 2ND DATA BYTE
        LD      (nn),A
        JP      EXIT            ;DONE
```

```
;       GET HERE IF ERROR IN OVERHEAD BYTES
CORR70    PUSH    DE
          LD      DE,nn             ;OVERHEAD BYTES BUFFER ADDRESS
          JP      CORR50            ;JOIN COMMON PATH
;
;
;
;
;      UNCORRECTABLE ERROR EXIT
;
UNCORR    NOP                       ;BRANCH TO ERR PATH IN MAIN PGM
;
;      CORRECTION COMPLETE EXIT
;
EXIT      NOP                       ;BRANCH BACK TO MAIN PGM
;
;
;
;      WORK STORAGE
;
ALGNFLG   DEFS    1                 ;ALIGNMENT FLAG - SEE FLOW CHART
RLBMO     DEFS    2                 ;RECORD LENGTH IN BYTES MINUS 1
BUFFADR   DEFS    2                 ;BUFFER ADDRESS - EITHER,
                                    ; -DATA BUFF ADDR MINUS
                                    ;   # OF OVERHEAD BYTES
                                    ; -HEADER BUFFER ADDRESS
FLDFLG    DEFS    1                 ;FIELD FLAG - SET BY CALLING PGM
                                    ; - ZERO FOR DATA FIELD
                                    ; - NONZERO FOR HEADER FIELD
```

```
;       POLYNOMIAL - '42402402105'  (OCTAL)
;       Z80 CODE FOR BYTE-SERIAL ALGORITHM
;       ---------------------------------
;
;       THIS ROUTINE PERFORMS ALL THE FUNCTIONS OF THE
;       BYTE-SERIAL ALGORITHM (SEE FLOWCHART)
;
;       TIMING IN THE 'SHIFT AND XOR' AREA OF THE CODE IS CRITICAL.
;
;       EXECUTE THIS ROUTINE AFTER ATTEMPTING
;       REREADS AND FINDING THE SAME SYNDROME ON 2
;       CONSECUTIVE READS.
;
;       FLOWCHART REGISTER ASSIGNMENTS --
;           R1=D
;           R2=E
;           R3=H
;           R4=L
;           'A'=C
;-----------------------------------------------------------------
;       IMPLEMENTATION CONSTANTS
;
;       DEFINE CONSTANTS K1 AND K2 (SEE FLOW CHART)
K1        EQU       ----            ;INSERT DATA FIELD CONSTANT K1
K2        EQU       ----            ;INSERT HEADER FIELD CONSTANT K2
;       DEFINE NUMBER OF OVERHEAD BYTES
OV        EQU       ----            ;INSERT # OF OVERHEAD BYTES
;       DEFINE CORRECTION SPAN MASK
CSM1      EQU       ----            ;INSERT APPROPRIATE MASK BELOW
                                    ; CORR SPAN  1  -  MASK '01111111'
                                    ;            2  -       '00111111'
                                    ;            3  -       '00011111'
                                    ;            4  -       '00001111'
                                    ;            5  -       '00000111'
                                    ;            6  -       '00000011'
                                    ;            7  -       '00000001'
                                    ;            8  -       '00000000'
                                    ;            9  -       '00000000'
                                    ;           10  -       '00000000'
                                    ;           11  -       '00000000'
CSM2     EQU      ----              ;INSERT APPROPRIATE MASK BELOW
                                    ; CORR SPAN  1  -  MASK '11111111'
                                    ;            2  -       '11111111'
                                    ;            3  -       '11111111'
                                    ;            4  -       '11111111'
                                    ;            5  -       '11111111'
                                    ;            6  -       '11111111'
                                    ;            7  -       '11111111'
                                    ;            8  -       '11111111'
                                    ;            9  -       '01111111'
                                    ;           10  -       '00111111'
                                    ;           11  -       '00011111'
```

```
;
;       INITIALIZE PSEUDO SHIFT REGS AND SHIFT COUNT (J)
;
INIT      LD      A,(FLDFLG)      ;LOAD FIELD FLAG
          OR      A
          JP      NZ,INIT20       ;JP TO INIT20 IF CORRECTING HEADER
;       INITIALIZE FOR DATA FIELD
INIT10    LD      HL,nn-OV        ;SAVE DATA BUFFER ADDRESS
          LD      (BUFFADR),HL    ;       - NUMBER OF OVERHEAD BYTES
          LD      HL,K1+3         ;SAVE
          LD         (RLBMO),HL        ;      DATA FIELD LENGTH IN BYTES
MINUS 1
          LD      HL,K1           ;LOAD J WITH K1 (CONST FOR DATA)
          JP      INIT30
;       INITIALIZE FOR HEADER FIELD
INIT20    LD      HL,nn           ;SAVE
          LD      (BUFFADR),HL    ;       HEADER BUFFER ADDRESS
          LD      HL,K2+3         ;SAVE
          LD      (RLBMO),HL      ;       HEADER LENGTH IN BYTES-1
          LD      HL,K2           ;LOAD J WITH K2 (CONST FOR HEADER)
INIT30    LD       BC,65535           ;CONSTANT  FOR DECREMENTING SHIFT
COUNT
          EXX
          LD      HL,(nn)         ;FETCH 1ST 2 SYNDROME BYTES
          LD      D,L             ;SYNDROME BITS X0-X7
          LD      E,H             ;SYNDROME BITS X8-X15
          LD      HL,(nn)         ;FETCH 2ND 2 SYND BYTES (X16-X31)
;
;       LEFT JUSTIFY FIRST NON-ZERO SYNDROME BYTE IN 'B'
;
JUST      XOR     A
          OR      D               ;TEST 'R1' FOR ZERO
          JP      NZ,SHIFT05      ;BRANCH ON NONZERO
          EXX
          LD      A,L
          ADD     1               ;J=J+1
          LD      L,A
          JP      NC,JUST9
          INC     H
JUST9     EXX
JUST10    LD      D,E
          LD      E,H
          LD      H,L
          LD      L,0
          JP      JUST
```

```
;
;       SHIFT PSEUDO SHIFT REG UNTIL CORRECTABLE PATTERN FOUND
;
SHIFT     EXX
SHIFT05   LD      B,0            ;INIT TO POINT TO TABLE (T4)
          LD      C,L            ;LOAD 'A' INDEX (SEE FLOW CHART)
;       R4=R3 'XOR' T4(A)   (SEE FLOW CHART)
          LD      A,(BC)         ;
          XOR     H              ;
          LD      L,A            ;
          INC     B              ;
;       R3=R2 'XOR' T3(A)   (SEE FLOW CHART)
          LD      A,(BC)         ;
          XOR     E              ;
          LD      H,A            ;
          INC     B              ;
;       R2=R1 'XOR' T2(A)   (SEE FLOW CHART)
          LD      A,(BC)         ;
          XOR     D              ;
          LD      E,A            ;
          INC     B              ;
;       R1=T1(A)            (SEE FLOW CHART)
          LD      A,(BC)         ;
          LD      D,A            ;
;       TEST LOW ORDER 8 BITS OF SHIFT REG FOR ZERO
          OR      A              ;
          JP      Z,PTRNTST      ;
;       DECREMENT SHIFT COUNT AND TEST FOR ZERO
SHIFT10   EXX                    ;
          ADD     HL,BC          ;BC='FFFF' FOR DECREMENTING HL BY 1
          JP      C,SHIFT        ;NO CARRY IF HL WAS 0 BEFORE ADD
          EXX                    ;
          JP      UNCORR         ;
```

```
;
;       TEST FOR CORRECTABLE PATTERN
;
PTRNTST  LD      A,L             ;
         JP      NZ,SHIFT10      ;BRANCH IF CORR PTRN NOT YET FOUND
;       SAVE SHIFT REG CONTENTS
PTRNTST2 LD      (nn),HL         ;SAVE HL
         EX      DE,HL           ;SAVE DE
PTRNTST3 LD      (nn),HL         ;
;       DETERMINE IF PTRN IN E,H AND L IS CORRECTABLE
PTRNTST4 BIT     7,E
         JP      NZ,PTRNTST5
         SLA     L
         RL      H
         RL      E
         JP      PTRNTST4
PTRNTST5 LD      A,H
         AND     CSM2            ;SEE DEFINITION OF CSM2 ABOVE
         JP      NZ,PTRNTST7     ;BRANCH IF CORR PTRN NOT YET FOUND
         LD      A,E
         AND     CSM1            ;SEE DEFINITION OF CSM1 ABOVE
         JP      Z,PTRNTST8      ;BRANCH IF CORR PTRN FOUND
;       CORR PTRN NOT YET FOUND, RESTORE S/R, RETURN TO SHIFTING
PTRNTST7 LD      HL,(nn)         ;
         EX      DE,HL           ;RESTORE DE (SAVED AT PTRNTST3)
         LD      HL,(nn)         ;RESTORE HL (SAVED AT PTRNTST2)
         JP      SHIFT10         ;
;       GET HERE IF CORR PTRN FOUND
PTRNTST8 LD      HL,(nn)         ;
         EX      DE,HL           ;RESTORE DE (SAVED AT PTRNTST3)
         LD      HL,(nn)         ;RESTORE HL (SAVED AT PTRNTST2)
         LD      C,E             ;PLACE PTRN IN REGS
         LD      D,H             ;    EXPECTED BY
         LD      E,L             ;        NEXT ROUTINE
         EXX
         LD      (nn),HL         ;SAVE HL
         EXX
         LD      HL,(nn)         ;RESTORE HL SAVED 2 STEPS UP
```

```
;
;       CORRECT BYTES IN ERROR
;
CORRECT  LD      B,C             ;MOVE
         LD      C,D             ;      PATTERN
         LD      D,E             ;
;       COMPUTE REV BYTE DISPLACEMENT
         SCF
         CCF
         PUSH    DE
         EX      DE,HL
         LD      HL,(RLBMO)
         SBC     HL,DE
         POP     DE
;       TEST REVERSE DISPLACEMENT CASES
         LD      A,H
         OR      A
         JP      NZ,CORR40       ;BR IF HI BYTE OF REV DISP NONZERO
         LD      A,L
         CP      6
         JP      NC,CORR40       ;BR IF REV DISP EQ OR GTH THAN 6
         CP      5
         JP      Z,CORR25        ;BR IF REV DISP EQ 5
         CP      4
         JP      Z,CORR30        ;BR IF REV DISP EQ 4
;       GET HERE IF ERROR IN CHECK BYTES
CORR20   JP      EXIT            ;IGNORE CORR ERR IN CHECK BYTES
;       GET HERE IF ERROR STARTS IN NEXT TO LAST DATA BYTE
CURR25   LD      A,(nn)
         XOR     B               ;CORRECT NEXT TO LAST DATA BYTE
         LD      (nn),A
         LD      B,C
;       GET HERE IF ERROR STARTS IN LAST DATA BYTE
CORR30   LD      A,(nn)
         XOR     B               ;CORRECT LAST DATA BYTE
         LD      (nn),A
         JP      EXIT            ;DONE
;       RECOMPUTE FWD BYTE DISPLACEMENT
CORR40   SCF
         CCF
         PUSH    DE
         EX      DE,HL
         LD      HL,(RLBMO)
         SBC     HL,DE
         POP     DE
```

```
;       TEST FWD DISPLACEMENT CASES
        LD      A,H
        OR      A
        JP      NZ,CORR45       ;BR IF HI BYTE OF FWD DISP NONZERO
        LD      A,L
        CP      4
        JP      NC,CORR45       ;BR IF FWD DISP EQ OR GTH THAN 4
        CP      3
        JP      Z,CORR60        ;BR IF FWD DISP EQ 3
        CP      2
        JP      Z,CORR55        ;BR IF FWD DISP EQ 2
        JP      CORR70
;
;       GET HERE IF ERROR IN DATA BYTES
CORR45  PUSH    DE
        LD      DE,(BUFFADR)    ;LOAD BUFFER ADDRESS
CORR50  ADD     HL,DE           ;ADD DATA BUFFER ADDR TO
                                ;  TO DISPLACEMENT
        POP     DE
        LD      A,(HL)
        XOR     B               ;CORRECT 1ST DATA BYTE IN ERROR
        LD      (HL),A
        INC     HL
        LD      A,(HL)
        XOR     C               ;CORRECT 2ND DATA BYTE IN ERROR
        LD      (HL),A
        INC     HL
        LD      A,(HL)
        XOR     D               ;CORRECT 3RD DATA BYTE IN ERROR
        LD      (HL),A
        JP      EXIT            ;DONE
;
;       ERROR STARTS IN NEXT TO LAST OVHD BYTE
CORR55  LD      A,(nn)
        XOR     B               ;CORRECT NEXT TO LAST OVHD BYTE
        LD      (nn),A
        LD      B,C
        LD      C,D
        LD      D,0
;
;       ERROR STARTS IN LAST OVHD BYTE
CORR60  LD      A,(nn)
        XOR     B               ;CORRECT LAST OVERHEAD BYTE
        LD      (nn),A
        LD      A,(nn)
        XOR     C               ;CORRECT FIRST DATA BYTE
        LD      (nn),A
        LD      A,(nn)
        XOR     D               ;CORRECT 2ND DATA BYTE
        LD      (nn),A
        JP      EXIT            ;DONE
```

```
;       GET HERE IF ERROR IN OVERHEAD BYTES
CORR70   PUSH    DE
         LD      DE,nn          ;OVERHEAD BYTES BUFFER ADDRESS
         JP      CORR50         ;JOIN COMMON PATH
;
;
;
;
;      UNCORRECTABLE ERROR EXIT
;
UNCORR   NOP                    ;BRANCH TO ERR PATH IN MAIN PGM
;
;      CORRECTION COMPLETE EXIT
;
EXIT     NOP                    ;BRANCH BACK TO MAIN PGM
;
;
;
;      WORK STORAGE
;
ALGNFLG  DEFS    1              ;ALIGNMENT FLAG - SEE FLOW CHART
RLBMO    DEFS    2              ;RECORD LENGTH IN BYTES MINUS 1
BUFFADR  DEFS    2              ;BUFFER ADDRESS - EITHER,
                                ; -DATA BUFF ADDR MINUS
                                ;    # OF OVERHEAD BYTES
                                ; -HEADER BUFFER ADDRESS
FLDFLG   DEFS    1              ;FIELD FLAG - SET BY CALLING PGM
                                ; - ZERO FOR DATA FIELD
                                ; - NONZERO FOR HEADER FIELD
;
;
;      RECIPROCAL POLYNOMIAL TABLES
;
;CONSTRUCT THE RECIPROCAL POLYNOMIAL TABLES AT THIS POINT.
;THE TABLES MUST BE ALIGNED TO AN ADDRESS BOUNDARY THAT
;IS DIVISIBLE BY 256.  THE TABLES MUST BE CONTIGUOUS IN THE
;FOLLOWING ORDER 'T4,T3,T2,T1'.  SEE SECTION 5.3.7 FOR A
;DEFINITION OF THE RECIPROCAL POLYNOMIAL TABLES.
```

5.3.4 *DIAGNOSTICS AND TESTING*

The diagnostic routines for the small-systems magnetic-disk code should be developed using the techniques of Chapter 6 TESTING OF ERROR-CONTROL SYSTEMS.

One of the diagnostic approaches described in Chapter 6, requires a test record that causes check bytes of zero to be generated. For the code described in this section such a record can be constructed as follows. Set the first four bytes to hex '0C 06 03 C3'. Set the last four bytes to hex 'F3 F9 FC 3C'. Clear the remaining bytes to zero.

For design debug, write the test record defined above. Debug the write path until the write check bytes written for this record are zero. Next, debug the read path until this record can be read without error. Finally, run diagnostics as defined in Chapter 6.

5.3.5 *PROTECTION FOR SYNC FRAMING ERRORS*

Protection for sync framing errors is built into circuits of Figures 5.3.2.1 and 5.3.2.2. First, the '1' state of each shift register stage is the low-voltage state. Therefore, the clear function sets the shift register to all ones. Secondly, degating the shift register input during ECC time forces '1's into the high order stage. This is equivalent to inverting certain groups of bits of the check bytes. Today's data integrity requirements dictate greater protection for sync framing errors than provided by the method discussed here. See Section 4.8.2 for a detailed discussion of sync framing errors.

5.3.6 *SIMULATION RUNS*

The following pages contain simulations of the hardware and software algorithms for several correctable errors. Each step of the algorithm, hardware, and software, is included in the simulation.

The test record for each simulation is the test record defined in Section 5.3.4.

Simulation run 1 is a dummy run that illustrates the first 40 shifts for each of the remaining simulation runs. Runs 2 through 4 simulate the bit-serial software algorithm. Runs 2 and 3 simulate error bursts in the data field, while run 4 simulates a single bit error in a check byte. Runs 5 through 7 simulate the byte-serial software algorithm and are similar to runs 2 through 4.

READ SIMULATION RUN # 1

THIS PAGE SHOWS FIRST 40 SHIFTS FOR EACH SIMULATION RUN

BEGIN HDW PART OF SIMULATION
(SHIFTING LEFT, EXTERNAL 'XOR' FORM OF SHIFT REG)

BIT NO.	DATA BITS	ERROR BURST	R1	R2	R3	R4	BYTE NO.
			0 X			31 X	
0			11111111	11111111	11111111	11111110	0
1			11111111	11111111	11111111	11111100	0
2			11111111	11111111	11111111	11111000	0
3			11111111	11111111	11111111	11110000	0
4	1		11111111	11111111	11111111	11100000	0
5	1		11111111	11111111	11111111	11000000	0
6			11111111	11111111	11111111	10000000	0
7			11111111	11111111	11111111	00000000	0
8			11111111	11111111	11111110	00000000	1
9			11111111	11111111	11111100	00000000	1
10			11111111	11111111	11111000	00000000	1
11			11111111	11111111	11110000	00000000	1
12			11111111	11111111	11100000	00000000	1
13	1		11111111	11111111	11000000	00000000	1
14	1		11111111	11111111	10000000	00000000	1
15			11111111	11111111	00000000	00000000	1
16			11111111	11111110	00000000	00000000	2
17			11111111	11111100	00000000	00000000	2
18			11111111	11111000	00000000	00000000	2
19			11111111	11110000	00000000	00000000	2
20			11111111	11100000	00000000	00000000	2
21			11111111	11000000	00000000	00000000	2
22	1		11111111	10000000	00000000	00000000	2
23	1		11111111	00000000	00000000	00000000	2
24	1		11111110	00000000	00000000	00000000	3
25	1		11111100	00000000	00000000	00000000	3
26			11111000	00000000	00000000	00000000	3
27			11110000	00000000	00000000	00000000	3
28			11100000	00000000	00000000	00000000	3
29			11000000	00000000	00000000	00000000	3
30	1		10000000	00000000	00000000	00000000	3
31	1		00000000	00000000	00000000	00000000	3
32			00000000	00000000	00000000	00000000	4
33			00000000	00000000	00000000	00000000	4
34			00000000	00000000	00000000	00000000	4
35			00000000	00000000	00000000	00000000	4
36			00000000	00000000	00000000	00000000	4
37			00000000	00000000	00000000	00000000	4
38			00000000	00000000	00000000	00000000	4
39			00000000	00000000	00000000	00000000	4

```
READ SIMULATION RUN #  2


SIMULATION OF HARDWARE AND SOFTWARE

BEGIN HDW PART OF SIMULATION
(SHIFTING LEFT, EXTERNAL XOR FORM OF SHIFT REG)

BIT     DATA    ERROR       R1          R2          R3          R4        BYTE
NO.     BITS    BURST                                                      NO.

0
,
,    (SEE SIMULATION RUN #1 FOR 1ST 40 SHIFTS)
,    (R IS RECORD LEN IN BITS INCLUDING CHK AND OVERHD)
,
                          0                                            31
                          X                                            X
R-96                      00000000 00000000 00000000 00000000   -11
R-95                      00000000 00000000 00000000 00000000   -11
R-94                      00000000 00000000 00000000 00000000   -11
R-93                      00000000 00000000 00000000 00000000   -11
R-92                      00000000 00000000 00000000 00000000   -11
R-91                      00000000 00000000 00000000 00000000   -11
R-90              1       00000000 00000000 00000000 00000001   -11
R-89              1       00000000 00000000 00000000 00000011   -11
R-88                      00000000 00000000 00000000 00000110   -10
R-87              1       00000000 00000000 00000000 00001101   -10
R-86                      00000000 00000000 00000000 00011011   -10
R-85                      00000000 00000000 00000000 00110111   -10
R-84                      00000000 00000000 00000000 01101111   -10
R-83                      00000000 00000000 00000000 11011110   -10
R-82                      00000000 00000000 00000001 10111101   -10
R-81                      00000000 00000000 00000011 01111010   -10
R-80                      00000000 00000000 00000110 11110100   -9
R-79                      00000000 00000000 00001101 11101001   -9
R-78                      00000000 00000000 00011011 11010010   -9
R-77                      00000000 00000000 00110111 10100101   -9
R-76                      00000000 00000000 01101111 01001010   -9
R-75                      00000000 00000000 11011110 10010100   -9
R-74                      00000000 00000001 10111101 00101000   -9
R-73                      00000000 00000011 01111010 01010001   -9
R-72                      00000000 00000110 11110100 10100010   -8
R-71                      00000000 00001101 11101001 01000101   -8
R-70                      00000000 00011011 11010010 10001011   -8
R-69                      00000000 00110111 10100101 00010111   -8
R-68                      00000000 01101111 01001010 00101111   -8
R-67                      00000000 11011110 10010100 01011110   -8
R-66                      00000001 10111101 00101000 10111100   -8
R-65                      00000011 01111010 01010001 01111001   -8
R-64      1               00000110 11110100 10100010 11110011   -7
R-63      1               00001101 11101001 01000101 11100110   -7
R-62      1               00011011 11010010 10001011 11001100   -7
R-61      1               00110111 10100101 00010111 10011001   -7
R-60                      01101111 01001010 00101111 00110011   -7
R-59                      11011110 10010100 01011110 01100111   -7
R-58      1               10111101 00101000 10111100 11001110   -7
R-57      1               01111010 01010001 01111001 10011100   -7
R-56      1               11110100 10100010 11110011 00111000   -6
R-55      1               11101001 01000101 11100110 01110000   -6
R-54      1               11010010 10001011 11001100 11100001   -6
```

```
SIMULATION RUN NO.  2 CONTINUED

R-53     1        10100101 00010111 10011001 11000011  -6
R-52     1        01001010 00101111 00110011 10000110  -6
R-51              10010100 01011110 01100111 00001101  -6
R-50              00101000 10111100 11001110 00011011  -6
R-49     1        01010001 01111001 10011100 00110111  -6
R-48     1        10100010 11110011 00111000 01101110  -5
R-47     1        01000101 11100110 01110000 11011100  -5
R-46     1        10001011 11001100 11100001 10111001  -5
R-45     1        00010111 10011001 11000011 01110010  -5
R-44     1        00101111 00110011 10000110 11100100  -5
R-43     1        01011110 01100111 00001101 11001001  -5
R-42              10111100 11001110 00011011 10010011  -5
R-41              01111001 10011100 00110111 00100111  -5
R-40              11110011 00111000 01101110 01001111  -4
R-39              11100110 01110000 11011100 10011110  -4
R-38     1        11001100 11100001 10111001 00111100  -4
R-37     1        10011001 11000011 01110010 01111000  -4
R-36     1        00110011 10000110 11100100 11110000  -4
R-35     1        01100111 00001101 11001001 11100001  -4
R-34              11001110 00011011 10010011 11000010  -4
R-33              10011100 00110111 00100111 10000101  -4

FINISHED READING DATA BYTES.   NOW READ CHECK BYTES.
INPUT TO SHIFT REGISTER NOW DEGATED.   PIN 9 OUTPUT
IS GATED TO DESERIALIZER TO BE STORED AS SYNDROME.

R-32              00111000 01101110 01001111 00001011  -3  PIN 9= 0
R-31              01110000 11011100 10011110 00010111  -3  PIN 9= 0
R-30              11100001 10111001 00111100 00101111  -3  PIN 9= 0
R-29              11000011 01110010 01111000 01011111  -3  PIN 9= 0
R-28              10000110 11100100 11110000 10111111  -3  PIN 9= 0
R-27              00001101 11001001 11100001 01111111  -3  PIN 9= 1
R-26              00011011 10010011 11000010 11111111  -3  PIN 9= 1
R-25              00110111 00100111 10000101 11111111  -3  PIN 9= 0
R-24              01101110 01001111 00001011 11111111  -2  PIN 9= 1
R-23              11011100 10011110 00010111 11111111  -2  PIN 9= 0
R-22              10111001 00111100 00101111 11111111  -2  PIN 9= 0
R-21              01110010 01111000 01011111 11111111  -2  PIN 9= 1
R-20              11100100 11110000 10111111 11111111  -2  PIN 9= 1
R-19              11001001 11100001 01111111 11111111  -2  PIN 9= 0
R-18              10010011 11000010 11111111 11111111  -2  PIN 9= 0
R-17              00100111 10000101 11111111 11111111  -2  PIN 9= 0
R-16              01001111 00001011 11111111 11111111  -1  PIN 9= 0
R-15              10011110 00010111 11111111 11111111  -1  PIN 9= 1
R-14              00111100 00101111 11111111 11111111  -1  PIN 9= 0
R-13              01111000 01011111 11111111 11111111  -1  PIN 9= 0
R-12              11110000 10111111 11111111 11111111  -1  PIN 9= 1
R-11              11100001 01111111 11111111 11111111  -1  PIN 9= 1
R-10              11000010 11111111 11111111 11111111  -1  PIN 9= 1
R -9              10000101 11111111 11111111 11111111  -1  PIN 9= 1
R -8              00001011 11111111 11111111 11111111  0   PIN 9= 0
R -7              00010111 11111111 11111111 11111111  0   PIN 9= 0
R -6              00101111 11111111 11111111 11111111  0   PIN 9= 0
R -5              01011111 11111111 11111111 11111111  0   PIN 9= 1
R -4              10111111 11111111 11111111 11111111  0   PIN 9= 0
R -3              01111111 11111111 11111111 11111111  0   PIN 9= 0
R -2              11111111 11111111 11111111 11111111  0   PIN 9= 1
R -1              11111111 11111111 11111111 11111111  0   PIN 9= 0

HDW PART NOW COMPLETE - SYNDOME HAS BEEN STORED.
```

SIMULATION RUN # 2 CONTINUED

SIMULATION OF CORRECTION PROCEDURE
BEGIN SHIFTING SYNDROME
THIS PART SIMULATES INTERNAL XOR FORM OF SHIFT REG
(SHIFTING RIGHT WITH SOFTWARE)

```
                    0                                     31
                   X                                     X
R-25               00000011 01001100 00100111 10001001   -3
R-26               10001011 10100011 00010001 11100110   -3
R-27               01000101 11010001 10001000 11110011   -3
R-28               10101000 11101101 11000110 01011011   -3
R-29               11011110 01110011 11100001 00001111   -3
R-30               11100101 00111100 11110010 10100101   -3
R-31               11111000 10011011 01111011 01110000   -3
R-32               01111100 01001101 10111101 10111000   -3
R-33               00111110 00100110 11011110 11011100   -4
R-34               00011111 00010011 01101111 01101110   -4
R-35               00001111 10001001 10110111 10110111   -4
R-36               10001101 11000001 11011001 11111001   -4
R-37               11001100 11100101 11101110 11011110   -4
R-38               01100110 01110010 11110111 01101111   -4
R-39               10111001 00111100 01111001 10010101   -4
R-40               11010110 10011011 00111110 11101000   -4
R-41               01101011 01001101 10011111 01110100   -5
R-42               00110101 10100110 11001111 10111010   -5
R-43               00011010 11010011 01100111 11011101   -5
R-44               10000111 01101100 10110001 11001100   -5
R-45               01000011 10110110 01011000 11100110   -5
R-46               00100001 11011011 00101100 01110011   -5
R-47               10011010 11101000 10010100 00011011   -5
R-48               11000111 01110001 01001000 00101111   -5
R-49               11101001 10111101 10100110 00110101   -6
R-50               11111110 11011011 11010001 00111000   -6
R-51               01111111 01101101 11101000 10011100   -6
R-52               00111111 10110110 11110100 01001110   -6
R-53               00011111 11011011 01111010 00100111   -6
R-54               10000101 11101000 10111111 00110001   -6
R-55               11001000 11110001 01011101 10111010   -6
R-56               01100100 01111000 10101110 11011101   -6
R-57               10111000 00111001 01010101 01001100   -7
R-58               01011100 00011100 10101010 10100110   -7
R-59               00101110 00001110 01010101 01010011   -7
R-60               10011101 00000010 00101000 10001011   -7
R-61               11000100 10000100 00010110 01100111   -7
R-62               11101000 01000111 00001001 00010001   -7
R-63               11111110 00100110 10000110 10101010   -7
R-64               01111111 00010011 01000011 01010101   -7
R-65               10110101 10001100 10100011 10001000   -8
R-66               01011010 11000110 01010001 11000100   -8
R-67               00101101 01100011 00101000 11100010   -8
R-68               00010110 10110001 10010100 01110001   -8
R-69               10000001 01011101 11001000 00011010   -8
R-70               01000000 10101110 11100100 00001101   -8
R-71               10101010 01010010 01110000 00100100   -8
R-72               01010101 00101001 00111000 00010010   -8
R-73               00101010 10010100 10011100 00001001   -9
R-74               10011111 01001111 01001100 00100110   -9
R-75               01001111 10100111 10100110 00010011   -9
R-76               10101101 11010110 11010001 00101011   -9
R-77               11011100 11101110 01101010 10110111   -9
```

SIMULATION RUN NO. 2 CONTINUED

```
R-78          11100100 01110010 00110111 01111001  -9
R-79          11111000 00111100 00011001 10011110  -9
R-80          01111100 00011110 00001100 11001111  -9
R-81          10110100 00001010 00000100 01000101  -10
R-82          11010000 00000000 00000000 00000000  -10
R-83          01101000 00000000 00000000 00000000  -10
R-84          00110100 00000000 00000000 00000000  -10
R-85          00011010 00000000 00000000 00000000  -10
R-86          00001101 00000000 00000000 00000000  -10
R-87          00000110 10000000 00000000 00000000  -10
R-88          00000011 01000000 00000000 00000000  -10
R-89          00000001 10100000 00000000 00000000  -11
R-90          00000000 11010000 00000000 00000000  -11
```

CORRECTABLE PATTERN FOUND, -90 IS BIT DISPLACEMENT.
NOW BEGIN BYTE ALIGNMENT.

```
R-91          00000000 01101000 00000000 00000000  -11
R-92          00000000 00110100 00000000 00000000  -11
R-93          00000000 00011010 00000000 00000000  -11
R-94          00000000 00001101 00000000 00000000  -11
R-95          00000000 00000110 10000000 00000000  -11
R-96          00000000 00000011 01000000 00000000  -11
```

BYTE ALIGNMENT COMPLETE - SIMULATION COMPLETE
BYTE DISPLACEMENT IS 11.
COUNTING FROM END OF RECORD. LAST BYTE IS ZERO.

READ SIMULATION RUN # 3

SIMULATION OF HARDWARE AND SOFTWARE

BEGIN HDW PART OF SIMULATION
(SHIFTING LEFT, EXTERNAL XOR FORM OF SHIFT REG)

```
BIT    DATA   ERROR     R1        R2        R3        R4      BYTE
NO     BITS   BURST                                           NO.

0
,
,   (SEE SIMULATION RUN #1 FOR 1ST 40 SHIFTS)
,   (R IS RECORD LEN IN BITS INCLUDING CHK AND OVERHD)
,
                     0                                  31
                     X                                  X
R-96                 00000000 00000000 00000000 00000000  -11
R-95                 00000000 00000000 00000000 00000000  -11
R-94                 00000000 00000000 00000000 00000000  -11
R-93                 00000000 00000000 00000000 00000000  -11
R-92                 00000000 00000000 00000000 00000000  -11
R-91                 00000000 00000000 00000000 00000000  -11
R-90                 00000000 00000000 00000000 00000000  -11
R-89                 00000000 00000000 00000000 00000000  -11
R-88                 00000000 00000000 00000000 00000000  -10
R-87                 00000000 00000000 00000000 00000000  -10
R-86                 00000000 00000000 00000000 00000000  -10
R-85                 00000000 00000000 00000000 00000000  -10
R-84           1     00000000 00000000 00000000 00000001  -10
R-83                 00000000 00000000 00000000 00000010  -10
R-82           1     00000000 00000000 00000000 00000101  -10
R-81           1     00000000 00000000 00000000 00001011  -10
R-80                 00000000 00000000 00000000 00010111  -9
R-79                 00000000 00000000 00000000 00101110  -9
R-78                 00000000 00000000 00000000 01011100  -9
R-77                 00000000 00000000 00000000 10111001  -9
R-76                 00000000 00000000 00000001 01110010  -9
R-75                 00000000 00000000 00000010 11100101  -9
R-74                 00000000 00000000 00000101 11001011  -9
R-73                 00000000 00000000 00001011 10010111  -9
R-72                 00000000 00000000 00010111 00101110  -8
R-71                 00000000 00000000 00101110 01011101  -8
R-70                 00000000 00000000 01011100 10111011  -8
R-69                 00000000 00000000 10111001 01110110  -8
R-68                 00000000 00000001 01110010 11101100  -8
R-67                 00000000 00000010 11100101 11011000  -8
R-66                 00000000 00000101 11001011 10110000  -8
R-65                 00000000 00001011 10010111 01100000  -8
R-64    1            00000000 00010111 00101110 11000001  -7
R-63    1            00000000 00101110 01011101 10000011  -7
R-62    1            00000000 01011100 10111011 00000110  -7
R-61    1            00000000 10111001 01110110 00001100  -7
R-60                 00000001 01110010 11101100 00011000  -7
R-59                 00000010 11100101 11011000 00110001  -7
R-58    1            00000101 11001011 10110000 01100010  -7
R-57    1            00001011 10010111 01100000 11000101  -7
R-56    1            00010111 00101110 11000001 10001011  -6
R-55    1            00101110 01011101 10000011 00010111  -6
R-54    1            01011100 10111011 00000110 00101111  -6
```

SIMULATION RUN NO.   3 CONTINUED

```
R-53      1     10111001 01110110 00001100 01011110   -6
R-52      1     01110010 11101100 00011000 10111101   -6
R-51            11100101 11011000 00110001 01111010   -6
R-50            11001011 10110000 01100010 11110101   -6
R-49      1     10010111 01100000 11000101 11101010   -6
R-48      1     00101110 11000001 10001011 11010101   -5
R-47      1     01011101 10000011 00010111 10101011   -5
R-46      1     10111011 00000110 00101111 01010110   -5
R-45      1     01110110 00001100 01011110 10101100   -5
R-44      1     11101100 00011000 10111101 01011001   -5
R-43      1     11011000 00110001 01111010 10110011   -5
R-42            10110000 01100010 11110101 01100111   -5
R-41            01100000 11000101 11101010 11001110   -5
R-40            11000001 10001011 11010101 10011101   -4
R-39            10000011 00010111 10101011 00111010   -4
R-38      1     00000110 00101111 01010110 01110101   -4
R-37      1     00001100 01011110 10101100 11101010   -4
R-36      1     00011000 10111101 01011001 11010101   -4
R-35      1     00110001 01111010 10110011 10101010   -4
R-34            01100010 11110101 01100111 01010101   -4
R-33            11000101 11101010 11001110 10101010   -4
```

FINISHED READING DATA BYTES.   NOW READ CHECK BYTES.
INPUT TO SHIFT REGISTER NOW DEGATED.   PIN 9 OUTPUT
IS GATED TO DESERIALIZER TO BE STORED AS SYNDROME.

```
R-32            10001011 11010101 10011101 01010101   -3   PIN 9= 1
R-31            00010111 10101011 00111010 10101011   -3   PIN 9= 1
R-30            00101111 01010110 01110101 01010111   -3   PIN 9= 1
R-29            01011110 10101100 11101010 10101111   -3   PIN 9= 0
R-28            10111101 01011001 11010101 01011111   -3   PIN 9= 1
R-27            01111010 10110011 10101010 10111111   -3   PIN 9= 1
R-26            11110101 01100111 01010101 01111111   -3   PIN 9= 1
R-25            11101010 11001110 10101010 11111111   -3   PIN 9= 1
R-24            11010101 10011101 01010101 11111111   -2   PIN 9= 1
R-23            10101011 00111010 10101011 11111111   -2   PIN 9= 1
R-22            01010110 01110101 01010111 11111111   -2   PIN 9= 0
R-21            10101100 11101010 10101111 11111111   -2   PIN 9= 0
R-20            01011001 11010101 01011111 11111111   -2   PIN 9= 1
R-19            10110011 10101010 10111111 11111111   -2   PIN 9= 0
R-18            01100111 01010101 01111111 11111111   -2   PIN 9= 1
R-17            11001110 10101010 11111111 11111111   -2   PIN 9= 0
R-16            10011101 01010101 11111111 11111111   -1   PIN 9= 1
R-15            00111010 10101011 11111111 11111111   -1   PIN 9= 1
R-14            01110101 01010111 11111111 11111111   -1   PIN 9= 1
R-13            11101010 10101111 11111111 11111111   -1   PIN 9= 1
R-12            11010101 01011111 11111111 11111111   -1   PIN 9= 0
R-11            10101010 10111111 11111111 11111111   -1   PIN 9= 1
R-10            01010101 01111111 11111111 11111111   -1   PIN 9= 0
R -9            10101010 11111111 11111111 11111111   -1   PIN 9= 1
R -8            01010101 11111111 11111111 11111111   0    PIN 9= 0
R -7            10101011 11111111 11111111 11111111   0    PIN 9= 1
R -6            01010111 11111111 11111111 11111111   0    PIN 9= 0
R -5            10101111 11111111 11111111 11111111   0    PIN 9= 0
R -4            01011111 11111111 11111111 11111111   0    PIN 9= 0
R -3            10111111 11111111 11111111 11111111   0    PIN 9= 0
R -2            01111111 11111111 11111111 11111111   0    PIN 9= 0
R -1            11111111 11111111 11111111 11111111   0    PIN 9= 1
```

HDW PART NOW COMPLETE - SYNDOME HAS BEEN STORED.

SIMULATION RUN # 3 CONTINUED

SIMULATION OF CORRECTION PROCEDURE
BEGIN SHIFTING SYNDROME
THIS PART SIMULATES INTERNAL XOR FORM OF SHIFT REG
(SHIFTING RIGHT WITH SOFTWARE)

	X^0			X^{31}	
R-25	11111101	11100000	01111000	10000010	-3
R-26	01111110	11110000	00111100	01000001	-3
R-27	10110101	01111101	00011100	00000010	-3
R-28	01011010	10111110	10001110	00000001	-3
R-29	10100111	01011010	01000101	00100010	-3
R-30	01010011	10101101	00100010	10010001	-3
R-31	10100011	11010011	10010011	01101010	-3
R-32	01010001	11101001	11001001	10110101	-3
R-33	10100010	11110001	11100110	11111000	-4
R-34	01010001	01111000	11110011	01111100	-4
R-35	00101000	10111100	01111001	10111110	-4
R-36	00010100	01011110	00111100	11011111	-4
R-37	10000000	00101010	00011100	01001101	-4
R-38	11001010	00010000	00001100	00000100	-4
R-39	01100101	00001000	00000110	00000010	-4
R-40	00110010	10000100	00000011	00000001	-4
R-41	10010011	01000111	00000011	10100010	-5
R-42	01001001	10100011	10000001	11010001	-5
R-43	10101110	11010100	11000010	11001010	-5
R-44	01010111	01101010	01100001	01100101	-5
R-45	10100001	10110000	00110010	10010000	-5
R-46	01010000	11011000	00011001	01001000	-5
R-47	00101000	01101100	00001100	10100100	-5
R-48	00010100	00110110	00000110	01010010	-5
R-49	00001010	00011011	00000011	00101001	-6
R-50	10001111	00001000	10000011	10110110	-6
R-51	01000111	10000100	01000001	11011011	-6
R-52	10101001	11000111	00100010	11001111	-6
R-53	11011110	11100110	10010011	01000101	-6
R-54	11100101	01110110	01001011	10000000	-6
R-55	01110010	10111011	00100101	11000000	-6
R-56	00111001	01011101	10010010	11100000	-6
R-57	00011100	10101110	11001001	01110000	-7
R-58	00001110	01010111	01100100	10111000	-7
R-59	00000111	00101011	10110010	01011100	-7
R-60	00000011	10010101	11011001	00101110	-7
R-61	00000001	11001010	11101100	10010111	-7
R-62	10001010	11100000	01110100	01101001	-7
R-63	11001111	01110101	00111000	00010110	-7
R-64	01100111	10111010	10011100	00001011	-7
R-65	10111001	11011000	01001100	00100111	-8
R-66	11010110	11101001	00100100	00110001	-8
R-67	11100001	01110001	10010000	00111010	-8
R-68	01110000	10111000	11001000	00011101	-8
R-69	10110010	01011001	01100110	00101100	-8
R-70	01011001	00101100	10110011	00010110	-8
R-71	00101100	10010110	01011001	10001011	-8
R-72	10011100	01001110	00101110	11100111	-8
R-73	11000100	00100010	00010101	01010001	-9
R-74	11101000	00010100	00001000	10001010	-9
R-75	01110100	00001010	00000100	01000101	-9
R-76	10110000	00000000	00000000	00000000	-9
R-77	01011000	00000000	00000000	00000000	-9

SIMULATION RUN NO. 3 CONTINUED

```
R-78      00101100 00000000 00000000 00000000   -9
R-79      00010110 00000000 00000000 00000000   -9
R-80      00001011 00000000 00000000 00000000   -9
R-81      00000101 10000000 00000000 00000000   -10
R-82      00000010 11000000 00000000 00000000   -10
R-83      00000001 01100000 00000000 00000000   -10
R-84      00000000 10110000 00000000 00000000   -10
```

CORRECTABLE PATTERN FOUND, -84 IS BIT DISPLACEMENT.
NOW BEGIN BYTE ALIGNMENT.

```
R-85      00000000 01011000 00000000 00000000   -10
R-86      00000000 00101100 00000000 00000000   -10
R-87      00000000 00010110 00000000 00000000   -10
R-88      00000000 00001011 00000000 00000000   -10
```

BYTE ALIGNMENT COMPLETE - SIMULATION COMPLETE
BYTE DISPLACEMENT IS 10.
COUNTING FROM END OF RECORD. LAST BYTE IS ZERO.

READ SIMULATION RUN # 4

SIMULATION OF HARDWARE AND SOFTWARE

BEGIN HDW PART OF SIMULATION
(SHIFTING LEFT, EXTERNAL XOR FORM OF SHIFT REG)

```
BIT   DATA   ERROR     R1         R2         R3         R4        BYTE
NO.   BITS   BURST                                                NO.

0
/
/   (SEE SIMULATION RUN #1 FOR 1ST 40 SHIFTS)
/   (R IS RECORD LEN IN BITS INCLUDING CHK AND OVERHD)
/                                                        31
                       0                                 X
                       X
R-96                   00000000 00000000 00000000 00000000  -11
R-95                   00000000 00000000 00000000 00000000  -11
R-94                   00000000 00000000 00000000 00000000  -11
R-93                   00000000 00000000 00000000 00000000  -11
R-92                   00000000 00000000 00000000 00000000  -11
R-91                   00000000 00000000 00000000 00000000  -11
R-90                   00000000 00000000 00000000 00000000  -11
R-89                   00000000 00000000 00000000 00000000  -11
R-88                   00000000 00000000 00000000 00000000  -10
R-87                   00000000 00000000 00000000 00000000  -10
R-86                   00000000 00000000 00000000 00000000  -10
R-85                   00000000 00000000 00000000 00000000  -10
R-84                   00000000 00000000 00000000 00000000  -10
R-83                   00000000 00000000 00000000 00000000  -10
R-82                   00000000 00000000 00000000 00000000  -10
R-81                   00000000 00000000 00000000 00000000  -10
R-80                   00000000 00000000 00000000 00000000  -9
R-79                   00000000 00000000 00000000 00000000  -9
R-78                   00000000 00000000 00000000 00000000  -9
R-77                   00000000 00000000 00000000 00000000  -9
R-76                   00000000 00000000 00000000 00000000  -9
R-75                   00000000 00000000 00000000 00000000  -9
R-74                   00000000 00000000 00000000 00000000  -9
R-73                   00000000 00000000 00000000 00000000  -9
R-72                   00000000 00000000 00000000 00000000  -8
R-71                   00000000 00000000 00000000 00000000  -8
R-70                   00000000 00000000 00000000 00000000  -8
R-69                   00000000 00000000 00000000 00000000  -8
R-68                   00000000 00000000 00000000 00000000  -8
R-67                   00000000 00000000 00000000 00000000  -8
R-66                   00000000 00000000 00000000 00000000  -8
R-65                   00000000 00000000 00000000 00000000  -8
R-64     1             00000000 00000000 00000000 00000001  -7
R-63     1             00000000 00000000 00000000 00000011  -7
R-62     1             00000000 00000000 00000000 00000111  -7
R-61     1             00000000 00000000 00000000 00001111  -7
R-60                   00000000 00000000 00000000 00011111  -7
R-59                   00000000 00000000 00000000 00111111  -7
R-58     1             00000000 00000000 00000000 01111111  -7
R-57     1             00000000 00000000 00000000 11111111  -7
R-56     1             00000000 00000000 00000001 11111111  -6
R-55     1             00000000 00000000 00000011 11111111  -6
R-54     1             00000000 00000000 00000111 11111111  -6
```

SIMULATION RUN NO.   4 CONTINUED

```
R-53      1          00000000 00000000 00001111 11111111   -6
R-52      1          00000000 00000000 00011111 11111111   -6
R-51                 00000000 00000000 00111111 11111111   -6
R-50                 00000000 00000000 01111111 11111111   -6
R-49      1          00000000 00000000 11111111 11111111   -6
R-48      1          00000000 00000001 11111111 11111111   -5
R-47      1          00000000 00000011 11111111 11111111   -5
R-46      1          00000000 00000111 11111111 11111111   -5
R-45      1          00000000 00001111 11111111 11111111   -5
R-44      1          00000000 00011111 11111111 11111111   -5
R-43      1          00000000 00111111 11111111 11111111   -5
R-42                 00000000 01111111 11111111 11111111   -5
R-41                 00000000 11111111 11111111 11111111   -5
R-40                 00000001 11111111 11111111 11111111   -4
R-39                 00000011 11111111 11111111 11111111   -4
R-38      1          00000111 11111111 11111111 11111111   -4
R-37      1          00001111 11111111 11111111 11111111   -4
R-36      1          00011111 11111111 11111111 11111111   -4
R-35      1          00111111 11111111 11111111 11111111   -4
R-34                 01111111 11111111 11111111 11111111   -4
R-33                 11111111 11111111 11111111 11111111   -4
```

FINISHED READING DATA BYTES. NOW READ CHECK BYTES.
INPUT TO SHIFT REGISTER NOW DEGATED. PIN 9 OUTPUT
IS GATED TO DESERIALIZER TO BE STORED AS SYNDROME.

```
R-32                 11111111 11111111 11111111 11111111   -3   PIN 9= 0
R-31                 11111111 11111111 11111111 11111111   -3   PIN 9= 0
R-30                 11111111 11111111 11111111 11111111   -3   PIN 9= 0
R-29                 11111111 11111111 11111111 11111111   -3   PIN 9= 0
R-28                 11111111 11111111 11111111 11111111   -3   PIN 9= 0
R-27                 11111111 11111111 11111111 11111111   -3   PIN 9= 0
R-26                 11111111 11111111 11111111 11111111   -3   PIN 9= 0
R-25                 11111111 11111111 11111111 11111111   -3   PIN 9= 0
R-24                 11111111 11111111 11111111 11111111   -2   PIN 9= 0
R-23                 11111111 11111111 11111111 11111111   -2   PIN 9= 0
R-22                 11111111 11111111 11111111 11111111   -2   PIN 9= 0
R-21                 11111111 11111111 11111111 11111111   -2   PIN 9= 0
R-20                 11111111 11111111 11111111 11111111   -2   PIN 9= 0
R-19                 11111111 11111111 11111111 11111111   -2   PIN 9= 0
R-18                 11111111 11111111 11111111 11111111   -2   PIN 9= 0
R-17                 11111111 11111111 11111111 11111111   -2   PIN 9= 0
R-16                 11111111 11111111 11111111 11111111   -1   PIN 9= 0
R-15                 11111111 11111111 11111111 11111111   -1   PIN 9= 0
R-14                 11111111 11111111 11111111 11111111   -1   PIN 9= 0
R-13           1     11111111 11111111 11111111 11111111   -1   PIN 9= 1
R-12                 11111111 11111111 11111111 11111111   -1   PIN 9= 0
R-11                 11111111 11111111 11111111 11111111   -1   PIN 9= 0
R-10                 11111111 11111111 11111111 11111111   -1   PIN 9= 0
R -9                 11111111 11111111 11111111 11111111   -1   PIN 9= 0
R -8                 11111111 11111111 11111111 11111111   0    PIN 9= 0
R -7                 11111111 11111111 11111111 11111111   0    PIN 9= 0
R -6                 11111111 11111111 11111111 11111111   0    PIN 9= 0
R -5                 11111111 11111111 11111111 11111111   0    PIN 9= 0
R -4                 11111111 11111111 11111111 11111111   0    PIN 9= 0
R -3                 11111111 11111111 11111111 11111111   0    PIN 9= 0
R -2                 11111111 11111111 11111111 11111111   0    PIN 9= 0
R -1                 11111111 11111111 11111111 11111111   0    PIN 9= 0
```

HDW PART NOW COMPLETE - SYNDOME HAS BEEN STORED.

```
SIMULATION RUN #  4 CONTINUED

SIMULATION OF CORRECTION PROCEDURE
BEGIN SHIFTING SYNDROME
THIS PART SIMULATES INTERNAL XOR FORM OF SHIFT REG
(SHIFTING RIGHT WITH SOFTWARE)

                         0                                      31
                         X                                      X
                         00001000 00000000 00000000 00000000   -1
R -9                     00000100 00000000 00000000 00000000   -1
R-10                     00000010 00000000 00000000 00000000   -1
R-11                     00000001 00000000 00000000 00000000   -1
R-12                     00000000 10000000 00000000 00000000   -1
R-13

CORRECTABLE PATTERN FOUND, -13 IS BIT DISPLACEMENT.
NOW BEGIN BYTE ALIGNMENT.

                         00000000 01000000 00000000 00000000   -1
R-14                     00000000 00100000 00000000 00000000   -1
R-15                     00000000 00010000 00000000 00000000   -1
R-16

BYTE ALIGNMENT COMPLETE - SIMULATION COMPLETE
BYTE DISPLACEMENT IS  1.
COUNTING FROM END OF RECORD.  LAST BYTE IS ZERO.
```

```
READ SIMULATION RUN #  5


SIMULATION OF HARDWARE AND SOFTWARE

BEGIN HDW PART OF SIMULATION
(SHIFTING LEFT, SERIAL EXTERNAL XOR FORM OF SHIFT REG)

BIT     DATA    ERROR    R1         R2         R3         R4        BYTE
NO.     BITS    BURST                                               NO.

0
/   (SEE SIMULATION RUN # 1 FOR FIRST 40 SHIFTS)
/   (R IS RECORD LEN IN BITS INCLUDING CHK AND OVERHD)
/
                         0                                    31
                         X                                    X
R-96                     00000000 00000000 00000000 00000000  -11
R-95                     00000000 00000000 00000000 00000000  -11
R-94                     00000000 00000000 00000000 00000000  -11
R-93                     00000000 00000000 00000000 00000000  -11
R-92                     00000000 00000000 00000000 00000000  -11
R-91                     00000000 00000000 00000000 00000000  -11
R-90                  1  00000000 00000000 00000000 00000001  -11
R-89                  1  00000000 00000000 00000000 00000011  -11
R-88                     00000000 00000000 00000000 00000110  -10
R-87                  1  00000000 00000000 00000000 00001101  -10
R-86                     00000000 00000000 00000000 00011011  -10
R-85                     00000000 00000000 00000000 00110111  -10
R-84                     00000000 00000000 00000000 01101111  -10
R-83                     00000000 00000000 00000000 11011110  -10
R-82                     00000000 00000000 00000001 10111101  -10
R-81                     00000000 00000000 00000011 01111010  -10
R-80                     00000000 00000000 00000110 11110100  -9
R-79                     00000000 00000000 00001101 11101001  -9
R-78                     00000000 00000000 00011011 11010010  -9
R-77                     00000000 00000000 00110111 10100101  -9
R-76                     00000000 00000000 01101111 01001010  -9
R-75                     00000000 00000000 11011110 10010100  -9
R-74                     00000000 00000001 10111101 00101000  -9
R-73                     00000000 00000011 01111010 01010001  -9
R-72                     00000000 00000110 11110100 10100010  -8
R-71                     00000000 00001101 11101001 01000101  -8
R-70                     00000000 00011011 11010010 10001011  -8
R-69                     00000000 00110111 10100101 00010111  -8
R-68                     00000000 01101111 01001010 00101111  -8
R-67                     00000000 11011110 10010100 01011110  -8
R-66                     00000001 10111101 00101000 10111100  -8
R-65                     00000011 01111010 01010001 01111001  -8
R-64     1               00000110 11110100 10100010 11110011  -7
R-63     1               00001101 11101001 01000101 11100110  -7
R-62     1               00011011 11010010 10001011 11001100  -7
R-61     1               00110111 10100101 00010111 10011001  -7
R-60                     01101111 01001010 00101111 00110011  -7
R-59                     11011110 10010100 01011110 01100111  -7
R-58     1               10111101 00101000 10111100 11001110  -7
R-57     1               01111010 01010001 01111001 10011100  -7
R-56     1               11110100 10100010 11110011 00111000  -6
R-55     1               11101001 01000101 11100110 01110000  -6
R-54     1               11010010 10001011 11001100 11100001  -6
R-53     1               10100101 00010111 10011001 11000011  -6
```

SIMULATION RUN NO. 5 CONTINUED

```
R-52    1    01001010 00101111 00110011 10000110  -6
R-51         10010100 01011110 01100111 00001101  -6
R-50         00101000 10111100 11001110 00011011  -6
R-49    1    01010001 01111001 10011100 00110111  -6
R-48    1    10100010 11110011 00111000 01101110  -5
R-47    1    01000101 11100110 01110000 11011100  -5
R-46    1    10001011 11001100 11100001 10111001  -5
R-45    1    00010111 10011001 11000011 01110010  -5
R-44    1    00101111 00110011 10000110 11100100  -5
R-43    1    01011110 01100111 00001101 11001001  -5
R-42         10111100 11001110 00011011 10010011  -5
R-41         01111001 10011100 00110111 00100111  -5
R-40         11110011 00111000 01101110 01001111  -4
R-39         11100110 01110000 11011100 10011110  -4
R-38    1    11001100 11100001 10111001 00111100  -4
R-37    1    10011001 11000011 01110010 01111000  -4
R-36    1    00110011 10000110 11100100 11110000  -4
R-35    1    01100111 00001101 11001001 11100001  -4
R-34         11001110 00011011 10010011 11000010  -4
R-33         10011100 00110111 00100111 10000101  -4
```

FINISHED READING DATA BYTES. NOW READ CHECK BYTES.
INPUT TO SHIFT REGISTER NOW DEGATED. PIN 9 OUTPUT
IS GATED TO DESERIALIZER TO BE STORED AS SYNDROME.

```
R-32    00111000 01101110 01001111 00001011  -3  PIN 9= 0
R-31    01110000 11011100 10011110 00010111  -3  PIN 9= 0
R-30    11100001 10111001 00111100 00101111  -3  PIN 9= 0
R-29    11000011 01110010 01111000 01011111  -3  PIN 9= 0
R-28    10000110 11100100 11110000 10111111  -3  PIN 9= 0
R-27    00001101 11001001 11100001 01111111  -3  PIN 9= 1
R-26    00011011 10010011 11000010 11111111  -3  PIN 9= 1
R-25    00110111 00100111 10000101 11111111  -3  PIN 9= 0
R-24    01101110 01001111 00001011 11111111  -2  PIN 9= 1
R-23    11011100 10011110 00010111 11111111  -2  PIN 9= 0
R-22    10111001 00111100 00101111 11111111  -2  PIN 9= 0
R-21    01110010 01111000 01011111 11111111  -2  PIN 9= 1
R-20    11100100 11110000 10111111 11111111  -2  PIN 9= 1
R-19    11001001 11100001 01111111 11111111  -2  PIN 9= 0
R-18    10010011 11000010 11111111 11111111  -2  PIN 9= 0
R-17    00100111 10000101 11111111 11111111  -2  PIN 9= 0
R-16    01001111 00001011 11111111 11111111  -1  PIN 9= 0
R-15    10011110 00010111 11111111 11111111  -1  PIN 9= 1
R-14    00111100 00101111 11111111 11111111  -1  PIN 9= 0
R-13    01111000 01011111 11111111 11111111  -1  PIN 9= 0
R-12    11110000 10111111 11111111 11111111  -1  PIN 9= 1
R-11    11100001 01111111 11111111 11111111  -1  PIN 9= 1
R-10    11000010 11111111 11111111 11111111  -1  PIN 9= 1
R -9    10000101 11111111 11111111 11111111  -1  PIN 9= 1
R -8    00001011 11111111 11111111 11111111  0   PIN 9= 0
R -7    00010111 11111111 11111111 11111111  0   PIN 9= 0
R -6    00101111 11111111 11111111 11111111  0   PIN 9= 0
R -5    01011111 11111111 11111111 11111111  0   PIN 9= 1
R -4    10111111 11111111 11111111 11111111  0   PIN 9= 0
R -3    01111111 11111111 11111111 11111111  0   PIN 9= 0
R -2    11111111 11111111 11111111 11111111  0   PIN 9= 1
R -1    11111111 11111111 11111111 11111111  0   PIN 9= 0
```

HDW PART NOW COMPLETE - SYNDOME HAS BEEN STORED.

SIMULATION RUN # 5 CONTINUED

SIMULATION OF CORRECTION PROCEDURE
BEGIN SHIFTING SYNDROME
THIS PART SIMULATES INTERNAL XOR FORM OF SHIFT REG
(SHIFTING RIGHT WITH SOFTWARE 8 BITS AT A TIME)

```
                    0                                  31
                    X                                  X
R-32                01111100 01001101 10111101 10111000  -3
R-40                11010110 10011011 00111110 11101000  -4
R-48                11000111 01110001 01001000 00101111  -5
R-56                01100100 01111000 10101110 11011101  -6
R-64                01111111 00010011 01000011 01010101  -7
R-72                01010101 00101001 00111000 00010010  -8
R-80                01111100 00011110 00001100 11001111  -9
R-88                00000011 01000000 00000000 00000000  -10
R-96                00000000 00000011 01000000 00000000  -11
```

CORRECTABLE PATTERN FOUND.

BYTE DISPLACEMENT IS 11.
COUNTING FROM END OF RECORD. LAST BYTE IS ZERO.

SIMULATION COMPLETE.

READ SIMULATION RUN # 6

SIMULATION OF HARDWARE AND SOFTWARE

BEGIN HDW PART OF SIMULATION
(SHIFTING LEFT, SERIAL EXTERNAL XOR FORM OF SHIFT REG)

BIT NO.	DATA BITS	ERROR BURST	R1	R2	R3	R4	BYTE NO.
0							
/	(SEE SIMULATION RUN # 1 FOR FIRST 40 SHIFTS)						
/	(R IS RECORD LEN IN BITS INCLUDING CHK AND OVERHD)						
/							
			0 X			31 X	
R-96			00000000	00000000	00000000	00000000	-11
R-95			00000000	00000000	00000000	00000000	-11
R-94			00000000	00000000	00000000	00000000	-11
R-93			00000000	00000000	00000000	00000000	-11
R-92			00000000	00000000	00000000	00000000	-11
R-91			00000000	00000000	00000000	00000000	-11
R-90			00000000	00000000	00000000	00000000	-11
R-89			00000000	00000000	00000000	00000000	-11
R-88			00000000	00000000	00000000	00000000	-10
R-87			00000000	00000000	00000000	00000000	-10
R-86			00000000	00000000	00000000	00000000	-10
R-85			00000000	00000000	00000000	00000000	-10
R-84		1	00000000	00000000	00000000	00000001	-10
R-83			00000000	00000000	00000000	00000010	-10
R-82		1	00000000	00000000	00000000	00000101	-10
R-81		1	00000000	00000000	00000000	00001011	-10
R-80			00000000	00000000	00000000	00010111	-9
R-79			00000000	00000000	00000000	00101110	-9
R-78			00000000	00000000	00000000	01011100	-9
R-77			00000000	00000000	00000000	10111001	-9
R-76			00000000	00000000	00000001	01110010	-9
R-75			00000000	00000000	00000010	11100101	-9
R-74			00000000	00000000	00000101	11001011	-9
R-73			00000000	00000000	00001011	10010111	-9
R-72			00000000	00000000	00010111	00101110	-8
R-71			00000000	00000000	00101110	01011101	-8
R-70			00000000	00000000	01011100	10111011	-8
R-69			00000000	00000000	10111001	01110110	-8
R-68			00000000	00000001	01110010	11101100	-8
R-67			00000000	00000010	11100101	11011000	-8
R-66			00000000	00000101	11001011	10110000	-8
R-65			00000000	00001011	10010111	01100000	-8
R-64	1		00000000	00010111	00101110	11000001	-7
R-63	1		00000000	00101110	01011101	10000011	-7
R-62	1		00000000	01011100	10111011	00000110	-7
R-61	1		00000000	10111001	01110110	00001100	-7
R-60			00000001	01110010	11101100	00011000	-7
R-59			00000010	11100101	11011000	00110001	-7
R-58	1		00000101	11001011	10110000	01100010	-7
R-57	1		00001011	10010111	01100000	11000101	-7
R-56	1		00010111	00101110	11000001	10001011	-6
R-55	1		00101110	01011101	10000011	00010111	-6
R-54	1		01011100	10111011	00000110	00101111	-6
R-53	1		10111001	01110110	00001100	01011110	-6

```
SIMULATION RUN NO.   6 CONTINUED

R-52      1        01110010 11101100 00011000 10111101   -6
R-51               11100101 11011000 00110001 01111010   -6
R-50               11001011 10110000 01100010 11110101   -6
R-49      1        10010111 01100000 11000101 11101010   -6
R-48      1        00101110 11000001 10001011 11010101   -5
R-47      1        01011101 10000011 00010111 10101011   -5
R-46      1        10111011 00000110 00101111 01010110   -5
R-45      1        01110110 00001100 01011110 10101100   -5
R-44      1        11101100 00011000 10111101 01011001   -5
R-43      1        11011000 00110001 01111010 10110011   -5
R-42               10110000 01100010 11110101 01100111   -5
R-41               01100000 11000101 11101010 11001110   -5
R-40               11000001 10001011 11010101 10011101   -4
R-39               10000011 00010111 10101011 00111010   -4
R-38      1        00000110 00101111 01010110 01110101   -4
R-37      1        00001100 01011110 10101100 11101010   -4
R-36      1        00011000 10111101 01011001 11010101   -4
R-35      1        00110001 01111010 10110011 10101010   -4
R-34               01100010 11110101 01100111 01010101   -4
R-33               11000101 11101010 11001110 10101010   -4

FINISHED READING DATA BYTES.   NOW READ CHECK BYTES.
INPUT TO SHIFT REGISTER NOW DEGATED.   PIN 9 OUTPUT
IS GATED TO DESERIALIZER TO BE STORED AS SYNDROME.

R-32               10001011 11010101 10011101 01010101   -3   PIN 9= 1
R-31               00010111 10101011 00111010 10101011   -3   PIN 9= 1
R-30               00101111 01010110 01110101 01010111   -3   PIN 9= 1
R-29               01011110 10101100 11101010 10101111   -3   PIN 9= 0
R-28               10111101 01011001 11010101 01011111   -3   PIN 9= 1
R-27               01111010 10110011 10101010 10111111   -3   PIN 9= 1
R-26               11110101 01100111 01010101 01111111   -3   PIN 9= 1
R-25               11101010 11001110 10101010 11111111   -3   PIN 9= 1
R-24               11010101 10011101 01010101 11111111   -2   PIN 9= 1
R-23               10101011 00111010 10101011 11111111   -2   PIN 9= 1
R-22               01010110 01110101 01010111 11111111   -2   PIN 9= 0
R-21               10101100 11101010 10101111 11111111   -2   PIN 9= 0
R-20               01011001 11010101 01011111 11111111   -2   PIN 9= 1
R-19               10110011 10101010 10111111 11111111   -2   PIN 9= 0
R-18               01100111 01010101 01111111 11111111   -2   PIN 9= 1
R-17               11001110 10101010 11111111 11111111   -2   PIN 9= 0
R-16               10011101 01010101 11111111 11111111   -1   PIN 9= 1
R-15               00111010 10101011 11111111 11111111   -1   PIN 9= 1
R-14               01110101 01010111 11111111 11111111   -1   PIN 9= 1
R-13               11101010 10101111 11111111 11111111   -1   PIN 9= 1
R-12               11010101 01011111 11111111 11111111   -1   PIN 9= 0
R-11               10101010 10111111 11111111 11111111   -1   PIN 9= 1
R-10               01010101 01111111 11111111 11111111   -1   PIN 9= 0
R -9               10101010 11111111 11111111 11111111   -1   PIN 9= 1
R -8               01010101 11111111 11111111 11111111   0    PIN 9= 0
R -7               10101011 11111111 11111111 11111111   0    PIN 9= 1
R -6               01010111 11111111 11111111 11111111   0    PIN 9= 0
R -5               10101111 11111111 11111111 11111111   0    PIN 9= 0
R -4               01011111 11111111 11111111 11111111   0    PIN 9= 0
R -3               10111111 11111111 11111111 11111111   0    PIN 9= 0
R -2               01111111 11111111 11111111 11111111   0    PIN 9= 0
R -1               11111111 11111111 11111111 11111111   0    PIN 9= 1

HDW PART NOW COMPLETE - SYNDOME HAS BEEN STORED.
```

```
SIMULATION RUN #  6 CONTINUED


SIMULATION OF CORRECTION PROCEDURE
BEGIN SHIFTING SYNDROME
THIS PART SIMULATES INTERNAL XOR FORM OF SHIFT REG
(SHIFTING RIGHT WITH SOFTWARE 8 BITS AT A TIME)

                         0                                   31
                         X                                   X
R-32                     01010001 11101001 11001001 10110101   -3
R-40                     00110010 10000100 00000011 00000001   -4
R-48                     00010100 00110110 00000110 01010010   -5
R-56                     00111001 01011101 10010010 11100000   -6
R-64                     01100111 10111010 10011100 00001011   -7
R-72                     10011100 01001110 00101110 11100111   -8
R-80                     00001011 00000000 00000000 00000000   -9
R-88                     00000000 00001011 00000000 00000000   -10

CORRECTABLE PATTERN FOUND.


BYTE DISPLACEMENT IS  10.
COUNTING FROM END OF RECORD.  LAST BYTE IS ZERO.

SIMULATION COMPLETE.
```

READ SIMULATION RUN # 7

SIMULATION OF HARDWARE AND SOFTWARE

BEGIN HDW PART OF SIMULATION
(SHIFTING LEFT, SERIAL EXTERNAL XOR FORM OF SHIFT REG)

BIT NO.	DATA BITS	ERROR BURST	R1	R2	R3	R4	BYTE NO.
0							
,	(SEE SIMULATION RUN # 1 FOR FIRST 40 SHIFTS)						
,	(R IS RECORD LEN IN BITS INCLUDING CHK AND OVERHD)						
,							
			0 X			31 X	
R-96			00000000	00000000	00000000	00000000	-11
R-95			00000000	00000000	00000000	00000000	-11
R-94			00000000	00000000	00000000	00000000	-11
R-93			00000000	00000000	00000000	00000000	-11
R-92			00000000	00000000	00000000	00000000	-11
R-91			00000000	00000000	00000000	00000000	-11
R-90			00000000	00000000	00000000	00000000	-11
R-89			00000000	00000000	00000000	00000000	-11
R-88			00000000	00000000	00000000	00000000	-10
R-87			00000000	00000000	00000000	00000000	-10
R-86			00000000	00000000	00000000	00000000	-10
R-85			00000000	00000000	00000000	00000000	-10
R-84			00000000	00000000	00000000	00000000	-10
R-83			00000000	00000000	00000000	00000000	-10
R-82			00000000	00000000	00000000	00000000	-10
R-81			00000000	00000000	00000000	00000000	-10
R-80			00000000	00000000	00000000	00000000	-9
R-79			00000000	00000000	00000000	00000000	-9
R-78			00000000	00000000	00000000	00000000	-9
R-77			00000000	00000000	00000000	00000000	-9
R-76			00000000	00000000	00000000	00000000	-9
R-75			00000000	00000000	00000000	00000000	-9
R-74			00000000	00000000	00000000	00000000	-9
R-73			00000000	00000000	00000000	00000000	-9
R-72			00000000	00000000	00000000	00000000	-8
R-71			00000000	00000000	00000000	00000000	-8
R-70			00000000	00000000	00000000	00000000	-8
R-69			00000000	00000000	00000000	00000000	-8
R-68			00000000	00000000	00000000	00000000	-8
R-67			00000000	00000000	00000000	00000000	-8
R-66			00000000	00000000	00000000	00000000	-8
R-65			00000000	00000000	00000000	00000000	-8
R-64	1		00000000	00000000	00000000	00000001	-7
R-63	1		00000000	00000000	00000000	00000011	-7
R-62	1		00000000	00000000	00000000	00000111	-7
R-61	1		00000000	00000000	00000000	00001111	-7
R-60			00000000	00000000	00000000	00011111	-7
R-59			00000000	00000000	00000000	00111111	-7
R-58	1		00000000	00000000	00000000	01111111	-7
R-57	1		00000000	00000000	00000000	11111111	-7
R-56	1		00000000	00000000	00000001	11111111	-6
R-55	1		00000000	00000000	00000011	11111111	-6
R-54	1		00000000	00000000	00000111	11111111	-6
R-53	1		00000000	00000000	00001111	11111111	-6

SIMULATION RUN NO.   7 CONTINUED

```
R-52      1          00000000 00000000 00011111 11111111  -6
R-51                 00000000 00000000 00111111 11111111  -6
R-50                 00000000 00000000 01111111 11111111  -6
R-49      1          00000000 00000000 11111111 11111111  -6
R-48      1          00000000 00000001 11111111 11111111  -5
R-47      1          00000000 00000011 11111111 11111111  -5
R-46      1          00000000 00000111 11111111 11111111  -5
R-45      1          00000000 00001111 11111111 11111111  -5
R-44      1          00000000 00011111 11111111 11111111  -5
R-43      1          00000000 00111111 11111111 11111111  -5
R-42                 00000000 01111111 11111111 11111111  -5
R-41                 00000000 11111111 11111111 11111111  -5
R-40                 00000001 11111111 11111111 11111111  -4
R-39                 00000011 11111111 11111111 11111111  -4
R-38      1          00000111 11111111 11111111 11111111  -4
R-37      1          00001111 11111111 11111111 11111111  -4
R-36      1          00011111 11111111 11111111 11111111  -4
R-35      1          00111111 11111111 11111111 11111111  -4
R-34                 01111111 11111111 11111111 11111111  -4
R-33                 11111111 11111111 11111111 11111111  -4
```

FINISHED READING DATA BYTES.   NOW READ CHECK BYTES.
INPUT TO SHIFT REGISTER NOW DEGATED.   PIN 9 OUTPUT
IS GATED TO DESERIALIZER TO BE STORED AS SYNDROME.

```
R-32                 11111111 11111111 11111111 11111111  -3   PIN 9= 0
R-31                 11111111 11111111 11111111 11111111  -3   PIN 9= 0
R-30                 11111111 11111111 11111111 11111111  -3   PIN 9= 0
R-29                 11111111 11111111 11111111 11111111  -3   PIN 9= 0
R-28                 11111111 11111111 11111111 11111111  -3   PIN 9= 0
R-27                 11111111 11111111 11111111 11111111  -3   PIN 9= 0
R-26                 11111111 11111111 11111111 11111111  -3   PIN 9= 0
R-25                 11111111 11111111 11111111 11111111  -3   PIN 9= 0
R-24                 11111111 11111111 11111111 11111111  -2   PIN 9= 0
R-23                 11111111 11111111 11111111 11111111  -2   PIN 9= 0
R-22                 11111111 11111111 11111111 11111111  -2   PIN 9= 0
R-21                 11111111 11111111 11111111 11111111  -2   PIN 9= 0
R-20                 11111111 11111111 11111111 11111111  -2   PIN 9= 0
R-19                 11111111 11111111 11111111 11111111  -2   PIN 9= 0
R-18                 11111111 11111111 11111111 11111111  -2   PIN 9= 0
R-17                 11111111 11111111 11111111 11111111  -2   PIN 9= 0
R-16                 11111111 11111111 11111111 11111111  -1   PIN 9= 0
R-15                 11111111 11111111 11111111 11111111  -1   PIN 9= 0
R-14                 11111111 11111111 11111111 11111111  -1   PIN 9= 0
R-13           1     11111111 11111111 11111111 11111111  -1   PIN 9= 1
R-12                 11111111 11111111 11111111 11111111  -1   PIN 9= 0
R-11                 11111111 11111111 11111111 11111111  -1   PIN 9= 0
R-10                 11111111 11111111 11111111 11111111  -1   PIN 9= 0
R -9                 11111111 11111111 11111111 11111111  -1   PIN 9= 0
R -8                 11111111 11111111 11111111 11111111  0    PIN 9= 0
R -7                 11111111 11111111 11111111 11111111  0    PIN 9= 0
R -6                 11111111 11111111 11111111 11111111  0    PIN 9= 0
R -5                 11111111 11111111 11111111 11111111  0    PIN 9= 0
R -4                 11111111 11111111 11111111 11111111  0    PIN 9= 0
R -3                 11111111 11111111 11111111 11111111  0    PIN 9= 0
R -2                 11111111 11111111 11111111 11111111  0    PIN 9= 0
R -1                 11111111 11111111 11111111 11111111  0    PIN 9= 0
```

HDW PART NOW COMPLETE - SYNDOME HAS BEEN STORED.

```
SIMULATION RUN #  7 CONTINUED

SIMULATION OF CORRECTION PROCEDURE
BEGIN SHIFTING SYNDROME
THIS PART SIMULATES INTERNAL XOR FORM OF SHIFT REG
(SHIFTING RIGHT WITH SOFTWARE 8 BITS AT A TIME)

                      0                                  31
                     X                                  X
R-16                 00000000 00010000 00000000 00000000  -1

CORRECTABLE PATTERN FOUND.

BYTE DISPLACEMENT IS  1.
COUNTING FROM END OF RECORD.  LAST BYTE IS ZERO.

SIMULATION COMPLETE.
```

5.3.7 *RECIPROCAL POLYNOMIAL TABLES*

The byte-serial software algorithm requires four, 256-byte tables. These tables are listed on the following pages. Since data entry is error prone, the tables should be regenerated by computer.

To regenerate the tables, implement a right-shifting internal-XOR serial shift register in software, using the reciprocal polynomial. For each address of the tables (0-255), place the address in the eight most significant (right-most) bits of the shift register and clear the remaining bits. Shift eight times, then store the four bytes of shift register contents in tables T1 through T4 at the location indexed by the current address. The coefficient of x^0 is stored as the high-order bit of T1; the coefficient of x^{31} is stored as the low-order bit of T4. Check the resulting tables against those on the following pages.

RECIPROCAL POLYNOMIAL TABLE T1

	0	1	2	3	4	5	6	7	8	9	A	B	C	D	E	F
00	00	14	28	3C	50	44	78	6C	A0	B4	88	9C	F0	E4	D8	CC
10	54	40	7C	68	04	10	2C	38	F4	E0	DC	C8	A4	B0	8C	98
20	A8	BC	80	94	F8	EC	D0	C4	08	1C	20	34	58	4C	70	64
30	FC	E8	D4	C0	AC	B8	84	90	5C	48	74	60	0C	18	24	30
40	45	51	6D	79	15	01	3D	29	E5	F1	CD	D9	B5	A1	9D	89
50	11	05	39	2D	41	55	69	7D	B1	A5	99	8D	E1	F5	C9	DD
60	ED	F9	C5	D1	BD	A9	95	81	4D	59	65	71	1D	09	35	21
70	B9	AD	91	85	E9	FD	C1	D5	19	0D	31	25	49	5D	61	75
80	8A	9E	A2	B6	DA	CE	F2	E6	2A	3E	02	16	7A	6E	52	46
90	DE	CA	F6	E2	8E	9A	A6	B2	7E	6A	56	42	2E	3A	06	12
A0	22	36	0A	1E	72	66	5A	4E	82	96	AA	BE	D2	C6	FA	EE
B0	76	62	5E	4A	26	32	0E	1A	D6	C2	FE	EA	86	92	AE	BA
C0	CF	DB	E7	F3	9F	8B	B7	A3	6F	7B	47	53	3F	2B	17	03
D0	9B	8F	B3	A7	CB	DF	E3	F7	3B	2F	13	07	6B	7F	43	57
E0	67	73	4F	5B	37	23	1F	0B	C7	D3	EF	FB	97	83	BF	AB
F0	33	27	1B	0F	63	77	4B	5F	93	87	BB	AF	C3	D7	EB	FF

RECIPROCAL POLYNOMIAL TABLE T2

	0	1	2	3	4	5	6	7	8	9	A	B	C	D	E	F
00	00	04	09	0D	12	16	1B	1F	24	20	2D	29	36	32	3F	3B
10	42	46	4B	4F	50	54	59	5D	66	62	6F	6B	74	70	7D	79
20	84	80	8D	89	96	92	9F	9B	A0	A4	A9	AD	B2	B6	BB	BF
30	C6	C2	CF	CB	D4	D0	DD	D9	E2	E6	EB	EF	F0	F4	F9	FD
40	02	06	0B	0F	10	14	19	1D	26	22	2F	2B	34	30	3D	39
50	40	44	49	4D	52	56	5B	5F	64	60	6D	69	76	72	7F	7B
60	86	82	8F	8B	94	90	9D	99	A2	A6	AB	AF	B0	B4	B9	BD
70	C4	C0	CD	C9	D6	D2	DF	DB	E0	E4	E9	ED	F2	F6	FB	FF
80	05	01	0C	08	17	13	1E	1A	21	25	28	2C	33	37	3A	3E
90	47	43	4E	4A	55	51	5C	58	63	67	6A	6E	71	75	78	7C
A0	81	85	88	8C	93	97	9A	9E	A5	A1	AC	A8	B7	B3	BE	BA
B0	C3	C7	CA	CE	D1	D5	D8	DC	E7	E3	EE	EA	F5	F1	FC	F8
C0	07	03	0E	0A	15	11	1C	18	23	27	2A	2E	31	35	38	3C
D0	45	41	4C	48	57	53	5E	5A	61	65	68	6C	73	77	7A	7E
E0	83	87	8A	8E	91	95	98	9C	A7	A3	AE	AA	B5	B1	BC	B8
F0	C1	C5	C8	CC	D3	D7	DA	DE	E5	E1	EC	E8	F7	F3	FE	FA

RECIPROCAL POLYNOMIAL TABLE T3

	0	1	2	3	4	5	6	7	8	9	A	B	C	D	E	F
00	00	82	04	86	09	8B	0D	8F	12	90	16	94	1B	99	1F	9D
10	21	A3	25	A7	28	AA	2C	AE	33	B1	37	B5	3A	B8	3E	BC
20	42	C0	46	C4	4B	C9	4F	CD	50	D2	54	D6	59	DB	5D	DF
30	63	E1	67	E5	6A	E8	6E	EC	71	F3	75	F7	78	FA	7C	FE
40	81	03	85	07	88	0A	8C	0E	93	11	97	15	9A	18	9E	1C
50	A0	22	A4	26	A9	2B	AD	2F	B2	30	B6	34	BB	39	BF	3D
60	C3	41	C7	45	CA	48	CE	4C	D1	53	D5	57	D8	5A	DC	5E
70	E2	60	E6	64	EB	69	EF	6D	F0	72	F4	76	F9	7B	FD	7F
80	02	80	06	84	0B	89	0F	8D	10	92	14	96	19	9B	1D	9F
90	23	A1	27	A5	2A	A8	2E	AC	31	B3	35	B7	38	BA	3C	BE
A0	40	C2	44	C6	49	CB	4D	CF	52	D0	56	D4	5B	D9	5F	DD
B0	61	E3	65	E7	68	EA	6C	EE	73	F1	77	F5	7A	F8	7E	FC
C0	83	01	87	05	8A	08	8E	0C	91	13	95	17	98	1A	9C	1E
D0	A2	20	A6	24	AB	29	AF	2D	B0	32	B4	36	B9	3B	BD	3F
E0	C1	43	C5	47	C8	4A	CC	4E	D3	51	D7	55	DA	58	DE	5C
F0	E0	62	E4	66	E9	6B	ED	6F	F2	70	F6	74	FB	79	FF	7D

RECIPROCAL POLYNOMIAL TABLE T4

	0	1	2	3	4	5	6	7	8	9	A	B	C	D	E	F
00	00	51	A2	F3	44	15	E6	B7	88	D9	2A	7B	CC	9D	6E	3F
10	55	04	F7	A6	11	40	B3	E2	DD	8C	7F	2E	99	C8	3B	6A
20	AA	FB	08	59	EE	BF	4C	1D	22	73	80	D1	66	37	C4	95
30	FF	AE	5D	0C	BB	EA	19	48	77	26	D5	84	33	62	91	C0
40	11	40	B3	E2	55	04	F7	A6	99	C8	3B	6A	DD	8C	7F	2E
50	44	15	E6	B7	00	51	A2	F3	CC	9D	6E	3F	88	D9	2A	7B
60	BB	EA	19	48	FF	AE	5D	0C	33	62	91	C0	77	26	D5	84
70	EE	BF	4C	1D	AA	FB	08	59	66	37	C4	95	22	73	80	D1
80	22	73	80	D1	66	37	C4	95	AA	FB	08	59	EE	BF	4C	1D
90	77	26	D5	84	33	62	91	C0	FF	AE	5D	0C	BB	EA	19	48
A0	88	D9	2A	7B	CC	9D	6E	3F	00	51	A2	F3	44	15	E6	B7
B0	DD	8C	7F	2E	99	C8	3B	6A	55	04	F7	A6	11	40	B3	E2
C0	33	62	91	C0	77	26	D5	84	BB	EA	19	48	FF	AE	5D	0C
D0	66	37	C4	95	22	73	80	D1	EE	BF	4C	1D	AA	FB	08	59
E0	99	C8	3B	6A	DD	8C	7F	2E	11	40	B3	E2	55	04	F7	A6
F0	CC	9D	6E	3F	88	D9	2A	7B	44	15	E6	B7	00	51	A2	F3

5.4 APPLICATION TO MASS STORAGE DEVICES

This section describes an interleaved Reed-Solomon code implementation that is suitable for many mass storage devices. It is a composite of several real world implementations, including the implementation described in U.S. Patent #4,142,174, Chen, et al. (1979).

The implementation has triple-symbol error-correction capability and is interleaved to depth 32. Symbols are one byte wide.

Key features of the implementation are:

- Corrects up to 3 random symbol errors in each interleave.
- Corrects a single burst up to 96 bytes in length.
- The data format includes a resync field after every 32 data bytes. This limits the length of an error burst resulting from synchronization loss.

The media data format is shown below. Data is transferred to and from the media one row at a time. Checking is performed in the column dimension.

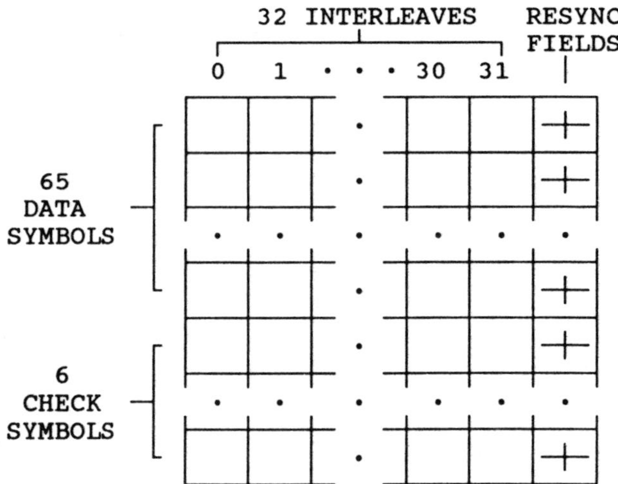

The following pages show, for the implementation:

- The write encoder circuit.
- The syndrome circuits.
- The finite field processor.
- An algorithm for determining the number of errors occurring and for generating coefficients of the error locator polynomial.
- Algorithms for finding the roots of the error locator polynomial in the single-, double-, and triple-error cases.
- Algorithms for determining error values for the single-, double-, and triple-error cases.
- ROM tables for taking logarithms and antilogarithms, for finding the roots of equation $y^2 + y + c = 0$, and for taking the cube root of a finite field element.

ENCODE POLYNOMIAL

$$(x + 1)\cdot(x + \alpha)\cdot(x + \alpha^2)\cdot(x + \alpha^3)\cdot(x + \alpha^4)\cdot(x + \alpha^5)$$

$$= x^6 + \alpha^{94}\cdot x^5 + \alpha^{10}\cdot x^4 + \alpha^{136}\cdot x^3 + \alpha^{15}\cdot x^2 + \alpha^{104}\cdot x + \alpha^{15}$$

WRITE ENCODER

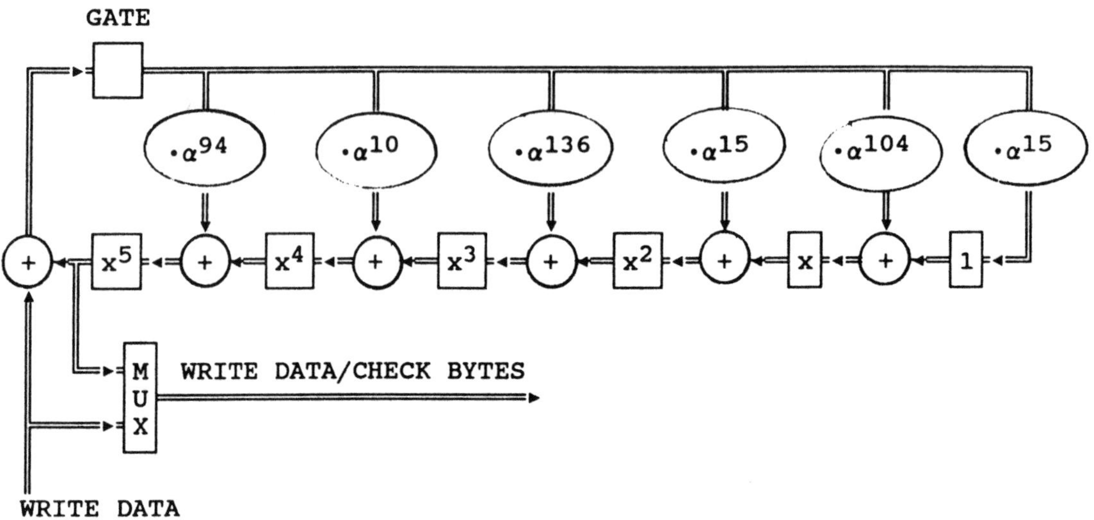

SYNDROME CIRCUITS

There are six circuits (i=0 to 5) and each circuit is interleaved to depth 32.

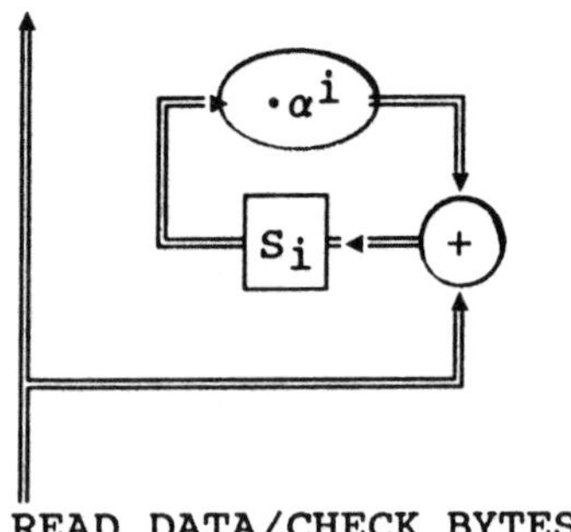

FINITE FIELD PROCESSOR

Except where noted, all paths are eight bits wide.

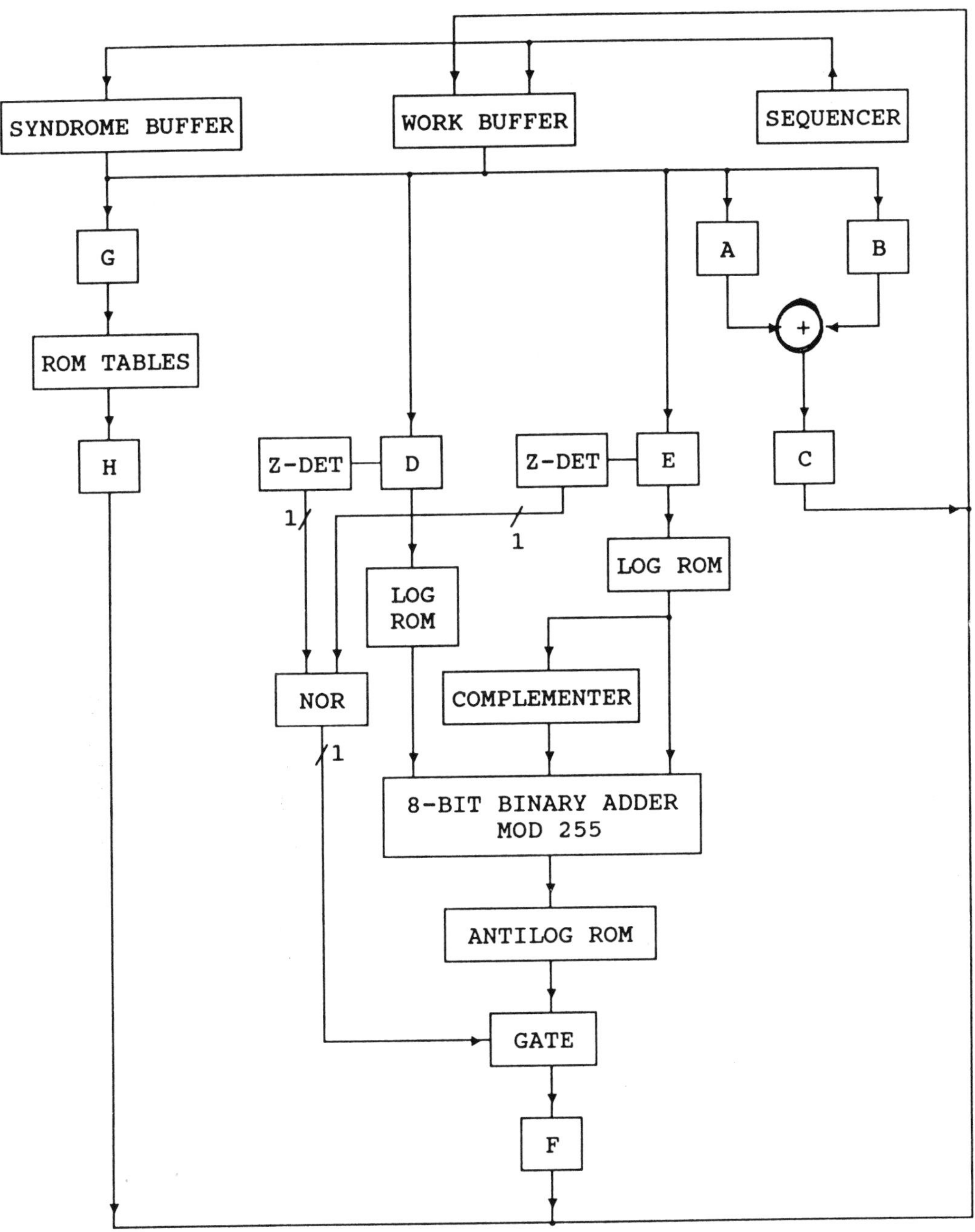

DETERMINE NUMBER OF ERRORS AND GENERATE ERROR LOCATOR POLYNOMIAL

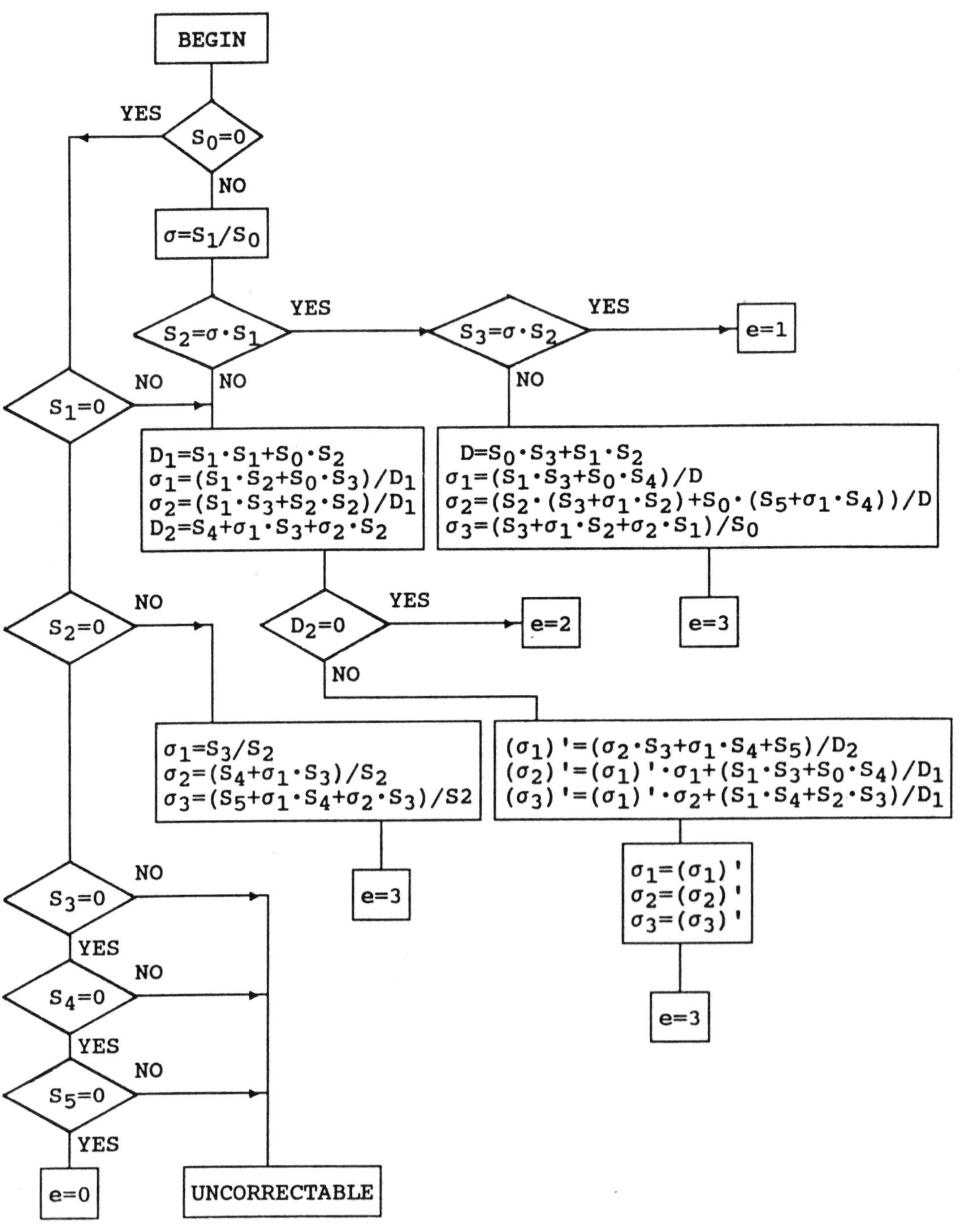

COMPUTE ERROR LOCATIONS AND ERROR VALUES

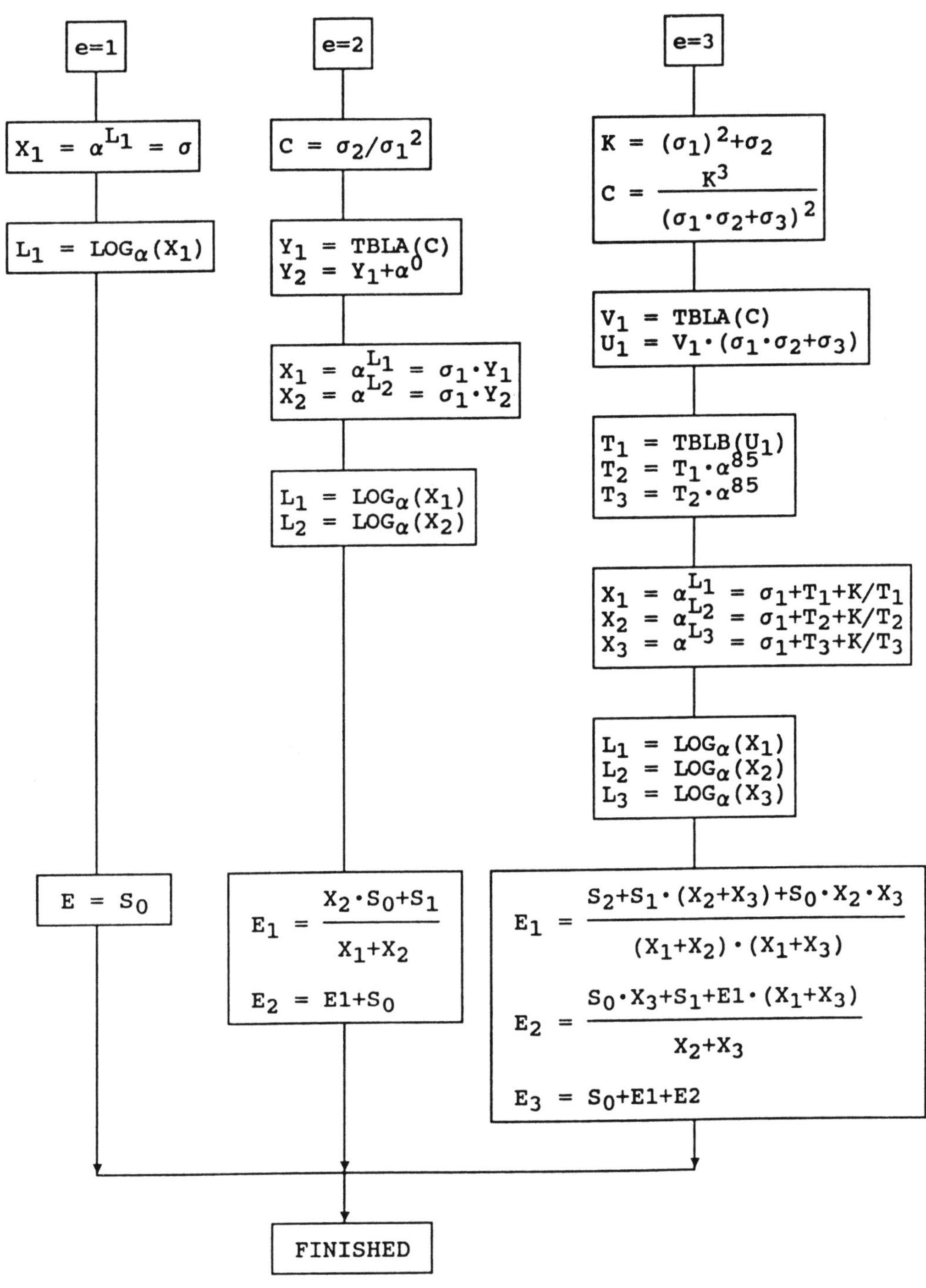

SOLVING THE THREE-ERROR LOCATOR POLYNOMIAL IN GF(2^8)

The three-error locator polynomial is

$$x^3 + \sigma_1 \cdot x^2 + \sigma_2 \cdot x + \sigma_3 = 0$$

First, substitute $w = x + \sigma_1$ to obtain

$$w^3 + ((\sigma_1)^2 + \sigma_2) \cdot w + (\sigma_1 \cdot \sigma_2 + \sigma_3) = 0$$

Second, apply the substitution

$$w = t + ((\sigma_1)^2 + \sigma_2)/t$$

to obtain

$$t^3 + (\sigma_1 \cdot \sigma_2 + \sigma_3) + ((\sigma_1)^2 + \sigma_2)^3/t^3 = 0$$

and thus

$$t^6 + (\sigma_1 \cdot \sigma_2 + \sigma_3) \cdot t^3 + ((\sigma_1)^2 + \sigma_2)^3 = 0$$

Third, substitute $u = t^3$ to obtain

$$u^2 + (\sigma_1 \cdot \sigma_2 + \sigma_3) \cdot u + ((\sigma_1)^2 + \sigma_2)^3 = 0$$

Finally, substitute

$$v = u/(\sigma_1 \cdot \sigma_2 + \sigma_3)$$

to obtain

$$v^2 + v + \frac{((\sigma_1)^2 + \sigma_2)^3}{(\sigma_1 \cdot \sigma_2 + \sigma_3)^2} = 0$$

Now fetch a root V_1 from the table developed for the two-error case:

$$V_1 = \text{TBLA}\left[\frac{((\sigma_1)^2 + \sigma_2)^3}{(\sigma_1 \cdot \sigma_2 + \sigma_3)^2} \right]$$

Next, apply the reverse substitution

$$u = v \cdot (\sigma_1 \cdot \sigma_2 + \sigma_3)$$

to obtain

$$U_1 = V_1 \cdot (\sigma_1 \cdot \sigma_2 + \sigma_3)$$

Apply the reverse substitution $t = (u)^{1/3}$ to obtain

$$T_1 = (V_1 \cdot (\sigma_1 \cdot \sigma_2 + \sigma_3))^{1/3}$$

T_1 may be fetched from a table of cube roots in GF(2^8):

$$T_1 = \text{TBLB}[\ V_1 \cdot (\sigma_1 \cdot \sigma_2 + \sigma_3)\ ]$$

Each element in GF(2^8) which has a cube root has three cube roots; the other two may be computed:

$$T_2 = T_1 \cdot \alpha^k$$

$$T_3 = T_2 \cdot \alpha^k$$

where $k = (2^8-1)/3 = 85$.

Now reverse the substitution

$$w = t + ((\sigma_1)^2 + \sigma_2)/t$$

to obtain

$$W_1 = T_1 + \frac{(\sigma_1)^2 + \sigma_2}{T_1}$$

$$W_2 = T_2 + \frac{(\sigma_1)^2 + \sigma_2}{T_2}$$

$$W_3 = T_3 + \frac{(\sigma_1)^2 + \sigma_2}{T_3}$$

And finally, apply the reverse substitution

$$x = w + \sigma_1$$

to obtain the roots of the original three-error locator polynomial:

$$X_1 = \alpha^{L1} = T_1 + \frac{(\sigma_1)^2 + \sigma_2}{T_1} + \sigma_1$$

$$X_2 = \alpha^{L2} = T_2 + \frac{(\sigma_1)^2 + \sigma_2}{T_2} + \sigma_1$$

$$X_3 = \alpha^{L3} = T_3 + \frac{(\sigma_1)^2 + \sigma_2}{T_3} + \sigma_1$$

The error locations L_1, L_2, and L_3 are the logs base α of X_1, X_2, and X_3, respectively.

ANTILOG TABLE
(INPUT IS n, OUTPUT IS α^n)

	0	1	2	3	4	5	6	7	8	9	A	B	C	D	E	F
00	01	02	04	08	10	20	40	80	71	E2	B5	1B	36	6C	D8	C1
10	F3	97	5F	BE	0D	1A	34	68	D0	D1	D3	D7	DF	CF	EF	AF
20	2F	5E	BC	09	12	24	48	90	51	A2	35	6A	D4	D9	C3	F7
30	9F	4F	9E	4D	9A	45	8A	65	CA	E5	BB	07	0E	1C	38	70
40	E0	B1	13	26	4C	98	41	82	75	EA	A5	3B	76	EC	A9	23
50	46	8C	69	D2	D5	DB	C7	FF	8F	6F	DE	CD	EB	A7	3F	7E
60	FC	89	63	C6	FD	8B	67	CE	ED	AB	27	4E	9C	49	92	55
70	AA	25	4A	94	59	B2	15	2A	54	A8	21	42	84	79	F2	95
80	5B	B6	1D	3A	74	E8	A1	33	66	CC	E9	A3	37	6E	DC	C9
90	E3	B7	1F	3E	7C	F8	81	73	E6	BD	0B	16	2C	58	B0	11
A0	22	44	88	61	C2	F5	9B	47	8E	6D	DA	C5	FB	87	7F	FE
B0	8D	6B	D6	DD	CB	E7	BF	0F	1E	3C	78	F0	91	53	A6	3D
C0	7A	F4	99	43	86	7D	FA	85	7B	F6	9D	4B	96	5D	BA	05
D0	0A	14	28	50	A0	31	62	C4	F9	83	77	EE	AD	2B	56	AC
E0	29	52	A4	39	72	E4	B9	03	06	0C	18	30	60	C0	F1	93
F0	57	AE	2D	5A	B4	19	32	64	C8	E1	B3	17	2E	5C	B8	01

LOG TABLE
(INPUT IS α^n, OUTPUT IS n)

	0	1	2	3	4	5	6	7	8	9	A	B	C	D	E	F
00	--	00	01	E7	02	CF	E8	3B	03	23	D0	9A	E9	14	3C	B7
10	04	9F	24	42	D1	76	9B	FB	EA	F5	15	0B	3D	82	B8	92
20	05	7A	A0	4F	25	71	43	6A	D2	E0	77	DD	9C	F2	FC	20
30	EB	D5	F6	87	16	2A	0C	8C	3E	E3	83	4B	B9	BF	93	5E
40	06	46	7B	C3	A1	35	50	A7	26	6D	72	CB	44	33	6B	31
50	D3	28	E1	BD	78	6F	DE	F0	9D	74	F3	80	FD	CD	21	12
60	EC	A3	D6	62	F7	37	88	66	17	52	2B	B1	0D	A9	8D	59
70	3F	08	E4	97	84	48	4C	DA	BA	7D	C0	C8	94	C5	5F	AE
80	07	96	47	D9	7C	C7	C4	AD	A2	61	36	65	51	B0	A8	58
90	27	BC	6E	EF	73	7F	CC	11	45	C2	34	A6	6C	CA	32	30
A0	D4	86	29	8B	E2	4A	BE	5D	79	4E	70	69	DF	DC	F1	1F
B0	9E	41	75	FA	F4	0A	81	91	FE	E6	CE	3A	22	99	13	B6
C0	ED	0F	A4	2E	D7	AB	63	56	F8	8F	38	B4	89	5B	67	1D
D0	18	19	53	1A	2C	54	B2	1B	0E	2D	AA	55	8E	B3	5A	1C
E0	40	F9	09	90	E5	39	98	B5	85	8A	49	5C	4D	68	DB	1E
F0	BB	EE	7E	10	C1	A5	C9	2F	95	D8	C6	AC	60	64	AF	57

QUADRATIC SOLUTION TABLE
FOR FINDING SOLUTION TO $y^2 + y + C = 0$
(INPUT IS C, OUTPUT IS Y1; Y1=0 => NO SOLUTION, ELSE Y2 = Y1 + α^0)

	0	1	2	3	4	5	6	7	8	9	A	B	C	D	E	F
00	01	DB	8F	55	8D	57	03	D9	00	00	00	00	00	00	00	00
10	89	53	07	DD	05	DF	8B	51	00	00	00	00	00	00	00	00
20	00	00	00	00	00	00	00	00	C3	19	4D	97	4F	95	C1	1B
30	00	00	00	00	00	00	00	00	4B	91	C5	1F	C7	1D	49	93
40	00	00	00	00	00	00	00	00	09	D3	87	5D	85	5F	0B	D1
50	00	00	00	00	00	00	00	00	81	5B	0F	D5	0D	D7	83	59
60	CB	11	45	9F	47	9D	C9	13	00	00	00	00	00	00	00	00
70	43	99	CD	17	CF	15	41	9B	00	00	00	00	00	00	00	00
80	FF	25	71	AB	73	A9	FD	27	00	00	00	00	00	00	00	00
90	77	AD	F9	23	FB	21	75	AF	00	00	00	00	00	00	00	00
A0	00	00	00	00	00	00	00	00	3D	E7	B3	69	B1	6B	3F	E5
B0	00	00	00	00	00	00	00	00	B5	6F	3B	E1	39	E3	B7	6D
C0	00	00	00	00	00	00	00	00	F7	2D	79	A3	7B	A1	F5	2F
D0	00	00	00	00	00	00	00	00	7F	A5	F1	2B	F3	29	7D	A7
E0	35	EF	BB	61	B9	63	37	ED	00	00	00	00	00	00	00	00
F0	BD	67	33	E9	31	EB	BF	65	00	00	00	00	00	00	00	00

CUBE ROOT TABLE
(INPUT IS α^n, OUTPUT IS $\alpha^{n/3}$; EXCEPT FOR α^0, OUTPUT=0 => NO ROOT)

	0	1	2	3	4	5	6	7	8	9	A	B	C	D	E	F
00	00	DB	00	EC	00	98	00	00	02	00	00	00	00	00	0D	1C
10	00	45	36	34	00	00	00	00	A9	00	80	00	00	00	00	00
20	00	00	00	00	00	00	00	00	41	00	00	00	9A	00	D5	00
30	00	82	69	D9	00	D8	10	00	00	00	00	D1	00	00	4F	00
40	04	00	A2	B1	00	00	00	00	00	00	48	00	00	97	00	00
50	00	00	3B	70	51	24	A5	46	00	00	8C	00	00	00	1B	40
60	00	00	00	00	00	00	00	BC	00	00	00	07	00	00	F7	00
70	1A	00	76	00	D4	D0	00	00	38	00	E0	00	00	00	00	BB
80	00	9E	00	00	00	00	00	00	8A	00	5F	00	D7	00	CA	00
90	6C	00	00	00	00	00	4C	00	68	00	00	00	12	00	00	F3
A0	00	00	00	00	00	00	00	AF	00	D3	00	09	00	00	00	00
B0	00	00	90	00	00	00	6A	00	00	00	00	00	00	4D	00	00
C0	23	20	00	00	00	E5	5E	00	00	00	00	0E	00	00	00	00
D0	71	00	00	00	00	DF	00	E2	00	C1	00	00	00	00	EF	00
E0	00	D2	08	9F	00	BE	00	00	00	C3	00	00	00	00	EA	B5
F0	00	00	35	00	00	65	26	00	00	75	13	00	2F	00	00	CF

AN ALTERNATIVE FINITE FIELD PROCESSOR DESIGN

The finite field processor shown below could be used instead of the one shown earlier in this section. It uses subfield multiplication; see Section 2.7 for more information. The timing for finite field multiplication includes only one ROM delay. This path for the other processor included two ROM delays and a binary adder delay. Inversion is accomplished with a ROM table.

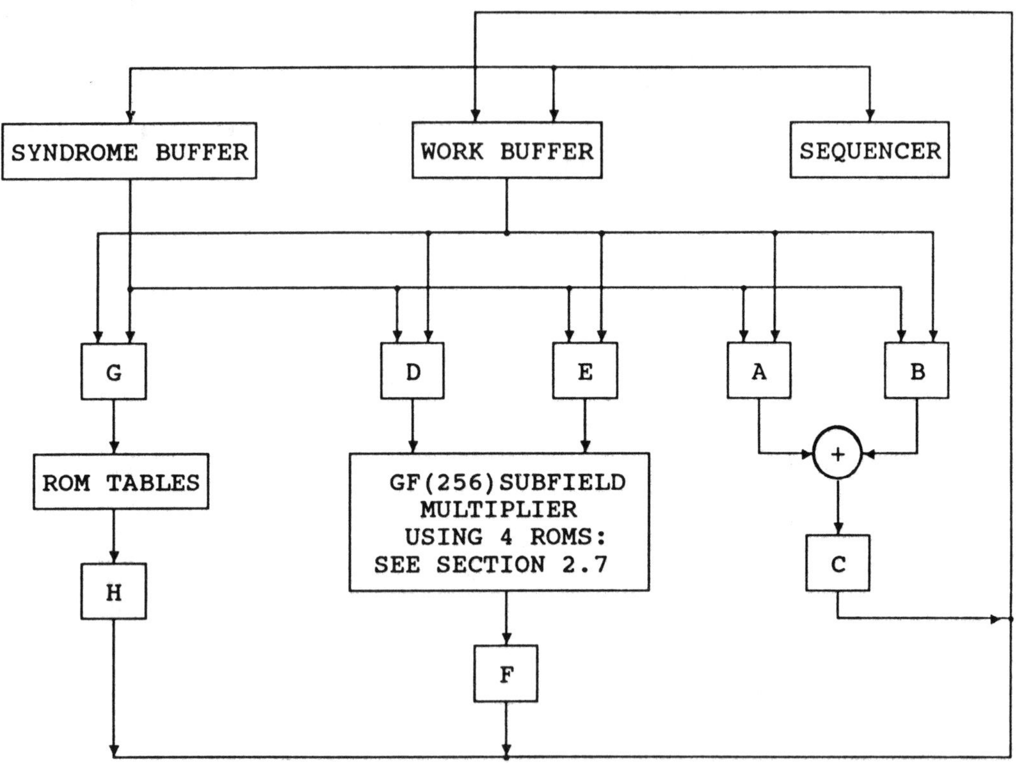

The following pages show, for this alternative finite field processor:

- A ROM table for the four multipliers comprising the GF(256) subfield multiplier.
- A ROM table for accomplishing inversion.
- ROM tables for taking logarithms and antilogarithms.
- A ROM table for finding roots of the finite field equation $y^2 + y + c = 0$.
- A ROM table for finding cube roots.

SUBFIELD MULTIPICATION TABLE
(INPUT IS TWO 4-BIT NIBBLES, OUTPUT IS ONE 4-BIT NIBBLE)

	0	1	2	3	4	5	6	7	8	9	A	B	C	D	E	F
0	0	0	0	0	0	0	0	0	0	0	0	0	0	0	0	0
1	0	1	2	3	4	5	6	7	8	9	A	B	C	D	E	F
2	0	2	4	6	8	A	C	E	9	B	D	F	1	3	5	7
3	0	3	6	5	C	F	A	9	1	2	7	4	D	E	B	8
4	0	4	8	C	9	D	1	5	B	F	3	7	2	6	A	E
5	0	5	A	F	D	8	7	2	3	6	9	C	E	B	4	1
6	0	6	C	A	1	7	D	B	2	4	E	8	3	5	F	9
7	0	7	E	9	5	2	B	C	A	D	4	3	F	8	1	6
8	0	8	9	1	B	3	2	A	F	7	6	E	4	C	D	5
9	0	9	B	2	F	6	4	D	7	E	C	5	8	1	3	A
A	0	A	D	7	3	9	E	4	6	C	B	1	5	F	8	2
B	0	B	F	4	7	C	8	3	E	5	1	A	9	2	6	D
C	0	C	1	D	2	E	3	F	4	8	5	9	6	A	7	B
D	0	D	3	E	6	B	5	8	C	1	F	2	A	7	9	4
E	0	E	5	B	A	4	F	1	D	3	8	6	7	9	2	C
F	0	F	7	8	E	1	9	6	5	A	2	D	B	4	C	3

INVERSE TABLE FOR ALTERNATIVE FINITE FIELD PROCESSOR
(INPUT IS α^n, OUTPUT IS $1/\alpha^n$)

	0	1	2	3	4	5	6	7	8	9	A	B	C	D	E	F
00	--	01	0C	08	06	0F	04	0E	03	0D	0B	0A	02	09	07	05
10	CC	C0	6A	6C	58	5D	D8	D5	FA	F5	8E	86	3E	3D	76	71
20	66	D7	60	DA	35	F3	36	FC	E4	B5	EA	BE	A4	47	AE	43
30	44	56	53	40	F2	24	26	FD	A8	C3	CF	A2	3F	1D	1C	3C
40	33	52	AF	2F	30	57	A5	2D	DE	72	BD	9E	D3	75	B6	97
50	BB	6F	41	32	69	B0	31	45	14	5C	92	84	59	15	8C	9B
60	22	DB	E3	CA	ED	C6	20	D6	B1	54	12	6D	13	6B	BA	51
70	77	1F	49	DF	D2	4D	1E	70	B9	F7	C8	9D	94	C4	F8	B2
80	DD	AD	E1	FE	5B	93	1B	8F	D0	A7	EF	F1	5E	9A	1A	87
90	AA	E7	5A	85	7C	C5	B7	4F	E9	A0	8D	5F	C9	7B	4B	BC
A0	99	E8	3B	CE	2C	46	D1	89	38	C2	90	E6	DC	81	2E	42
B0	55	68	7F	F9	E5	29	4E	96	F6	78	6E	50	9F	4A	2B	EB
C0	11	CD	A9	39	7D	95	65	EC	7A	9C	63	E2	10	C1	A3	3A
D0	88	A6	74	4C	D9	17	67	21	16	D4	23	61	AC	80	48	73
E0	FF	82	CB	62	28	B4	AB	91	A1	98	2A	BF	C7	64	F0	8A
F0	EE	8B	34	25	FB	19	B8	79	7E	B3	18	F4	27	37	83	E0

ANTILOG TABLE FOR ALTERNATIVE FINITE FIELD PROCESSOR
(INPUT IS n, OUTPUT IS α^n)

	0	1	2	3	4	5	6	7	8	9	A	B	C	D	E	F
00	01	10	12	32	16	72	5E	BA	1F	E2	C5	91	8B	39	A6	CD
10	11	02	20	24	64	2C	E4	A5	FD	27	54	1A	B2	9F	6B	DC
20	13	22	04	40	48	C8	41	58	DA	73	4E	A8	2D	F4	B7	CF
30	31	26	44	08	80	89	19	82	A9	3D	E6	85	D9	43	78	FE
40	17	62	4C	88	09	90	9B	2B	94	DB	63	5C	9A	3B	86	E9
50	75	2E	C4	81	99	0B	B0	BF	4F	B8	3F	C6	A1	BD	6F	9C
60	5B	EA	45	18	92	BB	0F	F0	F7	87	F9	67	1C	D2	F3	C7
70	B1	AF	5D	8A	29	B4	FF	07	70	7E	9E	7B	CE	21	34	76
80	1E	F2	D7	A3	9D	4B	F8	77	0E	E0	E5	B5	EF	15	42	68
90	EC	25	74	3E	D6	B3	8F	79	EE	05	50	5A	FA	57	2A	84
A0	C9	51	4A	E8	65	3C	F6	97	EB	55	0A	A0	AD	7D	AE	4D
B0	98	1B	A2	8D	59	CA	61	7C	BE	5F	AA	0D	D0	D3	E3	D5
C0	83	B9	2F	D4	93	AB	1D	C2	E1	F5	A7	DD	03	30	36	56
D0	3A	96	FB	47	38	B6	DF	23	14	52	7A	DE	33	06	60	6C
E0	AC	6D	BC	7F	8E	69	FC	37	46	28	A4	ED	35	66	0C	C0
F0	C1	D1	C3	F1	E7	95	CB	71	6E	8C	49	D8	53	6A	CC	01

LOG TABLE FOR ALTERNATIVE FINITE FIELD PROCESSOR
(INPUT IS α^n, OUTPUT IS n)

	0	1	2	3	4	5	6	7	8	9	A	B	C	D	E	F
00	--	00	11	CC	22	99	DD	77	33	44	AA	55	EE	BB	88	66
10	01	10	02	20	D8	8D	04	40	63	36	1B	B1	6C	C6	80	08
20	12	7D	21	D7	13	91	31	19	E9	74	9E	47	15	2C	51	C2
30	CD	30	03	DC	7E	EC	CE	E7	D4	0D	D0	4D	A5	39	93	5A
40	23	26	8E	3D	32	62	E8	D3	24	FA	A2	85	42	AF	2A	58
50	9A	A1	D9	FC	1A	A9	CF	9D	27	B4	9B	60	4B	72	06	B9
60	DE	B6	41	4A	14	A4	ED	6B	8F	E5	FD	1E	DF	E1	F8	5E
70	78	F7	05	29	92	50	7F	87	3E	97	DA	7B	B7	AD	79	E3
80	34	53	37	C0	9F	3B	4E	69	43	35	73	0C	F9	B3	E4	96
90	45	0B	64	C4	48	F5	D1	A7	B0	54	4C	46	5F	84	7A	1D
A0	AB	5C	B2	83	EA	17	0E	CA	2B	38	BA	C5	E0	AC	AE	71
B0	56	70	1C	95	75	8B	D5	2E	59	C1	07	65	E2	5D	B8	57
C0	EF	F0	C7	F2	52	0A	5B	6F	25	A0	B5	F6	FE	0F	7C	2F
D0	BC	F1	6D	BD	C3	BF	94	82	FB	3C	28	49	1F	CB	DB	D6
E0	89	C8	09	BE	16	8A	3A	F4	A3	4F	61	A8	90	EB	98	8C
F0	67	F3	81	6E	2D	C9	A6	68	86	6A	9C	D2	E6	18	3F	76

QUADRATIC SOLUTION TABLE FOR ALTERNATIVE FINITE FIELD PROCESSOR
FOR FINDING SOLUTION TO $y^2 + y + C = 0$
(INPUT IS C, OUTPUT IS Y1; Y1=0 => NO SOLUTION, ELSE Y2 = Y1 + α^0)

	0	1	2	3	4	5	6	7	8	9	A	B	C	D	E	F
00	01	0B	11	1B	13	19	03	09	1D	17	0D	07	0F	05	1F	15
10	B5	BF	A5	AF	A7	AD	B7	BD	A9	A3	B9	B3	BB	B1	AB	A1
20	00	00	00	00	00	00	00	00	00	00	00	00	00	00	00	00
30	00	00	00	00	00	00	00	00	00	00	00	00	00	00	00	00
40	00	00	00	00	00	00	00	00	00	00	00	00	00	00	00	00
50	00	00	00	00	00	00	00	00	00	00	00	00	00	00	00	00
60	3D	37	2D	27	2F	25	3F	35	21	2B	31	3B	33	39	23	29
70	89	83	99	93	9B	91	8B	81	95	9F	85	8F	87	8D	97	9D
80	00	00	00	00	00	00	00	00	00	00	00	00	00	00	00	00
90	00	00	00	00	00	00	00	00	00	00	00	00	00	00	00	00
A0	CF	C5	DF	D5	DD	D7	CD	C7	D3	D9	C3	C9	C1	CB	D1	DB
B0	7B	71	6B	61	69	63	79	73	67	6D	77	7D	75	7F	65	6F
C0	F3	F9	E3	E9	E1	EB	F1	FB	EF	E5	FF	F5	FD	F7	ED	E7
D0	47	4D	57	5D	55	5F	45	4F	5B	51	4B	41	49	43	59	53
E0	00	00	00	00	00	00	00	00	00	00	00	00	00	00	00	00
F0	00	00	00	00	00	00	00	00	00	00	00	00	00	00	00	00

CUBE ROOT TABLE FOR ALTERNATIVE FINITE FIELD PROCESSOR
(INPUT IS α^n, OUTPUT IS $\alpha^{n/3}$; EXCEPT FOR α^0, OUTPUT=0 => NO ROOT)

	0	1	2	3	4	5	6	7	8	9	A	B	C	D	E	F
00	00	0B	00	09	00	08	00	00	02	00	00	00	00	00	00	04
10	00	00	00	00	94	CF	00	00	22	20	E2	85	48	4C	00	00
20	5E	00	91	00	00	00	00	00	00	00	00	00	BA	00	1A	00
30	00	11	10	00	4E	00	00	3B	00	00	00	00	82	24	26	6B
40	00	00	00	00	00	00	00	00	8B	00	19	00	E4	00	A6	00
50	00	00	00	99	00	00	90	00	39	D9	00	13	27	41	12	00
60	63	00	00	00	00	00	E9	00	00	00	00	C5	00	5C	00	00
70	DA	00	00	00	00	00	00	F4	00	00	00	73	43	00	00	00
80	00	00	00	17	89	00	54	40	00	00	00	16	81	00	9A	44
90	A5	00	00	00	FD	00	00	00	00	B2	00	00	00	2D	00	00
A0	3D	00	00	00	86	00	00	00	00	00	78	00	00	00	E6	00
B0	00	00	00	00	58	00	2B	00	00	00	00	00	00	DC	00	9F
C0	00	75	00	00	00	00	00	C8	00	00	00	C4	00	72	00	00
D0	00	00	00	FE	62	00	00	00	00	64	00	00	00	00	DB	00
E0	00	00	32	00	00	B7	00	00	00	00	00	A9	31	00	00	00
F0	00	2E	A8	00	CD	88	00	00	00	00	80	9B	00	1F	2C	00

CHAPTER 6 - TESTING OF ERROR-CONTROL SYSTEMS

This chapter is concerned primarily with diagnostic capability for storage device applications. However, the techniques described are adaptable to semiconductor memory, communications, and other applications.

6.1 MICRODIAGNOSTICS

There are several approaches for implementing diagnostics for storage device error-correction circuits. Two approaches are discussed here. The first approach requires the implementation of "read long" and "write long" commands in the controller.

The "read long" command is identical to the normal read command except that check bytes are read as if they were data bytes. The "write long" command is identical to the normal write command except that check bytes to be written are supplied, not generated. They are supplied immediately behind the data bytes.

Use the "read long" command to read a known defect-free data record and its check bytes. XOR into the record a simulated error condition. Write the modified data record plus check bytes back to the storage device using the "write long" command. On read back, using the normal read command, an ECC error should be detected and the correction routines should generate the correct response for the error condition simulated. Repeat the test for several simulated error conditions, correctable and uncorrectable.

It is often desirable to reserve one or more diagnostic records for the testing of error-correction functions. It is important for any diagnostic routines testing these functions to first verify that the diagnostic record is error free.

In some cases, hardware computes syndromes but is not involved in the correction algorithm. The correction algorithm is totally contained in software. In this case, it is easy to get a breakdown between hardware and software failures by testing the software first. Supply syndromes to the software, for which proper responses have been recorded.

Using the second diagnostic approach, the hardware is designed so that, under diagnostic control, data records can be written with the check bytes forced to zero. A data record is selected that would normally cause all check bytes to be zero. Simulated error conditions are XOR'd into this record. The record is then written to the storage device under diagnostic control and check bytes are forced zero. On normal read back of this record, an error should be detected and the proper responses generated.

These techniques apply to error-control systems employing very complex codes as well as those employing simple codes. They apply to the interleaved Reed-Solomon code as well as the Fire code.

6.2 HOST SOFTWARE DIAGNOSTICS

Host testing of error-correction functions can be accomplished by implementing at the host software level either of the diagnostic approaches discussed in Section 6.1.

If the controller corrects data before it is transferred to the host, the host diagnostic software must check that the simulated error condition is corrected in the test record. The entire test record must be checked to verify that the error is corrected and that correct data is not altered. Alternatively, the controller could have a diagnostic status or sense command that transfers error pattern(s) and displacement(s) to the host for checking. However, this is not as protective as checking corrected data.

6.3 VERIFYING AN ECC IMPLEMENTATION

Error-correction implementations should be carefully verified to avoid incorrect operation and the transfer of undetected erroneous data under subtle circumstances. This verification should be performed at the host software level using host level diagnostic commands.

FORCING CORRECTABLE ERROR CONDITIONS

Use the "read long" command to read a known error free data record and its check bytes. XOR into this record a simulated error condition that is guaranteed to be correctable. Write the data record plus check bytes back to the storage device using the "write long" command.

Read back the record just written using the normal read command. Verify that the controller corrected the simulated error condition. Repeat, using many random guaranteed-correctable error conditions.

Some nonrandom error conditions should be forced as well. Select a set of error conditions that is known to test all paths of the error-correction implementation.

FORCING DETECTABLE ERROR CONDITIONS

Repeat the test defined under FORCING CORRECTABLE ERROR CONDITIONS, except use simulated error conditions that exceed guaranteed correction capability but not guaranteed detection capability. An uncorrectable error should be detected for each simulated error condition.

FORCING ERRORS THAT EXCEED DETECTION CAPABILITY

Repeat the test defined under FORCING CORRECTABLE ERROR CONDITIONS, except use simulated error conditions that far exceed both the guaranteed correction and guaranteed detection capabilities. Count the number of correctable and uncorrectable errors reported by the error-correction implementation. The ratio of counts should be approximately equal to the miscorrection probability of the code. Repeat for error conditions known to have a higher miscorrection probability.

6.4 ERROR LOGGING

For implementations where the data is actually corrected by the controller, it may be desirable to include an error-logging capability within the controller. A minimum error-logging capability would count the errors recovered by reread and the errors recovered by error correction. Logging requires the controller to have a method of signaling the host when the counters overflow and a command for offloading counts to the host.

A more sophisticated error log would also store information useful for:

- Reassigning areas of media for repeated errors.
- Retiring media when the number of reassignments exceeds a threshold.
- Isolation of devices writing marginal media. This may require that the physical address of the writing device be part of each record written.
- Hardware failure isolation.

It may be desirable to reserve space for error logging on each storage device.

6.5 SELF-CHECKING

HARDWARE SELF-CHECKING

Hardware self-checking can limit the amount of undetected erroneous data transferred when error-correction circuits fail.

Self-checking should be added to the design if the probability of error-correction circuit failure contributes significantly to the probability of transferring undetected erroneous data. One self-checking method duplicates the error-correction circuits and, on read, verifies that the error latches for both circuits agree. No circuits from the two sets of error-correction hardware share the same IC package. This concept can be extended by having separate sources and/or paths for clocks, power, and ground.

Another self-checking method is called parity predict. It is used for the self-checking of shift registers that are part of an error-correction implementation. On each clock, new parity for each shift register is predicted. The actual parity of each shift register is continuously monitored and at each clock, is compared to the predicted parity. If a difference is found, a hardware check flag is set.

The diagrams below define when parity is predicted to change for four shift-register configurations.

DIVIDE BY g(x), ODD NUMBER OF FEEDBACKS

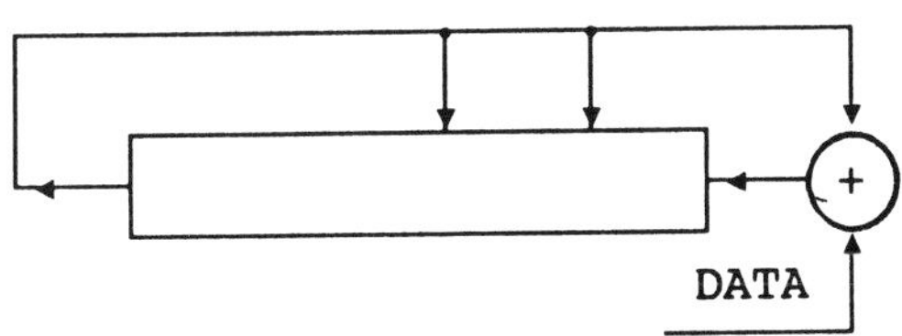

The parity of the shift register will flip each time the data bit is '1'.

DIVIDE BY g(x), EVEN NUMBER OF FEEDBACKS

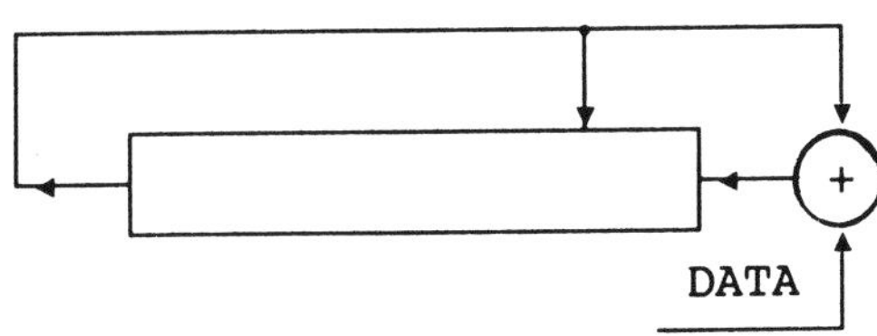

The parity of the shift register will flip if a '1' is shifted out of the shift register, or (exclusive) if the data bit is '1'.

MULTIPLY BY x^m AND DIVIDE BY g(x), ODD # OF FEEDBACKS

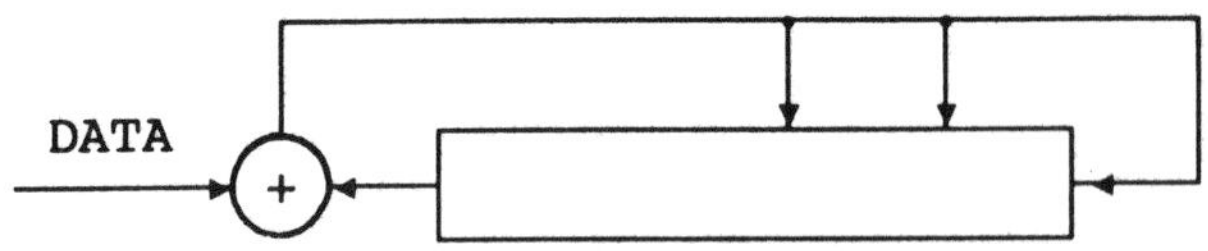

The parity of the shift register will flip if the data bit is '1'.

MULTIPLY BY x^m AND DIVIDE BY g(x), EVEN # OF FEEDBACKS

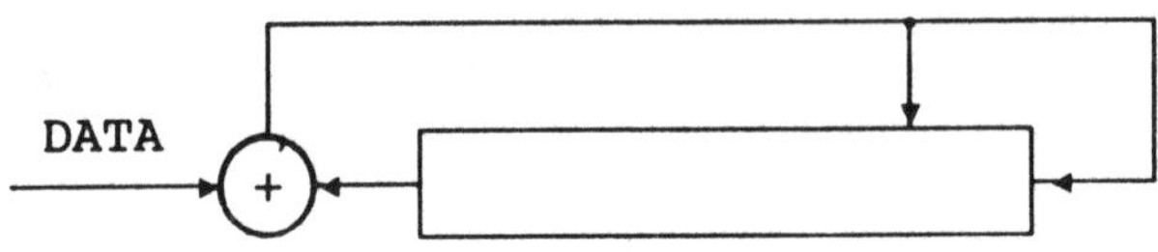

The parity of the shift register will flip if a '1' is shifted out of the shift register.

An m-bit shift register circuit using parity predict for self-checking is shown on the following page. An odd number of feedbacks and premultiplication by x^m is assumed. It is also assumed that the feedbacks are disabled during write check-bit time but not during read check-bit time. While writing data bits, reading data bits, and reading check bits, parity of the shift register is predicted to change for each data bit that is '1'. While writing check bits, parity is predicted to change for each '1' that is shifted out of the shift register.

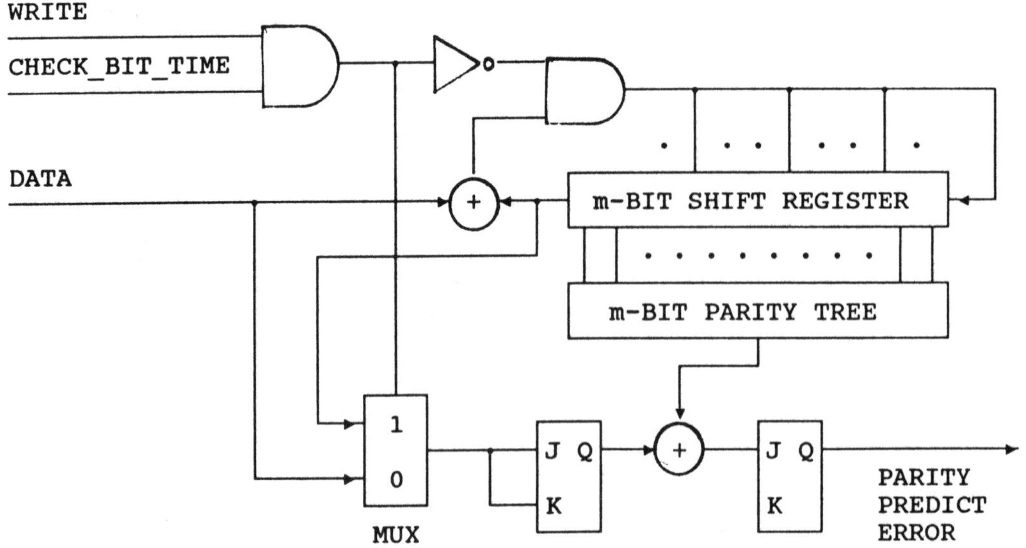

Another technique that aids the detection of error-correction hardware failures is to design the circuits so that nonzero check bytes result when the data is all zeros.

SELF-CHECKING WITH MICROCODE AND/OR SOFTWARE

Periodic microcode and/or software checking is another approach that can be used to limit the amount of undetected erroneous data transferred in case of an error-correction circuit failure. Diagnostic microcode or software could be run on a subsystem power-up and during idle times. These routines would force ECC errors and check for proper detection and correction. In some cases, this approach is the only form of self-checking incorporated in an implementation, even though it is not as protective as self-checking hardware. In other cases, this approach is used to supplement self-checking hardware.

SUPPLEMENTARY PROBLEMS

1. Write the syndrome equations for a three-error-correcting Reed-Solomon code.

2. Write out the error-locator polynomial for errors at locations 0, 3, and 5 for a Reed-Solomon code operating over $GF(2^4)$ defined by $x^4 + x + 1$.

3. Show a Chien search circuit to solve the error-locator polynomial from problem 2.

4. Once error locations for a Reed-Solomon code are known, the syndrome equations become a system of simultaneous linear equations with the error values as unknowns. The error-location vectors are coefficients of the unknown error values. Solve this set of simultaneous linear equations for the two error case.

5. Write out the encode polynomial for a two-error-correcting Reed-Solomon code using $GF(2^4)$ generated by $x^4 + x + 1$.

6. Given a small field generated by the rule $\beta^3 = \beta + 1$ and a large field generated by $\alpha^2 = \alpha + \beta$, develop the rule for accomplishing the square of any element in the large field by performing computation in the small field.

7. Show a complete decoder (on-the-fly, spaced data blocks) for a burst length 2 correcting, shortened cyclic code, using the polynomial $(x^4 + 1) \cdot (x^4 + x + 1)$. Record length is 20 bits, including check bits. Data and check bits are to be buffered in a 20-bit FIFO (first in first out) circuit.

8. Find a polynomial for a code of length 7 that has single-, double-, and triple-bit error detection.

9. For detection of random bit errors on a 32-bit memory word, would it be better to place parity on each byte or use a degree four error-detection polynomial across the entire 32-bit word?

10. A device using a 2048 bit record, including 16 check bits, has a random bit error rate of 1E-4. The 16 check bits are defined by the polynomial below. Can the device meet a 1E-15 specification for Pued (probability of undetected erroneous data)?

    $$x^{16} + x^{12} + x^5 + 1$$
    $$= (x + 1) \cdot (x^{15} + x^{14} + x^{13} + x^{12} + x^4 + x^3 + x^2 + x + 1)$$

11. Compute the probability for three or more error bursts in a block of 256 bytes when the raw burst error rate is 1E-7.

12. Compute the block error probability for a channel using a detection only code when the raw burst error rate is 1E-10.

13. Design a circuit to solve the equation $y^2 + y + C = 0$ for Y when C is given. The field is generated by $x^3 + x + 1$.

14. There is a Fire code in the industry defined by

$$x^{24} + x^{17} + x^{14} + x^{10} + x^3 + 1$$

a) For a correction span of four, determine the detection span using the inequalities for a Fire code.

b) Determine the miscorrection probability for correction span four and record length 259 bytes, (data plus check bytes.)

15. For an error-detection code using the shift register below for encoding and decoding of 2048 byte records:

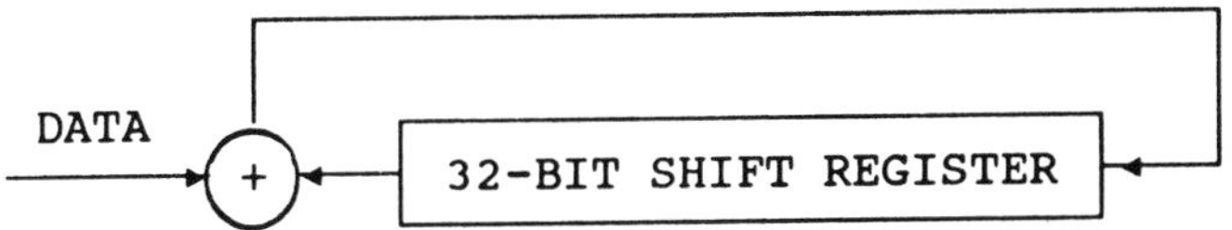

a) Determine the misdetection probability for all possible error bursts.

b) Determine the misdetection probability for all possible double-bit errors.

16. Which of the pairs of numbers below are relatively prime?

```
15 , 45
 9 , 31
 7 , 11
14 , 127
```

17. Write the integer 18 as residues of moduli 5 and 7.

18. Listed below are residues for several integers modulo 5 and 9. Compute the A_i and m_i of the Chinese Remainder Method. Then use the Chinese Remainder Method to determine the integers.

a) a MOD 5 = 4, a MOD 9 = 6, a = ?
b) a MOD 5 = 3, a MOD 9 = 5, a = ?
c) a MOD 5 = 0, a MOD 9 = 4, a = ?

What is the total number of unique integers that can be represented by residues modulo 5 and 9?

19. Define a fast division algorithm for dividing by 255 on an 8-bit processor that does not have a divide instruction. The dividend must be less than 65,536.

20. What is the total number of unique integers that can be represented by residues modulo 6 and 8?

21. Which of the finite field functions listed below are linear?

Log	Square	Antilog
Cube	Square Root	Cube Root
Sixth Power	Eight Root	Inverse
Modulo		

22. Determine the period of the following polynomials:

a) $x^4 + 1$
b) $x^3 + x^2 + x + 1$

23. Compute the reciprocal polynomial of $x^3 + x + 1$.

24. How many primitive polynomials are of degree eight?

25. Compute the residue of x^7 MOD $x^3 + x + 1$.

26. For a small-systems magnetic disk, list several factors influencing data accuracy.

27. Is it possible for a polynomial with an odd number of terms to have a factor of the form $(x^c + 1)$? Why?

28. Describe the difference between error locations and error-location vectors. Which are roots of an error-locator polynomial?

29. What method is used to solve error-locator polynomials of a high degree?

30. What is the difference between errata, errors, and erasures?

31. If g(x) divides $(x^{255} + 1)$, what can be said about the period of g(x)?

32. Given a field generated by $x^4 + x + 1$, show circuits to multiply an arbitrary field element by the following fixed field elements:

a) α^0
b) α^1
c) $\alpha2$

33. For a symbol-error-correcting code (symbol size eight bits) used with a 128 symbol (byte) record, what must the symbol-correcting capability be to have a block error rate less than 1E-8 for a raw symbol error rate of 1E-4? The block error rate is the ratio of block errors to blocks transferred.

34. Show a circuit to implement the equation below in GF(2^8).

$$R2 = R1 + \alpha^0$$

35. In a Reed-Solomon code implementation, it may be necessary to test an equality similar to the one below for true:

$$(S_0)^2 = (S_1) \cdot (S_{-1})$$

Suggest an equivalent test that would not require finite field multiplication or division.

36. Write log and antilog tables for the field generated by $x^3 + x + 1$.

37. Consider a Reed-Solomon code implementation where data is read from a storage device into a buffer. The data is corrected in the buffer and then transferred to a host. Define a way of loading and unloading the buffer such that the finite field processor does not have to take logs of error-location vectors before making corrections to the buffer.

38. Define an algorithm for computing the square root in a field of 15 elements using log and antilog tables.

39. Remember that miscorrection probability is the ratio of valid syndromes to all possible syndromes. Generate a miscorrection formula for a two-symbol-correcting Reed-Solomon code using GF(2^8). The symbol size is eight bits. The record length is 255 bytes, including check bytes.

40. List the first ten entries in an antilog table for a large field. The small field is generated by the rule $\beta^3 = \beta + 1$ and the large field is generated by the rule $\alpha^2 = \alpha + \beta$.

APPENDIX A. PRIME FACTORS OF 2^n-1

n	Factors of 2^n-1						
3	7						
4	3	5					
5	31						
6	3	3	7				
7	127						
8	3	5	17				
9	7	73					
10	3	11	31				
11	23	89					
12	3	3	5	7	13		
13	8191						
14	3	43	127				
15	7	31	151				
16	3	5	17	257			
17	131071						
18	3	3	3	7	19	73	
19	524287						
20	3	5	5	11	31	41	
21	7	7	127	337			
22	3	23	89	683			
23	47	178481					
24	3	3	5	7	13	17	241
25	31	601	1801				
26	3	2731	8191				
27	7	73	262657				
28	3	5	29	43	113	127	
29	233	1103	2089				
30	3	3	7	11	31	151	331
31	2147483647						
32	3	5	17	257	65537		

APPENDIX B

The following paper is included in slightly modified form.

METHODS FOR FINDING LOGARITHMS AND EXPONENTIALS OVER A FINITE FIELD

Neal Glover
DATA SYSTEMS TECHNOLOGY, CORP.
A Subsidiary of Cirrus Logic, Inc.
INTERLOCKEN BUSINESS PARK
100 Technology Drive, Suite 300
Broomfield, Colorado 80021
Phone (303) 466-5228 FAX (303) 466-5482

I.S. Reed, J.P. Huang
Department of Electrical Engineering
UNIVERSITY OF SOUTHERN CALIFORNIA
Los Angeles, California 90007

T.K. Truong
COMMUNICATIONS SYSTEM RESEARCH
Jet Propulsion Laboratory
Pasadena, California 91103

This work was supported in part by NASA contract No. NAS 7-100, in part by the U.S. Air Force Office of Scientific Research, under grant AFOSR-80-0151.

ABSTRACT

Let GF(q) be a finite field, where $q=p^m$ and p is prime. Multiplications are performed often using log and antilog tables of p^m-1 non-zero field elements. It is shown in this paper that for $q=p^{2n}$ and p^n+1 a prime, that the log and the antilog of a field element can be found with two substantially smaller tables of p^n+1 and p^n-1 elements, respectively. The method is based on a use of the Chinese Remainder theorem. This technique results in a significant reduction in the memory requirements of the problem. It is shown more generally that for:

$$q-1 = (p_1)^{r_1} \cdot (p_2)^{r_2} \cdots (p_k)^{r_k} = m_1 \cdot m_2 \cdots m_k,$$

where, $m_i=(p_i)^{r_i}$ for $1 \leq i \leq k$, tables of m_1 elements, m_2 elements, ... , and m_k elements also can be used to find logs and antilogs over GF(q). In the later method, further reductions in the memory requirements are achieved, however, at the expense of a greater number of operations.

I. *INTRODUCTION*

In order to efficiently encode and decode BCH and RS codes over a finite field GF(q), each symbol of GF(q) is representable as a power of a selected primitive element in GF(q), *i.e.*, $\alpha=\tau^i$ for $\alpha, \tau \in$ GF(q) where τ is primitive.

To multiply two field elements $\alpha,\beta \in$ GF(q), where $\alpha=\tau^i$ and $\beta=\tau^j$, one only needs to add i and j modulo (q-1) to obtain the resulting exponent k. That is,

$$\alpha\cdot\beta = \tau^i\cdot\tau^j = \tau^{(i+j) \bmod (q-1)} = \tau^k.$$

In the actual implementation of this multiplication process, a log table can be used to find the exponents. If the field elements are represented in the binary representation, binary addressing is used to locate a logarithm in the table. After the addition of the exponents modulo (q-1), an antilog table is used to find the binary representation of τ^k. The exponent k serves as the address of the antilog table. If q is large for many applications, such log and antilog tables may be prohibitively large.

In the next section it is shown that for a q of form p^{2n} where p^n+1 is a prime that substantially smaller tables of sizes p^n+1 and p^n-1 can be used to find the log and antilog of a field element. Since $q-1 = p^{2n}-1 = (p^n+1)\cdot(p^n-1)$ and $(p^n+1,p^n-1) = 1$, the Chinese Remainder theorem can be used to decompose the tables of $p^{2n}-1$ elements into smaller ones of p^n+1 elements and p^n-1 elements respectively. The results obtained from the tables of p^n+1 elements and p^n-1 elements can be recombined to yield the desired log table of $p^{2n}-1$ elements. A similar reduction can be made for the antilog table. The memory requirements of this new method for finding the log and antilog are reduced from $2(p^{2n}-1)$ to $2[p^n+1+p^n-1] = 4p^n$ memory elements.

In Section III a more involved method is developed that yields the logarithm with a minimum memory requirement but with a greater number of operations. Suppose:

$$q-1 = (p_1)^{r_1}\cdot(p_2)^{r_2}\cdots(p_k)^{r_k} = m_1\cdot m_2\cdots m_k,$$

where p_i is prime and $m_i=(p_i)^{r_i}$ for $1 \le i \le k$, $(m_i,m_j)=1$ for $i \neq j$. Then the Chinese Remainder theorem can be used to decompose tables of $p^{2n}-1$ elements into k smaller ones of m_1 elements, m_2 elements, ... and m_k elements, respectively. The log and antilog of a field element can be found by utilizing these k tables.

II. *A LOG AND ANTILOG ALGORITHM OVER $GF(p^{2n})$*

Let β be a primitive element in $GF(p^n)$ and $x \in GF(p^n)$. Also let m be the least integer such that $x=\beta^m$.

Definition 1. m is called the logarithm of x to base β, *i.e.*, $m=\log_\beta x$.

Theorem 1. Let β be a primitive element in $GF(p^n)$ such that the polynomial $p(x)=x^2+x+\beta$ is irreducible in this field. Also let $\alpha \in GF(p^{2n})$, where $GF(p^{2n})$ is an extension field of $GF(p^n)$. If α is a root of $p(x)$, *i.e.*, $p(\alpha)=0$, and p^n+1 a prime, then α is primitive in $GF(p^{2n})$.

Proof. If α is a root of p(x), its conjugate $\bar{\alpha}$ is also a root of p(x), where $\bar{\alpha}=\alpha^{p^n}$. Thus:

$$(x+\alpha)*(x+\bar{\alpha}) = x^2+(\alpha+\bar{\alpha})x+\alpha\cdot\bar{\alpha}$$

$$= x^2+x+\beta$$

It follows that:

$$\alpha+\bar{\alpha} = 1, \quad \alpha\cdot\bar{\alpha} = \beta \tag{1}$$

and:

$$\alpha^{p^n+1}=\beta \tag{2}$$

Now $(p^{2n}-1)=(p^n+1)\cdot(p^n-1)$ and p^n+1 is a prime by hypothesis. Hence, any number r such that $r|p^{2n}-1$ implies that $r|p^n-1$. Then, from (2):

$$\alpha^{(p^{2n}-1)/r} = (\alpha^{p^n+1})^{(p^n-1)/r}$$

$$= \beta^{(p^n-1)/r} \tag{3}$$

Since β is primitive over $GF(p^n)$, p^n-1 is least integer such that $\beta^{(p^n-1)/r}=1$. Hence:

$$\beta^{(p^n-1)/r} \neq 1 \text{ unless } r=1.$$

Thus, by (2) $\alpha^{(2p^n-1)/r} \neq 1$ unless $r=1$. Therefore, the order of α is $p^{2n}-1$ and α is primitive in $GF(p^{2n})$.

Q.E.D.

The above theorem guarantees the root α to be primitive in the extension field only when p^n+1 is a prime. To show that the theorem is not generally true for p^n+1 not a prime, consider the following counter example: Let $GF(11^2)$ be the extension field of GF(11). It is verified readily that $\beta=1/2 \in GF(11)$ and is a primitive element in this field. Also:

$$p(x)=x^2-x+1/2$$

is irreducible in GF(11). Suppose α is a solution to p(x) and $\alpha \in GF(11^2)$. Then $\alpha^2-\alpha+1/2=0$. From this equation it is seen that $\alpha^4=-1/4=1/7 \in GF(11)$. Since $\alpha^4 \in GF(11)$, $(\alpha^4)^{10}=1$. Thus, $\alpha^{40}=1$ and α is not a primitive element in $GF(11^2)$.

Definition 2. For $\alpha \in GF(p^{2n})$ and $a+\alpha b \in GF(p^{2n})$, where $a,b \in GF(p^n)$, the norm of $a+\alpha b$ is:

$$||a+\alpha b|| = (a+\alpha b)\cdot\overline{(a+\alpha b)}$$

Using the results of Theorem 1 and Definition 1 and Definition 2, the following theorem is demonstrated.

Theorem 2. Let β be a primitive element in $GF(p^n)$ such that the quadratic polynomial $x^2+x+\beta$ is irreducible over $GF(p^n)$. Suppose that p^n+1 is prime. Next let α be the root of this polynomial in the extension field $GF(p^{2n}) = \{a+\alpha b | a,b \in GF(p^n)\}$ of $GF(p^n)$. Suppose $\alpha^m= a+\alpha b \in GF(p^{2n})$. The following holds:

$$\log_\beta(a^2+ab+b^2p) \equiv m \bmod (p^n-1).$$

(The proof is given in Section V.)

By Theorem 2 one can construct a $\log_\beta$ table of $p^n - 1$ elements by storing the value $m_1 = m \bmod (p^n-1)$, where:

$$1 \leq m_1 \leq p^n-1,$$

at location $a^2+ab+b^2\beta$ such that $\alpha^m = a+\alpha b$. Then with a and b known, one can find m_1 using the $\log_\beta$ table. A $\log_\beta$ table is given in Section VI for $p^n-1=15$. Similarly, the antilog_β table is constructed by storing the binary representation of $a^2+ab+b^2\beta$ at location m_1 such that $\alpha^m = a+\alpha b$ and:

$$\text{antilog}_\beta(m_1) = a^2+ab+b^2\beta = x \tag{4}$$

An antilog_β table is also given in Section VI for $p^n-1=15$. Next, the constructions of tables of p^n+1 elements is shown.

Theorem 3. Let $\tau = \alpha^{p^n-1} \in GF(p^{2n})$, where α is primitive in $GF(p^{2n})$. Suppose $\alpha^m = a+\alpha b \in GF(p^{2n})$ for some $a,b \in GF(p^n)$. Then:

$$\log_\tau \left[\frac{a+\bar{\alpha}b}{a+\alpha b} \right] \equiv m \bmod (p^n+1)$$

(The proof is given in Section V.)

Using the results of Theorem 3, let:

$$f(a/b) = \tau^m = \frac{(a/b)+\bar{\alpha}}{(a/b)+\alpha} = \frac{a+\bar{\alpha}b}{a+\alpha b}$$

To construct the $\log_\tau$ table, notice that when a=0:

$$f(a/b) = \tau^m = \alpha^{p^n-1} = \tau$$

and m=1. For $m_2 \equiv \bmod (p^n+1)$, one has $m_2=1$ when a=0. When b=0:

$$f\ (a/b) = \frac{a+0}{a+0} = 1.$$

Thus, m=0 and $m_2=0$. The remaining part of the $\log_\tau$ table can then be constructed by storing the value $m_2 \equiv m \bmod (p^n+1)$ at location a/b for $a^m = a+\alpha b$, where $2 \le m_2 \le p^n$. A $\log_\tau$ table for $p^n+1=17$ is given in Section VII. Also, given there is an antilog_τ table for $p^n+1=17$. It is constructed by storing the binary representation of (a/b) ϵ $\{\beta^1, \beta^2, \cdots \beta^{15}\}$ at the corresponding location $i=m_2$ for $2 \le i \le 16$. Thus:

$$\text{Antilog}_\tau(m_2) = a/b = y. \tag{5}$$

From (4) and (5) the following two simultaneous equations need to be solved for a and b in order to reconstruct $\alpha^m = a+\alpha b$:

$$\left[\begin{array}{l} a^2+ab+b^2\beta = x \\ a/b = y \end{array}\right. \tag{6}$$

Relations (6) yield the following solution:

$$b = \left[\frac{x}{y^2+y+\beta} \right]^{1/2} \tag{7}$$

$$a = b \cdot y \tag{8}$$

For b ϵ GF(p^n) it is verified readily that:

$$b = \text{antilog}_\beta \left[\frac{\log_\beta z}{2} \right] \tag{9}$$

where:

$$z = \frac{x}{y^2+y+\beta}$$

Now, the logarithm of $\alpha^m = a+\alpha b$ ϵ GF(p^{2n}), where a,b ϵ GF(p^n) and α ϵ GF(p^{2n}) is primitive, can be found in terms of m_1 and m_2 by using the tables of p^n-1 elements and p^n+1 elements, respectively. Then the Chinese Remainder theorem warrants that:

$$m = m_1 \cdot n_1 \cdot (n_1)^{-1} + m_2 \cdot n_2 \cdot (n_2)^{-1} \tag{10}$$

where:

$$p^{2n}-1 = (p^n+1) \cdot (p^n-1)$$

$$= n_1 \cdot n_2$$

and $(n_1)^{-1}$ and $(n_2)^{-1}$ are the smallest numbers such that:

$$n_1 \cdot (n_1)^{-1} \equiv 1 \bmod n_2$$

$$n_2 \cdot (n_2)^{-1} \equiv 1 \bmod n_1$$

To recapitulate, the following algorithms for the log and antilog are given:

(a) THE LOG ALGORITHM

Given $a^m = a+\alpha b$ find m as follows:

1. Compute: $x = a^2+ab+b^2\beta$

   $y = a/b$

2. Use the $\log_\beta$ table to find $m_1 = \log_\beta(x)$ and the $\log_\tau$ table to find $m_2 = \log_\tau(y)$ for $a \neq 0$, $b \neq 0$.

3. By equation (10):

$$m = m_1 \cdot n_1 \cdot (n_1)^{-1} + m_2 \cdot n_2 (n_2)^{-1}.$$

(b) THE ANTILOG ALGORITHM

Given m, recover $\alpha^m = a + \alpha b$ as follows:

1. Compute: $m_1 = m \bmod (p^n - 1)$ and $m_2 = m \bmod (p^n + 1)$.

2. Use the antilog tables to find:

$$\text{antilog}_\beta(m_1) = x = a^2 + ab + b^2\beta, \text{ and}$$

$$\text{antilog}_\tau(m_2) = y = a/b, \text{ for } m_2 \neq 0,1.$$

3. Use the equation (9):

$$b = \text{antilog}_\beta \left[\frac{\log_\beta z}{2} \right]$$

where:

$$z = \frac{x}{y^2 + y + \beta}$$

Then:

$$a = b \cdot y$$

To illustrate the above procedures, the following examples are given over $GF(2^8)$.

Example 1: Given $\alpha^{127}=(0,1,1,0)+\alpha(1,1,1,0) \in GF(2^8)$. Then, $a=(0,1,1,0)$ and $b=(1,1,1,0)$. By the LOG algorithm:

$$x = a^2+ab+b^2\beta$$

$$= (1,1,1,0)$$

$$y = a/b = (1,1,1,1)$$

Now use Tables VI.1 and VI.3 to find m_1 and m_2, respectively. The results are $m_1=7$ and $m_2=8$. For this example, $n_1=17$, $n_2=15$, $(n_1)^{-1}=8$, $(n_2)^{-1}=8$, $n_1 \cdot (n_1)^{-1}=136$ and $n_2 \cdot (n_2)^{-1}=120$. By equation (10):

$$m = (136 \cdot m_1 + 120 \cdot m_2) \bmod (2^8-1)$$

$$= 127$$

Example 2. Given $m=127$, find $\alpha^{127}= a+\alpha b \in GF(2^8)$. Using the ANTILOG algorithm:

$$m_1 = m \bmod (p^n-1) = 7$$

$$m_2 = m \bmod (p^n+1) = 8$$

Then use Tables VI.2 and VI.4 to find x and y, respectively. The results are:

$x = (1,1,1,0)$ and $y = (1,1,1,1)$.

Thus:

$$z = \frac{x}{y^2+y+\beta} = (0,0,1,1)$$

By equation (9):

$$b = \text{antilog}_\beta - \left[\frac{\log_\beta(z)}{2} \right] = \text{antilog}_\beta$$

and:

$$b = (1,1,1,0)$$

Thus:

$$a = b \cdot y = (0,1,1,0)$$

Therefore:

$$\alpha^{127} = (0,1,1,0) + \alpha(1,1,1,0).$$

III. *A GENERAL ALGORITHM FOR FINDING LOG AND ANTILOG OVER GF(q)*

Consider a Galois field GF(q) and suppose that:

$$q-1 = (p_1)^{r_1} \cdot (p_2)^{r_2} \cdots (p_k)^{r_k} = n_1 \cdot n_2 \cdots n_k$$

where p_i is prime and $n_i=(p_i)^{r_i}$ for $1 \le i \le k$. Let $\alpha \in GF(q)$ be primitive. Then any field element of GF(q) can be represented by α^i for some i, where $1 \le i \le q-1$. By the Chinese Remainder theorem an exponent i is mapped onto ($i_1 \bmod n_1$, $i_2 \bmod n_2$, ..., $i_k \bmod n_k$). Then a primitive element α is expressed in the notation of the Chinese Remainder theorem as follows:

$$\begin{aligned} \alpha^1 &= \alpha^{(1 \bmod n_1,\ 1 \bmod n_2, \cdots, 1 \bmod n_k)} \\ &= \alpha^{(1,0,0,\ldots,0)}, \alpha^{(0,1,0\ldots,0)}, \ldots, \alpha^{(0,0,\ldots,0,1)} \end{aligned} \tag{11}$$

Here:

$$\alpha^{(0,0,\ldots,0,1,0,\ldots,0)} \equiv \tau_j \tag{12}$$

where the integer 1 in the exponent is in the location j. Element τ_j is an n_j-th root of unity. It follows from (11) that:

$$\alpha^m = (\tau_1)^m \cdot (\tau_2)^m, \ldots, (\tau_k)^m \tag{13}$$

for all integers m where τ_j is a primitive n_j-th root of unity.

By (12) and the reconstruction of the Chinese Remainder theorem:

$$\tau_j = \alpha^{m_j \cdot (m_j)^{-1}} \tag{14}$$

where:

$$m_j \cdot (m_j)^{-1} \equiv 1 \bmod n_j$$

and:

$$n_j \cdot m_j = q-1 \tag{15}$$

Now, suppose one computes $(\alpha^m)^{m_j \cdot (m_j)^{-1}}$ for any j such that $1 \le j \le k$. Observe by (15) that one has:

$$\begin{aligned} m \cdot m_j \cdot (m_j)^{-1} &\equiv (C_j + a \cdot n_j) \cdot m_j \cdot (m_j)^{-1} \bmod q-1 \\ &\equiv C_j \cdot m_j \cdot (m_j)^{-1} \bmod q-1 \end{aligned} \tag{16}$$

where $C_j \equiv m \bmod n_j$ for $1 \le j \le k$ and $m = C_j + a \cdot n_j$ for some integer a. Then, by (14) it follows that:

$$\begin{aligned} (\alpha^m)^{m_j \cdot (m_j)^{-1}} &= \left[\alpha^{m_j \cdot (m_j)^{-1}}\right]^{C_j} \bmod q-1 \\ &= (\tau_j)^{C_j} \end{aligned} \tag{17}$$

Therefore, by the use of equation (17) one can compute $(\tau_j)^{C_j}$ from α^m for $1 \le j \le k$.

Note that k small tables, each containing the value C_i, for $1 \le i \le n_j$, at location $(\tau_j)^{C_i}$, for $1 \le j \le k$, can be used to find the k exponents $C_1, C_2, \ldots, C_k$, respectively. Once the C_j are found, the Chinese Remainder theorem is used to compute the logarithm of α^m as follows:

$$m = \left[\sum_{i=1}^{k} C_i \cdot m_i \cdot (m_i)^{-1} \right] \bmod q-1 \tag{18}$$

By (13), (14), (16), and (17), the antilog of m is computed as:

$$\alpha^m = (\tau_1)^{C_1} \cdot (\tau_2)^{C_k}, \ldots, (\tau_k)^{C_k} \tag{19}$$

where $C_i \equiv m \bmod n_i$ for $1 \le i \le k$. Tables of $(\tau_j)^{C_i}$ for $1 \le j \le 3$ for $q-1 = 255 = 3x5x17 = n_1 \cdot n_2 \cdot n_3$ are given in Section VII.

Example 3: To demonstrate the general algorithm above, the logarithm of α^{20} is computed for $\alpha, \alpha^{20} \in GF(2^8)$ where α satisfies $x^8+x^4+x^3+x^2+1$. With an exponent of 20 given, the antilog α^{20} is recovered. In this case, $n_1=3$, $n_2=5$, $n_3=17$, $m_1 \cdot (m_1)^{-1} = 85$, $m_2 \cdot (m_2)^{-1} = 51$, and $m_3 \cdot (m_3)^{-1} = 120$.

LOGARITHM

Using the tables in Section VII, one finds C_1, C_2, and C_3 from the following computations:

$$(\alpha^{20})^{85} = (\alpha^{85})^{20} = (\tau_1)^{20 \bmod 3} = (\tau_1)^2$$

$$(\alpha^{20})^{51} = (\alpha^{51})^{20} = (\tau_2)^{20 \bmod 5} = (\tau_2)^0$$

$$(\alpha^{20})^{120} = (\alpha^{120})^{20} = (\tau_3)^{20 \bmod 17} = (\tau_3)^3$$

Thus, $C_1=2$, $C_2=0$, and $C_3=3$. The logarithm m is obtained by:

$$m = (2 \cdot 85 + 0 \cdot 51 + 3 \cdot 120) \bmod 255$$

$$= 20.$$

ANTILOG

From m=20, one computes:

$$C_1 \equiv 20 \bmod 3 = 2$$

$$C_2 \equiv 20 \bmod 5 = 0$$

$$C_3 \equiv 20 \bmod 17 = 3$$

Using the tables in Section VII gives:

$$(\tau_1)^2 = \alpha^{170}$$

$$(\tau_2)^0 = \alpha^0$$

$$(\tau_3)^3 = \alpha^{105}$$

Then the antilog of m=20 is recovered as:

$$(\tau_1)^2 \cdot (\tau_2)^0 \cdot (\tau_3)^3 = \alpha^{170} \cdot \alpha^0 \cdot \alpha^{105}$$

$$= \alpha^{(170+0+105) \bmod 255}$$

$$= \alpha^{20}$$

IV. *CONCLUSION*

To find the log and antilog of an element in a finite field GF(q), it $q=p^{2n}$ for some prime p, the technique shown in Section II can be used to reduce the table memory requirement from $2(p^{2n}-1)$ elements to $4p^n$ elements. A further memory reduction can be achieved, *i.e.* from (q-1) elements to:

$$\sum_{i=1}^{k} n_i$$

elements, by using the general method shown in Section III, however, at the expense of a greater number of operations. A comparison of the number of operations needed in these methods is given in Table IV.1. It is evident from Table IV.1 that the number of multiplications required in the general case can be prohibitively large in some situations. Thus, the technique shown in Section II has a better potential than the general method for many practical applications.

Table IV.1

A Complexity Comparison of the
Alternative Approaches for Computing
Logs and Antilogs over GF(q).

	when $q-1=p^{2n}-1$ and p^n+1 is prime		General Method for $q-1=n_1 n_2 \cdots n_k$	
No. of	LOG	ANTILOG	LOG	ANTILOG
Multiplication	7	5	$k+\sum_{j-1}^{k} m_j \cdot (m_j)^{-1*}$	k-1
Additions	4	2	k-1	0
Table Look-Ups	2	4	k	k
Modulus Operations	0	2	1	k

* $m_j \cdot (m_j)^{-1} \equiv 1 \bmod n_j$

V. PROOF OF THEOREMS

Proof of Theorem 2. Since $x^2+x+\beta$ is irreducible over $GF(p^n)$, it has roots α and $\bar{\alpha}=\alpha^{p^n}$ in the extension field $GF(p^{2n})$. By theorem 1, α is primitive in $GF(p^{2n})$. By Definition 2 and relations (1) and (2), one has the following:

$$
\begin{aligned}
||a+\alpha b|| &= (a+\alpha b)\,(\overline{a+\alpha b}) \\
&= (a+\alpha b)\,(a+\bar{\alpha} b) \\
&= a^2+ab+b^2\beta \qquad (V.1)
\end{aligned}
$$

If $c+\alpha d$ is any other element in $GF(p^{2n})$ and $c,d \in GF(p^n)$, then:

$$
\begin{aligned}
\overline{(a+\alpha b)\,(c+\alpha d)} &= (a+\alpha b)^{p^n}\;(c+\alpha d)^{p^n} \\
&= (\overline{a+\alpha b})\,(\overline{c+\alpha d}) \qquad (V.2)
\end{aligned}
$$

Thus, by (V.2) and the definition of the norm, one has:

$$
\begin{aligned}
||(a+\alpha b)\,(c+\alpha d)|| &= (a+\alpha b)\,(c+\alpha d)\,(\overline{a+\alpha b})\,(\overline{c+\alpha d}) \\
&= ||a+\alpha b|| \geq ||c+\alpha d|| \qquad (V.3)
\end{aligned}
$$

Observe next by (2) that:

$$
||\alpha|| = \alpha \geq \bar{\alpha} = \beta
$$

so that the theorem is true for $m=1$. For purposes of induction, assume that:

$$
||\alpha^k|| = \beta^k \qquad (V.4)
$$

for all k such that $1 \leq k \leq m$. Then by (V.3), for $k=m+1$:

$$||\alpha^{m+1}|| = ||\alpha^m|| \geq ||\alpha|| = \beta^{m+1}.$$

Hence, the induction is complete and (V.4) is true for all k.

Represent α^m by $a+\alpha b$ for some $a,b \in GF(p^n)$. Then, by (V.1) and (V.4):

$$||\alpha^m|| = \beta^m$$

The theorem follows by the definition of the logarithm and the fact that β has order p^n-1.

Q.E.D.

Proof of Theorem 3. Since α is primitive in $GF(p^{2n})$ and $\tau=\alpha^{p^n-1}$, the order of τ is p^n+1. By the definition of the norm, one has:

$$||\alpha|| = \alpha \geq \alpha^{p^n} = \tau\alpha^2 \qquad (V.6)$$

For purposes of induction, assume that:

$$||\alpha^k|| = \tau^k\alpha^{2k} \qquad (V.7)$$

for $1 \leq k \leq m$. Then, by (V.3) for $m=m+1$:

$$||\alpha^{m+1}|| = ||\alpha^m|| \geq ||\alpha||$$

$$= (\tau^m\alpha^{2m})(\tau\alpha^2)$$

$$= \tau^{m+1}\alpha^{2(m+1)}$$

Hence, the induction is complete and (V.7) is true for all k.

Representing α^m by $a+\alpha b$ for some $a,b \in GF(p^n)$, it follows from (V.6) that:

$$||a+\alpha b|| = \tau^m(a+\alpha b)^2 \qquad \text{(V.8)}$$

Multiplying both sides of (V.8) by $(a+\bar{\alpha}b)^2$ yields:

$$||a+\alpha b|| \; (a+\bar{\alpha}b)^2 = \tau^m(a+\alpha b)^2 \; (a+\bar{\alpha}b)^2$$

$$= \tau^m||a+\alpha b||^2$$

Therefore, from the definition of the norm:

$$\tau^m = \frac{||a+\alpha b|| \; (a+\bar{\alpha}b)^2}{||a+\alpha b||^2}$$

$$= \frac{a+\bar{\alpha}b}{a+\alpha b} \qquad \text{(V.9)}$$

The theorem follows by the definition of the logarithm and the fact that the order of τ is p^n+1.

Q.E.D.

VI. Let $p(x)=x^4+x^3+1$ be irreducible over GF(2) and $\beta \in GF(2_4)$ is a solution of $p(x)$. Then:

β^1

β^2

β^3

$\beta^4 = \beta^3+1$

$\beta^5 = \beta^3+\beta+1$

$\beta^6 = \beta^3+\beta^2+\beta+1$

$\beta^7 = \beta^2+\beta+1$

$\beta^8 = \beta^3+\beta^2+\beta$

$\beta^9 = \beta^2+1$

$\beta^{10} = \beta^3+\beta$

$\beta^{11} = \beta^3+\beta^2+1$

$\beta^{12} = \beta+1$

$\beta^{13} = \beta^2+\beta$

$\beta^{14} = \beta^3+\beta^2$

$\beta^{15} = 1$

Table VI.1
Log_{β}

Location	Content
0 0 0 1	3
0 0 1 0	2
0 0 1 1	14
0 1 0 0	1
0 1 0 1	10
0 1 1 0	13
0 1 1 1	8
1 0 0 0	15
1 0 0 1	4
1 0 1 0	9
1 0 1 1	11
1 1 0 0	12
1 1 0 1	5
1 1 1 0	7
1 1 1 1	6

Table VI.2
Antilog_β

Location	Content
0 0 0 0	0 0 0 0
0 0 0 1	0 1 0 0
0 0 1 0	0 0 1 0
0 0 1 1	0 0 0 1
0 1 0 0	1 0 0 1
0 1 0 1	1 1 0 1
0 1 1 0	1 1 1 1
0 1 1 1	1 1 1 0
1 0 0 0	0 1 1 1
1 0 0 1	1 0 1 0
1 0 1 0	0 1 0 1
1 0 1 1	1 0 1 1
1 1 0 0	1 1 0 0
1 1 0 1	0 1 1 0
1 1 1 0	0 0 1 1
1 1 1 1	1 0 0 0

For the τ table, note the following:

i) If a=0:

$$\tau^{m} = \frac{\alpha^{p^{n}}}{\alpha} = \alpha^{p^{n}-1} = \tau$$

Thus, m=1 and m_2=1.

ii) If b=0:

$$\tau^{m} = \frac{a}{a} = 1$$

Thus, m=0 and m_2=0.

iii) If a$\neq$0, b$\neq$0, for a,b $\in$ GF(2^4)

$$a,b \in \{\beta^{1},\beta^{2},\cdots,\beta^{15}\}$$

$$\text{and } a/b \in \{\beta^{1},\beta^{2},\cdots,\beta^{15}\}$$

Table VI.3
Antilog$_7$

Location	Content
0 0 0 1	14
0 0 1 0	12
0 0 1 1	6
0 1 0 0	2
0 1 0 1	10
0 1 1 0	4
0 1 1 1	9
1 0 0 0	16
1 0 0 1	3
1 0 1 0	5
1 0 1 1	11
1 1 0 0	15
1 1 0 1	7
1 1 1 0	13
1 1 1 1	8

Table VI.4
Antilog$_7$

Location	Content
0 0 1 0	0 1 0 0
0 0 1 1	1 0 0 1
0 1 0 0	0 1 1 0
0 1 0 1	1 0 1 0
0 1 1 0	0 0 1 1
0 1 1 1	1 1 0 1
1 0 0 0	1 1 1 1
1 0 0 1	0 1 1 1
1 0 1 0	0 1 0 1
1 0 1 1	1 0 1 1
1 1 0 0	0 0 1 0
1 1 0 1	1 1 1 0
1 1 1 0	0 0 0 1
1 0 0 0	1 1 0 0
1 0 0 0 0	1 0 0 0

VII. Tables for $m_1=3$, $m_2=5$, and $m_3=17$, where $m=255=m_1 \cdot m_2 \cdot m_3$, when $\alpha^8+\alpha^4+\alpha^3+\alpha^2+1=0$.

	$(\tau_1)^{m1}$	m_2
1	(0 0 0 0 0 0 0 1)	0
α^{85}	(0 1 1 0 1 0 1 1)	1
α^{170}	(1 1 1 0 0 1 0 1)	2

	$(\tau_2)^{m_2}$	m_2
1	(0 0 0 0 0 0 0 1)	0
α^{51}	(0 0 0 0 0 1 0 1)	1
α^{102}	(0 0 1 0 0 0 1 0)	2
α^{153}	(0 1 0 0 1 0 0 1)	3
α^{204}	(1 1 1 0 0 0 0 0)	4

	$(\tau_3)^{m_3}$	m_3
1	(0 0 0 0 0 0 0 1)	0
α^{120}	(1 0 0 1 0 0 1 1)	1
α^{240}	(0 0 0 1 0 1 1 0)	2
α^{105}	(0 0 0 0 1 1 0 1)	3
α^{225}	(0 1 1 1 0 1 1 0)	4
α^{90}	(1 1 1 0 0 0 0 1)	5
α^{210}	(1 0 1 0 0 0 1 0)	6
α^{75}	(1 0 0 0 1 0 0 1)	7
α^{195}	(0 0 1 1 0 0 1 0)	8
α^{60}	(1 1 0 1 0 0 1 0)	9
α^{180}	(0 1 0 0 1 0 1 1)	10
α^{45}	(1 1 1 0 1 1 1 0)	11
α^{165}	(1 1 0 0 0 1 1 0)	12
α^{30}	(0 0 1 1 0 0 0 0)	13
α^{150}	(1 0 1 0 0 1 0 0)	14
α^{15}	(0 0 1 0 0 1 1 0)	15
α^{135}	(1 1 0 1 1 0 1 0)	16

ABBREVIATIONS

BCH	Bose-Chaudhuri-Hocquenghem (code)
BER	Bit-Error Rate
CLK	Clock
CNT	Count
CRC	Cyclic Redundancy Check (code)
DBER	Decoded Bit-Error Rate
DEC	Double-Error Correction
DED	Double-Error Detection
ECC	Error Correcting Code
EDAC	Error Detection And Correction
FBSR	Feedback Shift Register
FEC	Forward Error Correction
FWD	Forward
GF	Galois Field
LFSR	Linear Feedback Shift Register
LSC	Linear Sequential Circuit
LSR	Linear Shift Register
LRC	Longitudinal Redundancy check
RS	Reed-Solomon (code)
REV	Reverse
SR	Shift Register
SEC	Single-Error Correction
TED	Triple-Error Detection
VRC	Vertical Redundancy Check

GLOSSARY

ALARM

Any condition detected by a correction algorithm that prevents correction, such as error-correction capability exceeded. In some cases, alarms will cause the error-control system to try another approach, for example using a different set of pointers.

BINARY SYMMETRIC CHANNEL

A channel in which there is equal probability for an information bit being 1 or 0.

BLOCK CODE

A block code is a code in which the check bits cover only the immediately preceding block of information bits.

BURST ERROR RATE

The number of burst-error occurrences divided by total bits transferred.

BURST LENGTH

The number of bits between and including the first and last bits in error; not all of the bits in between are necessarily in error.

CATASTROPHIC ERROR PROBABILITY (P_c)

The probability that a given defect event causes an error burst which exceeds the correction capability of a code.

CHARACTERISTIC

See Ground Field.

CODE POLYNOMIAL

See Codeword.

CODE RATE

See Rate.

CODE VECTOR

See Codeword.

CODEWORD

A set of data symbols (*i.e.* information symbols or message symbols) together with its associated redundancy symbols; also called a code vector or a code polynomial.

CONCATENATION

A method of combining an inner code and an outer code, to form a larger code. The inner code is decoded first. An example would be a convolutional inner code and a Reed-Solomon outer code.

CONVOLUTIONAL CODE

A code in which the check bits check information bits of prior blocks as well as the immediately preceding block.

CORRECTABLE ERROR

One that can be corrected without rereading.

CORRECTED ERROR RATE

Error rate after correction.

CORRECTION SPAN

The maximum length of an error burst which is guaranteed to be corrected by a burst-correcting code.

CYCLIC CODE

A linear code with the property that each cyclic (end-around) shift of each codeword is also a codeword.

CYCLIC REDUNDANCY CHECK (CRC)

An error-detection method in which check bits are generated by taking the remainder after dividing the data bits by a cyclic code polynomial.

DEFECT

A permanent fault on the media which causes an error burst.

DEFECT EVENT

A single occurrence of a defect regardless of the number of bits in error caused by the defect.

DEFECT EVENT RATE (Pe)

The ratio of total defect events to total bits, having the units of defect events per bit.

DETECTION SPAN

For a single-burst detection code, the single-burst detection span is the maximum length of an error burst which is guaranteed to be detected.

For a single-burst correction code, the single-burst detection span is the maximum length of an error burst which is guaranteed to be detected without possibility of miscorrection.

If a correction code has a double-burst detection span, then each of two bursts is guaranteed to be detected without possibility of miscorrection, provided neither burst exceeds the double-burst detection span.

DISCRETE MEMORYLESS CHANNEL

A channel for which noise affects each transmitted symbol independently, for example, the binary symmetric channel (BSC).

DISTANCE

See Hamming Distance.

ELEMENTARY SYMMETRIC FUNCTIONS

Elementary symmetric functions are the coefficients of the error locator polynomial.

ERASURE

An errata for which location information is known. An erasure has a known location, but an unknown value.

ERASURE CORRECTION

The process of correcting errata when erasure pointers are available. A Reed-Solomon code can correct more errata when erasure pointers are available. It is not necessary for erasure pointers to be available for all errata when erasure correction is employed.

ERASURE LOCATOR POLYNOMIAL

A polynomial whose roots provide erasure-location information.

ERASURE POINTER

Information giving the location of an erasure. Internal erasure pointers might be derived from adjacent interleave error locations. External erasure pointers might be derived from run-length violations, amplitude sensing, timing sensing, etc.

ERRATA LOCATOR POLYNOMIAL

A polynomial whose roots provide errata-location information.

ERRATUM

Either an error or an erasure.

ERROR

An errata for which location information is not known. In general, an error represents two unknowns, error location and value. In the binary case, the only unknown is the location.

ERROR BURST

A clustered group of bits in error.

ERROR LOCATION OR DISPLACEMENT

The distance by some measure (*e.g.*, bits or bytes) from a reference point (*e.g.*, beginning or end of sector or interleave) to the burst. For Reed-Solomon codes, the error location is the log of the error-location vector and is the symbol displacement of the error from the end of the codeword.

ERROR LOCATION VECTOR

Vector form of error location (antilog of error location).

ERROR LOCATOR POLYNOMIAL

A polynomial whose roots provide error-location information.

ERROR VALUE

The error value is the bit pattern which must be exclusive-or-ed (XOR-ed) against the data at the burst location in order to correct the error.

EXPONENT

See Period.

EXTENSION FIELD

See Ground Field.

FIELD

Refer to Section 2.8 for the definition of a field.

FINITE FIELD

A field with a finite number of elements; also called a Galois field and denoted as GF(n) where n is the number of elements in the field.

FORWARD-ACTING CODE

An error-control code that contains sufficient redundancy for correcting one or more symbol errors at the receiver.

FORWARD POLYNOMIAL

A polynomial is called the forward polynomial when it is necessary to distinguish it from its reciprocal polynomial.

GROUND FIELD

A finite field with q elements, GF(q), exists if, and only if, q is a power of a prime. Let $q=p^n$ where p is a prime and n is an integer, then GF(p) is referred to as the ground field and $GF(p^n)$ as the extension field of GF(p).

The prime P is called the characteristic of the field.

GROUP CODE

See Linear Code.

HAMMING DISTANCE

The Hamming distance between two vectors is the number of corresponding symbol positions in which the two vectors differ.

HAMMING WEIGHT

The Hamming weight of a vector is the number of nonzero symbols in the vector.

HARD ERROR

An error condition that persists on re-read; a hard error is assumed to be caused by a defect on the media.

IRREDUCIBLE

A polynomial of degree n is said to be irreducible if it is not divisible by any polynomial of degree greater than zero but less than n.

ISOMORPHIC

If two fields are isomorphic they have the same structure. That is, one can be obtained from the other by some appropriate one-to-one mapping of elements and operations.

LINEAR (GROUP) CODE

A code wherein the EXCLUSIVE-OR sum of every pair of codewords is also a codeword.

LINEAR FUNCTION

A function is said to be linear if the properties below hold:

a. Linearity: $f(a \cdot x) = a \cdot f(x)$
b. Superposition: $f(x+y) = f(x)+f(y)$

LINEARLY DEPENDENT

A set of n vectors is linearly dependent if, and only if, there exists a set of n scalars C_i, not all zero, such that:

$$C_1{*}v_1 + C_2{*}v_2 + \cdots + Cn{*}v_n = 0$$

LINEARLY INDEPENDENT

A set of vectors is linearly independent if they are not linearly dependent. See Linearly Dependent.

LONGITUDINAL REDUNDANCY CHECK (LRC)

A check byte or check word at the end of a block of data bytes or words, selected to make the parity of each column of bits odd or even.

MAJORITY LOGIC

A majority logic gate has an output of one if, and only if, more than half its inputs are ones.

MAJORITY LOGIC DECODABLE CODE

A code that can be decoded with majority logic gates. See Majority Logic.

MINIMUM DISTANCE OF A CODE

The minimum Hamming distance between all possible pairs of codewords. The minimum distance of a linear code is equal to its minimum weight.

MINIMUM FUNCTION

See Minimum Polynomial.

MINIMUM POLYNOMIAL OF α^i

The monic polynomial m(x) of smallest degree with coefficients in a ground field such that $m(\alpha^i)=0$, where α^i is any element of an extension field. The minimum polynomial of α^i is also called the minimum function of α^i.

MINIMUM WEIGHT OF A CODE

The minimum weight of a linear (group) code's non-zero codewords.

MISCORRECTION PROBABILITY (Pmc)

The probability that an error burst which exceeds the guaranteed capabilities of a code will appear correctable to a decoder. In this case, the decoder actually increases the number of errors by changing correct data. Miscorrection probability is determined by record length, total redundancy, and correction capability of the code.

Pmc usually represents the miscorrection probability for all possible error bursts, assuming all errors are possible and equally probable. Some codes, such as the Fire Code, have a higher miscorrection probability for particular error bursts than for all possible error bursts.

MISDETECTION PROBABILITY (Pmd)

The probability that an error burst which exceeds the correction and detection capabilities of a code will cause all syndromes to be zero and thereby go undetected. Misdetection probability is determined by the total number of redundancy bits, assuming that all errors are possible and equally probable.

MONIC POLYNOMIAL

A polynomial is said to be monic if the coefficient of the highest degree term is one.

(n,k) CODE

A block code with k information symbols, n-k check symbols, and n total symbols (information plus check symbols).

A convolutional code with constant length n, code rate R (efficiency), and information symbols k=Rn.

$\begin{bmatrix} n \\ r \end{bmatrix}$

Number of combinations of n objects taken r at a time, without regard to order.

$$\begin{bmatrix} n \\ r \end{bmatrix} = \frac{n!}{r!(n-r)!}$$

n-TUPLE

An ordered set of n field elements a_i, denoted by $(a_1, a_2, \cdots, a_n)$.

ORDER OF A FIELD

The order of a field is the number of elements in the field. The number of elements may be infinite (infinite field) or finite (finite field).

ORDER OF A FIELD ELEMENT

The order e of a field element β is the least positive integer for which $\beta^e=1$. Elements of order 2^n-1 in $GF(2^n)$ are called primitive elements.

PARITY

The property of being odd or even. The parity of a binary vector is the parity of the number of ones the vector contains. Parity may be computed by summing modulo-2 the bits of the vector.

PARITY CHECK CODE

A code in which the encoder accepts a block of information bits and computes for transmission, a set of modulo-2 sums (XOR) across various of these information bits and possibly information bits in prior blocks. A decoder at the receiving point reconstructs the original information bits from the set of modulo-2 sums. Every binary parity-check code is also a linear, or group code. See also Block Code and Convolutional Code.

PERFECT CODE

An e error correcting code over GF(q) is said to be *perfect* if every vector is distance no greater than e from the nearest codeword. Examples are Hamming and Golay codes.

PERIOD

The period of a polynomial P(x) is the least positive integer e such that x^e+1 is divisible by P(x).

POINTER

Location information for an erasure. This information is normally provided by special hardware.

POLYNOMIAL CODE

A linear block code whose codewords can be expressed in polynomial form and are divisible by a generator polynomial. This class of codes includes the cyclic and shortened cyclic codes.

POWER SUM SYMMETRIC FUNCTIONS

The power sum symmetric functions are the syndromes.

PRIME FIELD

A field is called prime if it possesses no subfields except that consisting of the whole field.

PRIME SUBFIELD

The prime subfield of a field is the intersection of all subfields of the field.

PRIME POLYNOMIAL

See Irreducible.

PRIMITIVE POLYNOMIAL

A polynomial is said to be primitive if its period is 2^m-1, where m is the degree of the polynomial.

RANDOM ERRORS

For the purposes of this book, the term 'random errors' refers to an error distribution in which error bursts (defect events) occur at random intervals and each burst affects only a single symbol, usually one bit or one byte.

RATE

The code rate, or rate (R) of a code is the ratio of information bits (k) to total bits (n); information bits plus redundancy. It is a measure of code efficiency.

$$R = \frac{k}{n}$$

RAW BURST ERROR RATE

Burst error rate before correction.

READABLE ERASURE

A suspected erasure that contains no errors.

RECIPROCAL POLYNOMIAL

The reciprocal of a polynomial F(x) is defined as

$$x^m \cdot F(1/x)$$

where m is the degree of F(x).

RECURRENT CODE

See Convolutional Code.

REDUCIBLE

A polynomial of degree n is said to be reducible if it is divisible by some polynomial of a degree greater than 0 but less than n.

RELATIVELY PRIME

If the greatest common divisor of two polynomials is 1, they are said to be relatively prime.

SELF-RECIPROCAL POLYNOMIAL

A polynomial which is equal to its reciprocal polynomial.

SHORTENED CYCLIC CODE

A linear code formed by deleting leading information digits from the code words of a cyclic code. Shortened cyclic codes are not cyclic.

SOFT ERROR

An error that disappears or becomes correctable on re-read; a soft error is assumed to be due, at least in part, to a transient cause such as electrical noise.

SUBFIELD

A subset of a field which satisfies the definition of a field. See Section 2.8 for the definition of a field.

SYNC FRAMING ERROR

When synchronization occurs early or late by one or more bits.

SYNDROME

A syndrome is a vector (a symbol or set of symbols) containing information about an error or errors. Some codes use a single syndrome while others use multiple syndromes. A syndrome is usually generated by taking the EXCLUSIVE-OR sum of two sets of redundant bits or symbols, one set generated on write and one set generated on read.

SYSTEMATIC CODE

A code in which the codewords are separated into two parts, with all information symbols occurring first and all redundancy symbols following.

UNCORRECTABLE ERROR

An error situation which exceeds the correction capability of a code. An uncorrectable error caused by a soft error on read will become correctable on re-read.

UNCORRECTABLE SECTOR

A sector which contains an uncorrectable error.

UNCORRECTABLE SECTOR EVENT RATE

The ratio of total uncorrectable sectors to total bits, having the units of uncorrectable sector events per bit.

UNDETECTED ERRONEOUS DATA PROBABILITY (Pued)

The probability that erroneous data will be transferred and not detected, having the units of undetected erroneous data events per bit. Pued for a code that does not have pattern sensitivity is the product of miscorrection probability (Pmc) of the error correcting code (if present), the misdetection probability (Pmd) of the error detecting code (if present), and the probability of having an error that exceeds guaranteed capabilities of the code (Pe*Pc).

A code with pattern sensitivity will have two undetected erroneous data rates: one for all possible error bursts, and a higher one for the sensitive patterns.

UNREADABLE ERASURE

A suspected erasure that actually contains an error.

UNRECOVERABLE ERROR

Same as hard error.

VERTICAL REDUNDANCY CHECK (VRC)

Check bit(s) on a byte or word selected to make total byte or word parity odd or even.

WEIGHT

The weight of a codeword is the number of non-zero symbols it contains.

BIBLIOGRAPHY

BOOKS

Abramson, N., *Information Theory and Coding*, McGraw-Hill, New York, 1963.

Aho, A., et. al., *The Design and Analysis of Computer Algorithms*, Addison-Wesley, Massachusetts, 1974.

Albert, A., *Fundamental Concepts of Higher Algebra*, 1st ed., University of Chicago Press, Chicago, 1956.

Artin, E., *Galois Theory*, 2nd ed., University of Notre Dame Press, Notre Dame, 1944.

Ash, R., *Information Theory*, Wiley-Interscience, New York, 1965.

Berlekamp, E. R., *Algebraic Coding Theory*, McGraw-Hill, New York, 1968.

Berlekamp, E. R., *A Survey of Algebraic Coding Theory*, Springer-Verlag, New York, 1970.

Berlekamp, E. R., *Key Papers in The Development of Coding Theory*, IEEE Press, New York, 1974.

Bhargava, V., et. al., *Digital Communications by Satellite*, Wiley, New York, 1981.

Birkhoff, G. and T. C. Bartee., *Modern Applied Algebra*, McGraw-Hill, New York, 1970.

Birkhoff, G. and S. MacLane, *A Survey of Modern Algebra*, 4th ed., Macmillan, New York, 1977.

Blake, I. F., *Algebraic Coding Theory: History and Development*, Dowden, Hutchinson & Ross, Pennsylvania, 1973.

Blake, I. F. and R. C. Mullin, *The Mathematical Theory of Coding*, Academic Press, New York, 1975.

Burton, D. M., *Elementary Number Theory*, Allyn & Bacon, Boston, 1980.

Cameron, P. and J. H. Van Lint, *Graphs, Codes and Designs*, Cambridge University Press, Cambridge, 1980.

Campbell, H. G., *Linear Algebra With Applications*, 2nd ed., Prentice-Hall, New Jersey, 1980.

Carlson, A. B., *Communication Systems*, 2nd ed., McGraw-Hill, New York, 1968.

Clark, Jr., G. C. and J. B. Cain, *Error-Correction Coding for Digital Communications*, Plenum Press, New York, 1981.

Cohen, D. I. A., *Basic Techniques of Combinatorial Theory*, Wiley, New York, 1978.

Crouch, R. and E. Walker, *Introduction to Modern Algebra and Analysis*, Holt, Rinehart & Winston, New York, 1962.

Davies, D. W. and D. L. A. Barber, *Communication Networks for Computers*, Wiley, New York, 1973.

Davisson, L. D. and R. M. Gray, *Data Compression*, Dowden, Hutchinson & Ross, Pennsylvania, 1976.

Doll, D. R., *Data Communications: Facilities, Networks, and Systems Design*, Wiley, New York, 1978.

Durbin, J. B., *Modern Algebra: An Introduction*, Wiley, New York, 1979.

Feller, W., *An Introduction to Probability Theory and Its Applications*, 2nd ed., Wiley, New York, 1971.

Fisher, J. L., *Application-Oriented Algebra*, Harper & Row, New York, 1977.

Folts, H. C., *Data Communications Standards*, 2nd ed., McGraw-Hill, New York, 1982.

Forney, Jr., G. D., *Concatenated Codes*, M.I.T. Press, Massachusetts, 1966.

Gallager, R. G., *Information Theory and Reliable Communication*, Wiley, New York, 1968.

Gere, J. M. and W. W. Williams, Jr., *Matrix Algebra for Engineers*, Van Nostrand, New York, 1965.

Gill, A., *Linear Sequential Circuits*, McGraw-Hill, New York, 1967.

Gill, A., *Applied Algebra for the Computer Sciences*, Prentice-Hall, New Jersey, 1976.

Golomb, S., et. al., *Digital Communications with Space Applications*, Peninsula Publishing, Los Altos, California, 1964.

Golomb, S., et. al., *Shift Register Sequences*, Aegean Park Press, Laguna Hills, California, 1982.

Gregg, W. D., *Analog and Digital Communication*, Wiley, New York, 1977.

Hamming, R. W., *Coding and Information Theory*, Prentice-Hall, New Jersey, 1980.

Hardy, G. and E. M. Wright, *An Introduction to the Theory of Numbers*, 5th ed., Clarendon Press, Oxford, 1979.

Herstein, I. N., *Topics in Algebra*, 2nd ed., Wiley, New York, 1975.

Jayant, N. S., *Waveform Quantization and Coding*, IEEE Press, New York, 1976.

Jones, D. S., *Elementary Information Theory*, Clarendon Press, Oxford, 1979.

Kaplansky, I., *Fields and Rings*, 2nd ed., The University of Chicago Press, Chicago, 1965.

Khinchin, A. I., *Mathematical Foundations of Information Theory*, Dover, New York, 1957.

Knuth, D. E., *The Art of Computer Programming*, Vol. 1, 2nd ed., Addison-Wesley, Massachusetts, 1973.

Knuth, D. E., *The Art of Computer Programming*, Vol. 2, Addison-Wesley, Massachusetts, 1969.

Knuth, D. E., *The Art of Computer Programming*, Vol. 3, Addison-Wesley, Massachusetts, 1973.

Kuo, F. F., *Protocols and Techniques for Data Communication Networks*, Prentice-Hall, New Jersey, 1981.

Lathi, B. P., *An Introduction to Random Signals and Communication Theory*, International Textbook Company, Pennsylvania, 1968.

Lin, S., *An Introduction to Error-Correcting Codes*, Prentice-Hall, New Jersey, 1970.

Lipson, J. D., *Elements of Algebra and Algebraic Computing*, Addison-Wesley, Massachusetts, 1981.

Lucky, R. W., et. al., *Principles of Data Communication*, McGraw-Hill, New York, 1968.

MacWilliams, F. J. and N. J. A. Sloane, *The Theory of Error-Correcting*, Vol. 16, North-Holland, Amsterdam, 1977.

Martin, J., *Communications Satellite Systems*, Prentice-Hall, New Jersey, 1978.

Martin, J., *Telecommunications and the Computer*, Prentice-Hall, New Jersey, 1969.

McEliece, R. J., *The Theory of Information and Coding*, Addison-Wesley, Massachusetts, 1977.

McNamara, J. E., *Technical Aspects of Data Communication*, Digital Press, Massachusetts, 1978.

Niven, I., *An Introduction to the Theory of Numbers*, 4th ed., Wiley , New York, 1960.

Owen, F. E., *PCM and Digital Transmission Systems*, McGraw-Hill, New York, 1982.

Peterson, W. W., and E. J. Weldon, Jr., *Error-Correcting Codes*, 2nd ed., MIT Press, Massachusetts, 1972.

Pless, V., *Introduction to the Theory of Error-Correcting Codes*, Wiley, New York, 1982.

Rao, T. R. N., *Error Coding for Arithmetic Processors*, Academic Press, New York, 1974.

Sawyer, W. W., *A Concrete Approach to Abstract Algebra*, Dover, New York, 1959.

Sellers, Jr., F. F., et. al., *Error Detecting Logic for Digital Computers*, McGraw-Hill, New York, 1968.

Shanmugam, K. S., *Digital and Analog Communication Systems*, Wiley, New York, 1979.

Shannon, C. E. and W. Weaver, *The Mathematical Theory of Communication*, University of Illinois Press, Chicago, 1980.

Slepian, D., *Key Papers in The Development of Information Theory*, IEEE Press, New York, 1974.

Spencer, D. D., *Computers in Number Theory*, Computer Science Press, Maryland, 1982.

Stafford, R. H., *Digital Television: Bandwidth Reduction and Communication Aspects*, Wiley-Interscience, New York, 1980.

Stark, H. M, *An Introduction to Number Theory*, The MIT Press, Cambridge, 1970.

Tanenbaum, A. S., *Computer Networks*, Prentice-Hall, New Jersey, 1981.

Viterbi, A. J., *Principles of Digital Communication and Coding*, McGraw-Hill, New York, 1979.

Wakerly, J., *Error Detecting Codes, Self-Checking Circuits and Applications*, North-Holland, New York, 1978.

Wiggert, D., *Error-Control Coding and Applications*, Artech House, Massachusetts, 1978.

Ziemer, R. E., and W. H. Tranter, *Principles of Communications: Systems, Modulation, and Noise*, Houghton Mifflin, Boston, 1976.

IBM TECHNICAL DISCLOSURE BULLETIN

(Chronologically Ordered)

D. C. Bossen, et. al., "Intermittent Error Isolation in a Double-Error Environment." 15 (12), 3853 (May 1973).

W. C. Carter, "Totally Self-Checking Error for K Out of N Coded Data." 15 (12), 3867-3870 (May 1973).

L. R. Bahl and D. T. Tang, "Shortened Cyclic Code With Burst Error Detection and Synchronization Recovery Capability." 16 (6), 2026-2027 (Nov. 1973).

L. R. Bahl and D. T. Tang, "Shortened Cyclic Code With Burst Error Detection and Synchronization Recovery Capability." 16 (6), 2028-2030 (Nov. 1973).

P. Hodges, "Error Detecting Code With Enhanced Error Detecting Capability." 16 (11), 3749-3751 (Apr. 1974).

D. M. Oldham and A. M. Patel, "Cyclical Redundancy Check With a Nonself-Reciprocal Polynomial." 16 (11), 3501-3503 (Apr. 1974).

G. H. Thompson, "Error Detection and Correction Apparatus." 17 (1), 7-8 (June 1974).

R. A. Healey, "Error Checking and Correction of Microprogram Control Words With a Late Branch Field." 17 (2), 374-381 (July 1974).

A. M. Patel, "Coding Scheme for Multiple Selections Error Correction." 17 (2), 473-475 (July 1974).

D. C. Bossen and M. Y. Hsiao, "Serial Processing of Interleaved Codes." 17 (3), 809-810 (Aug. 1974).

K. B. Day and H. C. Hinz, "Error Pointing in Digital Signal Recording." 17 (4), 977-978 (Sept. 1974).

W. C. Carter and A. B. Wadia, "Contracted Reed-Solomon Codes With Combinational Decoding." 17 (5), 1505-1507 (Oct. 1974).

W. D. Brodd and R. A. Donnan, "Cyclic Redundancy Check for Variable Bit Code Widths." 17 (6), 1708-1709 (Nov. 1974).

T. A. Adams, et. al., "Alternate Sector Assignment." 17 (6), 1738-1739 (Nov. 1974).

W. C. Carter, et. al., "Practical Length Single-Bit Error Correction/Double-Bit Error Detection Codes for Small Values of b." 17 (7), 2174-2176 (Dec. 1974).

J. E. Dohermann, "Defect Skipping Among Fixed Length Records in Direct Access Storage Devices." 19 (4), 1424-1426 (Sept. 1976).

R. E. Cummins, "Displacement Calculation of Error Correcting Syndrome Bytes by Table Lookup." 22 (8b), 3809-3810 (Jan. 1980).

R. C. Cocking, et. al., "Self-Checking Number Verification and Repair Techniques." 22 (10), 4673-4676 (Mar. 1980).

P. Hodges, "Error-Detecting Code for Buffered Disk." 22 (12), 5441- 5443 (May 1980).

F. G. Gustavson and D. Y. Y. Yun, "Fast Computation of Polynomial Remainder Sequences." 22 (12), 5580-5581 (May 1980).

V. Goetze, et, al., "Single Error Correction in CCD Memories." 23 (1), 215-216 (June 1980).

J. W. Barrs and J. C. Leininger, "Modified Gray Code Counters." 23 (2), 460-462 (July 1980).

J. L. Rivero, "Program for Calculating Error Correction Code." 23 (3), 986-988 (Aug. 1980).

N. N. Nguyen, "Error Correction Coding for Binary Data." 23 (4), 1525-1527 (Sept. 1980).

J. C. Mears, Jr., "High-Speed Error Correcting Encoder/Decoder." 23 (4), 2135-2136 (Oct. 1980).

G. W. Kurtz, et. al., "Odd-Weight Error Correcting Code for 32 Data Bits and 13 Check Bits." 23 (6), 2338 (Nov.1980).

J. R. Calva, et. al., "Distributed Parity Check Function." 23 (6), 2451-2456 (Nov. 1980).

J. R. Calva and B. J. Good, "Fail-Safe Error Detection With Improved Isolation of I/0 Faults." 23 (6), 2457-2460 (Nov. 1980).

S. G. Katsafouros and D. A. Kluga, "Memory With Selective Use of Error Detection and Correction Circuits." 23 (7a), 2866-2867 (Dec. 1980).

R. A. Forsberg, et. al., "Error Detection for Memory With Partially Good Chips." 23 (7b), 3272-3273 (Dec. 1980).

R. H. Linton, "Detection of Single Bit Failures in Memories Using Longitudinal Parity." 23 (8), 3603-3604 (Jan. 1981).

C. L. Chen, "Error Correcting Code for Multiple Package Error Detection." 23 (8), 3808-3810 (Jan. 1981).

D. C. Bossen, et. al., "Separation of Error Correcting Code Errors and Addressing Errors." 23 (9), 4224 (Feb. 1981).

G. S. Sager and A. J. Sutton, "System Correction of Alpha-Particle- Induced Uncorrectable Error Conditions by a Service Processor." 23 (9), 4225-4227 (Feb. 1981).

W. G. Bliss, et. al., "Error Correction Code." 23 (10), 4629-4632 (Mar.1981).

W. G. Bliss, "Circuitry for Performing Error Correction Calculations on Baseband Encoded Data to Eliminate Error Propagation." 23 (10), 4633-4634 (Mar. 1981).

P. A. Franaszek, "Efficient Code for Digital Magnetic Recording." 23 (11), 5229-5232 (Apr. 1981).

C. L. Chen, "Error Checking of ECC Generation Circuitry." 23 (11), 5055-5057 (Apr. 1981).

C. L. Chen and B. L. Chu, "Extended Error Correction With an Error Correction Code." 23 (11), 5058-5060 (Apr. 1981).

G. G. Langdon, Jr., "Table-Driven Decoder Involving Prefix Codes." 23 (12), 5559-5562 (May 1981).

D. F. Kelleher, "Error Detection for All Errors in a 9-Bit Memory Chip." 23 (12), 5441 (May 1981).

S. W. Hinkel, "Utilization of CRC Bytes for Error Correction on Multiple Formatted Data Strings." 24 (1b), 639-643 (June 1981).

D. A. Gourneau and S. W. Hinkel, "Error Correction as an Extension of Error Recovery on Information Strings." 24 (1b), 651-652 (June 1981).

J. D. Dixon, et. al., "Parity Mechanism for Detecting Both Address and Data Errors." 24 (1b), 794 (June 1981).

A. M. Patel, "Dual-Function ECC Employing Two Check Bytes Per Word." 24,(2), 1002-1004 (July 1981).

D. Meltzer, "CCD Error Correction System." 24 (3), 1392-1396 (Aug. 1981).

I. Jones, "Variable-Length Code-Word Encoder/Decoder." 24 (3), 1514-1515 (Aug. 1981).

D. B. Convis, et. al., "Sliding Window Cross-Hatch Match Algorithm for Spelling Error Correction." 24 (3), 1607-1609 (Aug. 1981).

N. N. Heise and W. G. Verdoorn, "Serial Implementation of b-Adjacent Codes." 24 (5), 2366-2370 (Oct. 1981).

S. R. McBean, "Error Correction at a Display Terminal During Data Verification." 24 (5), 2426-2427 (Oct. 1981).

D. T. Tang and P. S. Yu, "Error Detection With Imbedded Forward Error Correction." 24 (5), 2469-2472 (Oct. 1981).

R. W. Alexander and J. L. Mitchell, "Uncompressed Mode Trigger." 24 (5), 2476-2480 (Oct. 1981).

V. A. Albaugh, et. al., "Sequencer for Converting Any Shift Register Into a Shift Register Having a Lesser Number of Bit Positions." (Oct. 1981).

A. R. Barsness, W. H. Cochran, W. A. Lopour and L. P. Segar, "Longitudinal Parity Generation for Single- Bit Error Correction." 24 (6), 2769-2770 (Nov. 1981).

S. Lin and P. S. Yu, "Preventive Error Control Scheme." 24 (6), 2886-2891 (Nov. 1981).

D. T. Tang and P. S. Yu, "Hybrid Go-Back-N ARQ With Extended Code Block." 24 (6), 2892-2896 (Nov. 1981).

F. Neves and A. K. Uht, "Memory Error Correction Without ECC." 24 (7a), 3471 (Dec. 1981).

E. S. Anolick, et. al., "Alpha Particle Error Correcting Device." 24 (8), 4386 (Jan. 1982).

W. H. McAnney, "Technique for Test and Diagnosis of Shift-Register Strings." 24 (8), 4387-4389 (Jan. 1982).

F. J. Aichelmann, Jr. and L. K. Lange, "Paging Error Correction for Intermittent Errors." 24 (9), 4782-4783 (Feb. 1982).

R. E. Starbuck, "Self-Correcting DASD." 24 (10), 4916 (Mar. 1982).

W. H. Cochran and W. A. Lopour, "Optimized Error Correction/ Detection for Chips Organized Other Than By-1." 24 (10), 5275-5276 (Mar. 1982).

S. Bederman, et. al., "Codes for Accumulated- Error Channels." 24 (11a), 5744-5748 (Apr. 1982).

M. P. Deuser, et. al., "Correcting Errors in Cached Storage Subsystems." 24 (11a), 5347-6214 (Apr. 1982).

P. T. Burton, "Method for Enhancement of Correctability of Recording Data Errors in Computer Direct-Access Storage Devices." 24 (11b), 6213 (Apr. 1982).

A. R. Barsness, et. al., "ECC Memory Card With Built-in Diagnostic Aids and Multiple Usage." 24 (11b), 6173 (Apr. 1982).

NATIONAL TECHNICAL INFORMATION SERVICE

Altman, F. J., et. al., "Satellite Communications Reference Data Handbook." AD-746 165, (July 1972).

Assmus, Jr., E. F. and H. F. Mattson, Jr., "Research to Develop the Algebraic Theory of Codes." AD-678 108, (Sept. 1968).

Assmus, Jr., E. F., et. al., "Error-Correcting Codes." AD-754 234, (Aug. 1972).

Assmus, Jr., E. F., et. al., "Cyclic Codes." AD-634 989, (Apr. 1966).

Assmus, Jr., E. F., et. al., "Research to Develop the Algebraic Theory of Codes." AD-656 783, (June 1967).

Bahl, L. R., "Correction of Single and Multiple Bursts of Error." AD-679 877, (Oct. 1968).

Benelli, G., "Multiple-Burst-Error-Correcting-Codes." N78-28316, (Apr. 1977).

Benice, R. J., et. al., "Adaptive Modulation and Error Control Techniques." AD-484 188, (May 1966).

Brayer, K., "Error Patterns and Block Coding for the Digital High-Speed Autovon Channel." AD-A022 489, (Feb. 1976).

Bussgang, J. J. and H. Gish, "Analog Coding." AD-721 228, (Mar. 1971).

Cetinyilmaz, N., "Application of the Computer for Real Time Encoding and Decoding of Cyclic Block Codes." AD/A-O21 818, (Dec. 1975).

Chase, D., et. al., "Troposcatter Interleaver Study Report." AD/A-008 523, (Feb. 1975).

Chase, D., et. al., "Coding/MUX Overhead Study." AD/A-009 174, (Mar. 1975.

Chase, Dr., D., et. al., "Multi-Sample Error Protection Modulation Study." AD/A-028 985, (May 1976).

Chase, Dr. D., et. al., "Demod/Decoder Integration." AD/A-053 685, (Apr. 1978).

Chien, R. T. and S. W. Ng., "L-Step Majority Logic Decoding." AD-707 877, (June 1970).

Chien, R. T., et. al., "Hardware and Software Error Correction Coding." AD/A-017 377, (Aug. 1975).

Choy, D. M-H., "Application of Fourier Transform Over Finite Fields to Error-Correcting Codes." AD-778 102, (Apr. 1974).

Covitt, A. L., "Performance Analysis of a Frequency Hopping Modem." AD-756 840, (Dec. 1972).

Donnally, W., "Error Probability in Binary Digital FM Transmission Systems." AD/A-056 237, (Feb. 1978).

Ellison, J. T., "Universal Function Theory and Galois Logic Studies." AD-740 849, (Mar. 1972).

Ellison, J. T. and B. Kolman, "Galois Logic Design." AD-717 205, (Oct. 1970).

Forney, Jr,, G., "Study of Correlation Coding." AD-822 106, (Sept. 1967).

Gilhousen, K. S., et. al., "Coding Systems Study for High Data Rate Telemetry Links." N71-27786, (Jan. 1971).

Gish, H., "Digital Modulation Enhancement Study." AD-755 939, (Jan. 1973).

Hamalaninen, J. R. and E. N. Skoog, "Error Correction Coding With NMOS Microprocessors: a 6800-Based 7,3 Reed-Solomon Decoder." AD/A-073 088, (May 1979).

Horn, F. M., "Design Study of Error-Detecting and Error-Correcting Shift Register." N65-21302, (Apr. 1965).

Janoff, N. S., "Computer Simulation of Error-Correcting Codes." AD-777 198, (Sept. 1973).

Kindle, J. T., "Map Error Bit Decoding of Convolutional Codes." AD/A-061 639, (Aug. 1977).

Lee, L., "Concatenated Coding Systems Employing a Unit-Memory Convolutional Code and a Byte-Oriented Decoding Algorithm." N76-31932, (July 1976).

Liu, K. Y., et. al., "The Fast Decoding of Reed-Solomon Codes Using High-Radix Fermat Theoretic Transforms." N77-14057.

Martin, A. F., "Investigation of Bit Interleaving Techniques for Use with Viterbi Decoding Over Differentially Coded Satellite Channels." AD/A-003 807, (July 1974).

Marver, J. M., "Complexity Reduction in Galois Logic Design." AD/A-056 190, "(Dec. 1977).

Massey, J. L., "Joint Source and Channel Coding." AD/A-045 938, (Sept. 1977).

Mitchell, M. E., "Coding for Turbulent Channels." AD-869-973, (Apr. 1970).

Mitchell, M. E. and Colley, L. E., "Coding for Turbulent Channels." AD-869 942, (Apr. 1970).

Mitchell, M. E., et. al., "Coding for Turbulent Channels." AD-869 941, (Apr. 1970).

Morakis, J. C., "Shift Register Generators and Applications to Coding." X-520-68-133, (Apr. 1963).

Muggia, A., "Effect of the Reduction of the Prandtl in the Stagnation Region Past an Axisymmetric Blunt Body in Hypersonic Flow." AD-676 388, (July 1968).

McEliece, R. J., et. al., "Synchronization Strategies for RFI Channels." N77-21123.

Nesenbergs, M. "Study of Error Control Coding for the U. S. Postal Service Electronic Message System." PB-252 689, (May 1975).

Oderwalder, J. P., et. al., "Hybrid Coding Systems Study Final Report." N72-32206, (Sept. 1972).

Paschburg, R. H., "Software Implementation of Error-Correcting Codes." AD-786 542, (Aug. 1974).

Pierce, J. N., "Air Force Cambridge Research Laboratories." AD-744 069, (Mar. 1972).

Reed, I. S., "kth-Order Near-Orthogonal Codes." AD-725 901, (1971).

Reed, I. S. and T. K. Truong, "A Simplified Algorithm for Correcting Both Errors and Erasures of R-S Codes." N79-16012, (Sept./Oct. 1978).

Roome, T. F., "Generalized Cyclic Codes Finite Field Arithmetic." AD/A-070 673, (May 1979).

Rudolph, L. D., "Decoding Complexity Study." AD/A-002 155, (Nov. 1974).

Rudolph, L. D., "Decoding Complexity Study II." AD/A-039 023, (Mar. 1977).

Sarwate, D. V., "A Semi-Fast Fourier Transform Algorithm Over GF(2^m)." AD/A-034 982, (Sept. 1976).

Schmandt, F. D., "The Application of Sequential Code Reduction." AD-771 587, (Oct. 1973).

Sewards, A., et. al., "Forward Error-Correction for the Aeronautical Satellite Communications Channel." N79-19193, (Feb. 1979).

Skoog, E. N., "Error Correction Coding with NMOS Microprocessors: Concepts." AD/A-072 982, (May 1979).

Solomon, G., "Error Correcting Codes for the English Alphabet and Generalizations." AD-774 850, (July 1972).

Solomon G. and D. J. Spencer, "Error Correction/Multiplex for Megabit Data Channels." AD-731 567, (Sept. 1971).

Solomon, G., et. al., "Error Correction. Multiplex for Megabit Data Channels." AD-731 568, (Sept. 1971).

Stutt, C. A., "Coding for Turbulent Channels." AD-869 979, (Apr. 1970).

Tomlinson, M. and B. H. Davies, "Low Rate Error Correction Coding for Channels with Phase Jitter." AD/A-044 658, (Feb. 1977).

Viterbi, A. J., et. al., "Concatenation of Convolutional and Block Codes" N71-32505, (June 1971).

Welch, L. R., et. al., "The Fast Decoding of Reed-Solomon Codes Using Fermat Theoretic Transforms and Continued Fractions." N77-14056.

Wong, J. S. L., et. al., "Review of Finite Fields: Applications to Discrete Fourier Transforms and Reed-Solomon Coding." N77-33875, (July 1977).

. . . . . ,"Coding Investigation for Time Division Multiple Access Communications." AD-766 540, (July 1973).

. . . . . ,"Feedback Communications." AD/A-002 284, (Oct. 1974).

. . . . . ,"Majority Decoding Apparatus for Geometric Codes.", AD-D003 369, (Oct. 1976).

AUDIO ENGINEERING SOCIETY PREPRINTS

Adams, R. W., "Filtering in the Log Domain." 1470 (B-5), (May 1979).

Doi, T. T., "Channel Codings for Digital Audio Recordings." 1856 (I-1), (Oct./Nov. 1981).

Doi, T. T., "A Design of Professional Digital Audio Recorder." 1885 (G-2), (Mar. 1982).

Doi, T. T., et. al., "Cross Interleave Code for Error Correction of Digital Audio Systems." 1559 (H-4), (Nov. 1979).

Doi, Dr. T. T., et. al., "A Long Play Digital Audio Disc System." 1442 (G-4), (Mar. 1979).

Doi, T. T., et. al., "A Format of Stationary-Head Digital Audio Recorder Covering Wide Range of Application." 1677 (H-6), (Oct./Nov. 1980).

Engberg, E. W., "A Digital Audio Recorder Format for Professional Applications." 1413 (F-1), (Nov. 1978).

Fukuda, G. and T. Doi, "On Dropout Compensation of PCM Systems-Computer Simulation Method and a New Error-Correcting Code (Cross Word Code)." 1354 (E-7), (May 1978).

Fukuda, G., et. al., "On Error Correctability of EIAJ-Format of Home Use Digital Tape Recorders." 1560 (G-5), (Nov. 1979).

Furukawa, T., et. al., "A New Run Length Limited Code." 1839 (I-2), (Oct./Nov. 1981).

Inoue, T., et. al., "Comparison of Performances Between IPC Code and RSC Code When Applied to PCM Tape Recorder." 1541 (H-5), (Nov. 1979).

Ishida, Y., et. al., "A PCM Digital Audio Processor for Home Use VTR'S." 1528 (G-6), (Nov. 1979).

Kosaka, M., et. al., "A Digital Audio System Based on a PCM Standard Format." 1520 (G-4), (Nov. 1979).

Lagadec, Dr. R., et. al., "A Digital Interface for the Interconnection of Professional Digital Audio Equipment." 1883 (G-6), (Mar. 1982).

Locanthi, B. N. and M. Komamura, "Computer Simulation for Digital Audio Systems." 1653 (K-4), (May 1980).

Muraoka, T., et. al., "A Group Delay Analysis of Magnetic Recording Systems." 1466 (A-5), (May 1979).

Nakajima, H., et. al., "A New PCM Audio System as an Adapter of Video Tape Recorders." 1352 (E-11), (May 1978).

Nakajima, H., et. al., "Satellite Broadcasting System for Digital Audio." 1855 (L-8), (Oct./Nov. 1981).

Odaka, K., et. al., "LSIs for Digital Signal Processing to be Used in "Compact Disc Digital Audio" Players." 1860 (G-5), (Mar. 1982).

Sadashige, K. and H. Matsushima, "Recent Advances in Digital Audio Recording Technique." 1652 (K-5), (May 1980).

Seno, K., et. al., "A Consideration of the Error Correcting Codes for PCM Recording System." 1397 (H-4), (Nov. 1978).

Tanaka, K., et. al., "2-Channel PCM Tape Recorder for Professional Use." 1408 (F-3), (Nov. 1978).

Tanaka, K., et. al., "Improved Two Channel PCM Tape Recorder for Professional Use." 1533 (G-3), (Nov. 1979).

Tanaka, K., et. al., "On a Tape Format for Reliable PCM Multi-Channel Tape Recorders." 1669 (K-1), (May 1980).

Tanaka, K., et. al., "On PCM Multi-Channel Tape Recorder Using Powerful Code Format." 1690 (H-5), (Oct./Nov. 1980).

Tsuchiya, Y., et. al., "A 24-Channel Stationary-Head Digital Audio Recorder." 1412 (F-2), (Nov. 1978).

Van Gestel, W. J, et. al., " A Multi-Track Digital Audio Recorder for Consumer Applications." 1832 (I-4), (Oct. 1981).

Vries, L. B., "The Error Control System of Philips Compact Disc." 1548 (G-8), (Nov. 1979).

Vries, L. B., et. al., "The Compact Disc Digital Audio System: Mudulation and Error-Correction." 1674 (H-8), (Oct. 1980).

White, L., et. al., "Refinements of the Threshold Error Correcting Algorithm." 1790 (B-5), (May 1981).

Yamada, Y., et. al., "Professional-Use PCM Audio Processor With a High Efficiency Error Correction System." 1628 (G-7), (May 1980).

PATENTS

2,864,078, "Phased, Timed Pulse Generator," Seader, (1958).

2,957,947, "Pulse Code Transmission System," Bowers, (1960).

3,051,784, "Error-Correcting System," Neumann, (1962).

3,162,837, "Error Correcting Code Device With Modulo-2 Adder and Feedback Means," Meggitt, (1964).

3,163,848, "Double Error Correcting System," Abramson, (1964).

3,183,483, "Error Detection Apparatus," Lisowski, (1965).

3,226,685, "Digital Recording Systems Utilizing Ternary, N Bit Binary and Other Self-Clocking Forms," Potter, et al., (1965).

3,227,999, "Continuous Digital Error-Correcting System," Hagelbarger, (1966).

3,242,461, "Error Detection System," Silberg, et al., (1966).

3,264,623, "High Density Dual Track Redundant Recording System," Gabor, (1966).

3,278,729, "Apparatus For Correcting Error-Bursts In Binary Code," Chien, (1966).

3,281,804, "Redundant Digital Data Storage System," Dirks, (1966).

3,281,806, "Pulse Width Modulation Representation of Paired Binary Digits," Lawrance, et al., (1966).

3,291,972, "Digital Error Correcting Systems," Helm, (1966).

3,319,223, "Error Correcting System," Helm, (1967).

3,372,376, "Error Control Apparatus," Helm, (1968).

3,374,475, "High Density Recording System," Gabor, (1968).

3,387,261, "Circuit Arrangement for Detection and Correction of Errors Occurring in the Transmission of Digital Data," Betz, (1968).

3,389,375, "Error Control System," (1968).

3,398,400, "Method and Arrangement for Transmitting and Receiving Data Without Errors," Rupp, et al., (1968).

3,402,390, "System for Encoding and Decoding Information Which Provides Correction of Random Double-Bit and Triple-Bit Errors," Tsimbidis, et al., (1968).

3,411,135, "Error Control Decoding System," Watts, (1968).

3,413,599, "Handling Of Information With Coset Codes," Freiman, (1968).

3,416,132, "Group Parity Handling," MacSorley, (1968).

3,418,629, "Decoders For Cyclic Error-Correcting Codes," (1968).

3,421,147, "Buffer Arrangement," Burton, et al., (1969).

3,421,148, "Data Processing Equipment," (1969).

3,423,729, "Anti-Fading Error Correction System," Heller, (1969).

3,437,995, "Error Control Decoding System," Watts, (1969).

3,452,328, "Error Correction Device For Parallel Data Transmission System," Hsiao, et al., (1969).

3,457,562, "Error Correcting Sequential Decoder," (1969).

3,458,860, "Error Detection By Redundancy Checks," Shimabukuro, (1969).

3,465,287, "Burst Error Detector," (1969).

3,475,723, "Error Control System," Burton, et al., (1969).

3,475,724, "Error Control System," Townsend, et al., (1969).

3,475,725, "Encoding Transmission System," Frey, Jr., (1969).

3,478,313, "System For Automatic Correction Of Burst-Errors," Srinivasan, (1969).

3,504,340, "Triple Error Correction Circuit," Allen, (1970).

3,506,961, "Adaptively Coded Data Communications System," Abramson, et al., (1970).

3,508,194, "Error Detection and Correction System," Brown, (1970).

3,508,195, "Error Detection and Corrections Means," Sellers, Jr., (1970).

3,508,196, "Error Detection and Correction Features," Sellers, Jr., et al., (1970).

3,508,197, "Single Character Error and Burst-Error Correcting Systems Utilizing Convolution Codes," (1970).

3,508,228, "Digital Coding Scheme Providing Indicium AT Cell Boundaries Under Prescribed Circumstances to Facilitate Self-Clocking," Bishop, (1970).

3,519,988, "Error Checking Arrangement for Data Processing Apparatus," Grossman, (1970).

3,533,067, "Error Correcting Digital Coding and Decoding Apparatus," (1970).

3,534,331, "Encoding-Decoding Array," Kautz, (1970).

3,542,756, "Error Correcting," Gallager, (1970).

3,557,356, "Pseudo-Random 4-Level m-Sequences Generators," Balza, et al., (1971).

3,559,167, "Self-Checking Error Checker for Two-Rail Coded Data," (1971).

3,559,168, "Self-Checking Error Checker for k-Out-of-n Coded Data," (1971).

3,560,925, "Detection and Correction of Errors in Binary Code Words," Ohnsorge, (1971).

3,560,942, "Clock for Overlapped Memories With Error Correction," Enright, Jr., (1971).

3,562,711, "Apparatus for Detecting Circuit Malfunctions," Davis, et al., (1971).

3,568,148, "Decoder for Error Correcting Codes," Clark, Jr., (1971).

3,573,728, "Memory With Error Correction for Partial Store Operation," Kolankowsky, et al., (1971).

3,576,952, "Forward Error Correcting Code Telecommunicating System," VanDuuren, et al., (1971).

3,577,186, "Inversion-Tolerant Random Error Correcting Digital Data Transmission System," Mitchell, (1971).

3,582,878, "Mltiple Random Error Correcting System," (1971).

3,582,881, "Burst-Error Correcting Systems," Burton, (1971).

3,585,586, "Facsimile Transmission System," Harmon, et al., (1971).

3,587,090, "Great Rapidity Data Transmission System," Labeyrie, (1971).

3,601,798, "Error Correcting and Detecting Systems," Haize, (1971).

3,601,800, "Error Correcting Code Device for Parallel-Serial Transmissions," Lee, (1971).

3,662,337, "Mod 2 Sequential Function Generator for Multibit Binary Sequence," (1972).

3,622,982, "Method and Apparatus for Triple Error Correction," Clark, Jr., et al., (1971).

3,622,984, "Error Correcting System and Method," (1971).

3,622,985, "Optimum Error-Correcting Code Device for Parallel-Serial Transmissions in Shortened Cyclic Codes," Ayling, et al., (1971).

3,622,986, "Error-Detecting Technique for Multilevel Precoded Transmission," Tang, et al., (1971).

3,623,155, "Optimum Apparatus and Method for Check Bit Generation and Error Detection, Location and Correction," (1971).

3,624,637, "Digital Code to Digital Code Conversions," Irwin, (1971).

3,629,824, "Apparatus for Multiple-Error Correcting Codes," Bossen, (1971).

3,631,428, "Quarter-Half Cycle Coding for Rotating Magnetic Memory System," King, (1971).

3,634,821, "Error Correcting System," (1972).

3,638,182, "Random and Burst Error-Correcting Arrangment with Guard Space Error Correction," Burton, et al., (1972)

3,639,900, "Enhanced Error Detection and Correction for Data Systems," Hinz, Jr., (1972).

3,641,525, "Self-Clocking Five Bit Record-Playback System," Milligan, (1972).

3,641,526, "Intra-Record Resynchronization," Bailey, et al. (1972).

3,648,236, "Decoding Method and Apparatus for Bose-Chaudhuri- Hocquenghem Codes," Burton, (1972).

3,648,239, "System for Translating to and From Single Error Correction-Double Error Detection Hamming Code and Byte Parity Code," (1972).

3,649,915, "Digital Data Scrambler-Descrambler Apparatus for Improved Error Performance," Mildonian, Jr., (1972).

3,662,337, "Mod 2 Sequential Function Generator for Multibit Binary Sequence," Low, et al., (1972).

3,662,338, "Modified Threshold Decoder for Convolutional Codes," (1972).

3,665,430, "Digital Tape Error Recognition Method Utilizing Complementary Information," Hinrichs, et al., (1972).

3,668,631, "Error Detection and Correction System with Statistically Optimized Data Recovery," Griffith, et al., (1972).

3,668,632, "Fast Decode Character Error Detection and Correction System," Oldham III, (1972).

3,671,947, "Error Correcting Decoder," (1972).

3,675,200, "System for Expanded Detection and Correction of Errors in Parallel Binary Data Produced by Data Tracks," (1972).

3,675,202, "Device for Checking a Group of Symbols to Which a Checking Symbol is Joined and for Determining This Checking Symbol," Verhoeff, (1972).

3,685,014, "Automatic Double Error Detection and Correction Device," (1972).

3,685,016, "Array Method and Apparatus for Encoding, Detecting, and/or Correcting Data," (1972).

3,688,265, "Error-Free Decoding for Failure-Tolerant Memories," (1972).

3,689,899, "Run-Length-Limited Variable-Length Coding with Error Propagation Limitation," Franaszek, (1972).

3,697,947, "Character Correcting Coding System and Method for Deriving the Same," (1972).

3,697,948, "Apparatus for Correcting Two Groups of Multiple Errors," Bossen, (1972).

3,697,949, "Error Correction System for Use With a Rotational Single-Error Correction, Double-Error Detection Hamming Code," (1972).

3,697,950, "Versatile Arithmetic Unit for High Speed Sequential Decoder," (1972).

3,699,516, "Forward-Acting Error Control System," Mecklenburg, (1972).

3,701,094, "Error Control Arrangement for Information Comparison," Howell, (1972).

3,714,629, "Double Error Correcting Method and System," (1973).

3,718,903, "Circuit Arrangement for Checking Stored Information," Oiso, et al., (1973).

3,725,859, "Burst Error Detection and Correction System," Blair, et al., (1973).

3,728,678, "Error-Correcting Systems Utilizing Rate 1/2 Diffuse Codes," (1973).

3,742,449, "Burst and Single Error Detection and Correction System," Blair, (1973).

3,745,525, "Error Correcting System," (1973).

3,745,526, "Shift Register Error Correcting System," Hong, et al., (1973).

3,745,528, "Error Correction for Two Tracks in a Multi-Track System," Patel, (1973).

3,753,227, "Parity Check Logic for a Code Reading System," Patel, (1973).

3,753,228, "Synchronizing Arrangement for Digital Data Transmission Systems," Nickolas, et al., (1973).

3,753,230, "Methods and Apparatus for Unit-Distance Counting and Error-Detection," Hoffner II, (1973).

3,755,779, "Error Correction System for Single-Error Correction, Related-Double-Error Correction and Unrelated- Double-Error Detection," (1973).

3,764,998, "Methods and Apparatus for Removing Parity Bits from Binary Words," Spencer, (1973).

3,766,521, "Multiple B-Adjacent Group Error Correction and Detection Codes and Self-Checking Translators Therefor," (1973).

3,768,071, "Compensation for Defective Storage Positions," Knauft, et al., (1973).

3,771,126, "Error Correction for Self-Synchronized Scramblers," (1973).

3,771,143, "Method and Apparatus for Providing Alternate Storage Areas on a Magnetic Disk Pack," Taylor, (1973).

3,774,154, "Error Control Circuits and Methods," Devore, et al., (1973).

3,775,746, "Method and Apparatus for Detecting Odd Numbers of Errors and Burst Errors of Less Than a Predetermined Length in Scrambled Digital Sequences," Boudreau, et al., (1973).

3,777,066, "Method and System for Synchronizing the Transmission of Digital Data While Providing Variable Length Filler Code," Nicholas, (1973).

3,780,271, "Error Checking Code and Apparatus for an Optical Reader," (1973).

3,780,278, "Binary Squaring Circuit," Way, (1973).

3,781,109, "Data Encoding and Decoding Apparatus and Method," Mayer, Jr., et al., (1973).

3,781,791, "Method and Apparatus for Decoding BCH Codes," Sullivan, (1973).

3,786,201, "Audio-Digital Recording System," (1974).

3,786,439, "Error Detection Systems," McDonald, et al., (1974).

3,794,819, "Error Correction Method and Apparatus," Berding, (1974).

3,794,821, "Memory Protection Arrangements for Data Processing Systems," (1974).

3,798,597, "System and Method for Effecting Cyclic Redundancy Checking," Frambs, et al., (1974).

3,800,281, "Error Detection and Correction Systems," Devore, et al., (1974).

3,801,955, "Cyclic Code Encoder/Decoder," Howell, (1974).

3,810,111, "Data Coding With Stable Base Line for Recording and Transmitting Binary Data," (1974).

3,814,921, "Apparatus and Method for a Memory Partial-Write of Error Correcting Encoded Data," Nibby, et al., (1974).

3,818,442, "Error-Correcting Decoder for Group Codes," (1974).

3,820,083, "Coded Data Enhancer, Synchronizer, and Parity Remover System," Way, (1974).

3,825,893, "Modular Distributed Error Detection and Correction Apparatus and Method," (1974).

3,828,130, "Data Transmitting System," Yamaguchi, (1974).

3,831,142, "Method and Apparatus for Decoding Compatible Convolutional Codes," (1974).

3,831,143, "Concatenated Burst-Trapping Codes," Trafton, (1974).

3,832,684, "Apparatus for Detecting Data Bits and Error Bits In Phase Encoded Data," Besenfelder, (1974).

3,842,400, "Method and Circuit Arrangement for Decoding and Correcting Information Transmitted in a Convolutional Code," (1974).

3,843,952, "Method and Device for Measuring the Relative Displacement Between Binary Signals Corresponding to Information Recorded on the Different Tracks of a Kinematic Magnetic Storage Device," Husson, (1974).

3,851,306, "Triple Track Error Correction," (1974).

3,858,179, "Error Detection Recording Technique," (1974).

3,859,630, "Apparatus for Detecting and Correcting Errors in Digital Information Organized into a Parallel Format by Use of Cyclic Polynomial Error Detecting and Correcting Codes," Bennett, (1975).

3,863,228, "Apparatus for Detecting and Eliminating a Transfer of Noise Records to a Data Processing Apparatus," Taylor, (1975).

3,866,170, "Binary Transmission System Using Error-Correcting Code," Verzocchi, (1975).

3,868,632, "Plural Channel Error Correcting Apparatus and Methods," Hong, et al., (1975).

3,872,431, "Apparatus for Detecting Data Bits and Error Bits in Phase Encoded Data," Besenfelder, et al., (1975).

3,876,978, "Archival Data Protection," (1975).

3,878,333, "Simplex ARQ System," Shimizu, et al., (1975).

3,882,457, "Burst Error Correction Code," En, (1975).

3,891,959, "Coding System for Differential Phase Modulation," Tsuji, et al., (1975).

3,891,969, "Syndrome Logic Checker for an Error Correcting Code Decoder," Christensen, (1975).

3,893,070, "Error Correction and Detection Circuit With Modular Coding Unit," (1975).

3,893,071, "Multi Level Error Correction System for High Density Memory," (1975).

3,893,078, "Method and Apparatus for Calculating the Cyclic Code of a Binary Message," Finet, (1975).

3,895,349, "Pseudo-Random Binary Sequence Error Counters," Robson, (1975).

3,896,416, "Digital Teleccommunications Apparatus Having Error-Correcting Facilities," Barrett, et al., (1975).

3,903,474, "Periodic Pulse Check Circuit," (1975).

3,909,784, "Information Coding With Error Tolerant Code," Raymond, (1975).

3,913,068, "Error Correction of Serial Data Using a Subfield Code," (1975).

3,920,976, "Information Storage Security System," Christensen, et al., (1975).

3,921,210, "High Density Data Processing System," Halpern, (1975).

3,925,760, "Method of and Apparatus for Optical Character Recognition, Reading and Reproduction," Mason, et al., (1975).

3,928,823, "Code Translation Arrangement," (1975).

3,930,239, "Integrated Memory," Salters, et al., (1975).

3,938,085, "Transmitting Station and Receiving Station for Operating With a Systematic Recurrent Code," (1976).

3,944,973, "Error Syndrome and Correction Code Forming Devices," Masson, (1976).

3,949,380, "Peripheral Device Reassignment Control Technique," Barbour, et al., (1976).

3,958,110, "Logic Array with Testing Circuitry," Hong, et al., (1976).

3,958,220, "Enhanced Error Correction," (1976).

3,982,226, "Means and Method for Error Detection and Correction of Digital Data," (1976).

3,983,536, "Data Signal Handling Arrangements," Telfer, (1976).

3,988,677, "Space Communication System for Compressed Data With a Concatenated Reed-Solomon-Viterbi Coding Channel," Fletcher, et al., (1976).

3,996,565, "Programmable Sequence Controller," Nakao, et al., (1976).

3,997,876, "Apparatus and Method for Avoiding Defects in the Recording Medium within a Peripheral Storage System," Frush, (1976).

4,001,779, "Digital Error Correcting Decoder," Schiff, (1977).

4,009,469, "Loop Communications System with Method and Apparatus for Switch to Secondary Loop," Boudreau, et al., (1977).

4,013,997, "Error Detection/Correction System," Treadwell III, (1977).

4,015,238, "Metric Updater for Maximum Likelihood Decoder," (1977).

4,020,461, "Method of and Apparatus for Transmitting and Receiving Coded Digital Signals," Adams, et al., (1977).

4,024,498, "Apparatus for Dead Track Recovery," (1977).

4,030,067, "Table Lookup Direct Decoder for Double-Error Correcting Correcting (DEC) BCH Codes Using a Pair of Syndromes," Howell, et al., (1977).

4,030,129, "Pulse Code Modulated Digital Audio System," (1977).

4,032,886, "Concatentation Technique for Burst-Error Correction and Synchronization," En, et al., (1977).

4,035,767, "Error Correction Code and Apparatus for the Correction of Differentially Encoded Quadrature Phase Shift Keyed Data (DQPSK)," Chen, et al., (1977).

4,037,091, "Error Correction Circuit Utilizing Multiple Parity Bits," (1977).

4,037,093, "Matrix Multiplier in GF(2^m)," (1977).

4,044,328, "Data Coding and Error Correcting Methods and Apparatus," Herff, (1977).

4,044,329, "Variable Cyclic Redundancy Character Detector," (1977).

4,047,151, "Adaptive Error Correcting Transmission System," Rydbeck, et al., (1977).

4,052,698, "Multi-Parallel-Channel Error Checking," (1977).

4,054,921, "Automatic Time-Base Error Correction System," (1977).

4,055,832, "One-Error Correction Convolutional Coding System," (1977).

4,058,851, "Conditional Bypass of Error Correction for Dual Memory Access Time Selection," Scheuneman, (1977).

4,063,038, "Error Coding Communication Terminal Interface," Kaul, et al., (1977).

4,064,483, "Error Correcting Circuit Arrangement Using Cube Circuits," (1977).

4,072,853, "Apparatus and Method for Storing Parity Encoded Data from a Plurality of Input/Output Sources," Barlow, et al., (1978).

4,074,228, "Error Correction of Digital Signals," (1978).

4,077,028, "Error Checking and Correcting Device," Lui, et al., (1978).

4,081,789, "Switching Arrangement for Correcting the Polarity of a Data Signal Transmitted With a Recurrent Code," (1978).

4,087,787, "Decoder for Implementing an Approximation of the Viterbi Algorithm Using Analog Processing Techniques," (1978).

4,092,713, "Post-Write Address Word Correction in Cache Memory System," Scheuneman, (1978).

4,099,160, "Error Location Apparatus and Methods," (1978).

4,105,999, "Parallel-Processing Error Correction System," Nakamura, (1978).

4,107,650, "Error Correction Encoder and Decoder," Luke, et al., (1978).

4,107,652, "Error Correcting and Controlling System," (1978).

4,110,735, "Error Detection and Correction," Maxemchuk, (1978).

4,112,502, "Conditional Bypass of Error Correction for Dual Memory Access Time Selection," Scheuneman, (1978).

4,115,768, "Sequential Encoding and Decoding of Variable Word Length, Fixed Rate Data Codes," Eggenberger, et al., (1978).

4,117,458, "High Speed Double Error Correction Plus Triple Error Detection System," (1978).

4,119,945, "Error Detection and Correction," Lewis, Jr., et al., (1978).

4,129,355, "Light Beam Scanner with Parallelism Error Correction," Noguchi, (1978).

4,138,694, "Video Signal Recorder/Reproducer for Recording and Reproducing Pulse Signals," (1979).

4,139,148, "Double Bit Error Correction Using Single Bit Error Correction, Double Bit Error Detection Logic and Syndrome Bit Memory," (1979).

4,141,039, "Recorder Memory With Variable Read and Write Rates," Yamamoto, (1979).

4,142,174, "High Speed Decoding of Reed-Solomon Codes," Chen, et al., (1979).

4,145,683, "Single Track Audio-Digital Recorder and Circuit for Use Therein Having Error Correction," Brookhart, (1979).

4,146,099, "Signal Recording Method and Apparatus," Matsushima, et al. (1979).

4,146,909, "Sync Pattern Encoding System for Run-Length Limited Codes," Beckenhauer, et al., (1979).

4,151,510, "Method and Apparatus for an Efficient Error Detection and Correction System," Howell, et al., (1979).

4,151,565, "Discrimination During Reading of Data Prerecorded in Different Codes," Mazzola, (1979).

4,156,867, "Data Communication System With Random and Burst Error Protection and Correction," Bench, et al., (1979).

4,157,573, "Digital Data Encoding and Reconstruction Circuit," (1979).

4,159,468, "Communications Line Authentication Device," Barnes, et al., (1979).

4,159,469, "Method and Apparatus for the Coding and Decoding of Digital Information," (1979).

4,160,236, "Feedback Shift Register," Oka, et al., (1979).

4,162,480, "Galois Field Computer," (1979).

4,163,147, "Double Bit Error Correction Using Double Bit Complementing," (1979).

4,167,701, "Decoder for Error-Correcting Code Data," Kuki, et al., (1979).

4,168,468, "Radio Motor Control System," Mabuchi, et al., (1979).

4,168,486, "Segmented Error-Correction System," (1979).

4,175,692, "Error Correction and Detection Systems," (1979).

4,181,934, "Microprocessor Architecture with Integrated Interrupts and Cycle Steals Prioritized Channel," Marenin, (1980).

4,183,463, "Ram Error Correction Using Two Dimensional Parity Checking," (1980).

4,185,269, "Error Correcting System for Serial by Byte Data," (1980).

4,186,375, "Magnetic Storage Systems for Coded Numerical Data with Reversible Transcoding into High Density Bipolar Code of Order N," Castellani, et al., (1980).

4,188,616, "Method and System for Transmitting and Receiving Blocks of Encoded Data Words to Minimize Error Distortion in the Recovery of Said Data Words," Kazami, et al., (1980).

4,189,710, "Method and Apparatus for Detecting Errors in a Transmitted Code," Iga, (1980).

4,191,970, "Interframe Coder for Video Signals," Witsenhausen, et al., (1980).

4,193,062, "Triple Random Error Correcting Convolutional Code," En, (1980).

4,196,445, "Time-Base Error Correction," (1980).

4,201,337, "Data Processing System Having Error Detection and Correction Circuits," Lewis, et al., (1980).

4,201,976, "Plural Channel Error Correcting Methods and Means Using Adaptive Reallocation of Redundant Channels Among Groups of Channels," (1980).

4,202,018, "Apparatus and Method for Providing Error Recognition and Correction of Recorded Digital Information," Stockham, Jr., (1980).

4,204,199, "Method and Means for Encoding and Decoding Digital Data," Isailovic, (1980).

4,204,634, "Storing Partial Words in Memory," (1980).

4,205,324, "Methods and Means for Simultaneously Correcting Several Channels in Error in a Parallel Multi Channel Data System Using Continuously Modifiable Syndromes and Selective Generation of Internal Channel Pointers," (1980).

4,205,352, "Device for Encoding and Recording Information with Peak Shift Compensation," Tomada, (1980).

4,206,440, "Encoding for Error Correction of Recorded Digital Signals," Doi, et al., (1980).

4,209,809, "Apparatus and Method for Record Reorientation Following Error Detection in a Data Storage Subsystem," Chang, et al., (1980).

4,209,846, "Memory Error Logger Which Sorts Transient Errors From Solid Errors," Seppa, (1980).

4,211,996, "Error Correction System for Differential Phase-Shift-Keying," Nakamura, (1980).

4,211,997, "Method and Apparatus Employing an Improved Format for Recording and Reproducing Digital Audio," Rudnick, et al., (1980).

4,213,163, "Video-Tape Recording," Lemelson, (1980).

4,214,228, "Error-Correcting and Error-Detecting System," (1980).

4,215,402, "Hash Index Table Hash Generator Apparatus," Mitchell, et al., (1980).

4,216,532, "Self-Correcting Solid-State Mass Memory Organized by Words for a Stored-Program Control System," Garetti, et al., (1980).

4,216,540, "Programmable Polynomial Generator," McSpadden, (1980).

4,216,541, "Error Repairing Method and Apparatus for Bubble Memories," Clover, et al., (1980).

4,223,382, "Closed Loop Error Correct," Thorsrud, (1980).

4,225,959, "Tri-State Bussing System," (1980).

4,234,804, "Signal Correction for Electrical Gain Control Systems," Bergstrom, (1980).

4,236,247, "Apparatus for Correcting Multiple Errors in Data Words Read From a Memory," Sundberg, (1980).

4,238,852, "Error Correcting System," Iga, et al., (1980).

4,240,156, "Concatenated Error Correcting System," (1980).

4,241,446, "Apparatus for Performing Single Error Correction and Double Error Detection," Trubisky, (1980).

4,242,752, "Circuit Arrangement for Coding or Decoding of Binary Data," Herkert, (1980).

4,249,253, "Memory With Selective Intervention Error Checking and Correcting Device," Gentili, et al., (1981).

4,253,182, "Optimization of Error Detection and Correction Circuit," (1981).

4,254,500, "Single Track Digital Recorder and Circuit for Use Therein Having Error Correction," Brookhart, (1981).

4,255,809, "Dual Redundant Error Detection System for Counters," Hillman, (1981)

4,261,019, "Compatible Digital Magnetic Recording System," McClelland, (1981).

4,271,520, "Synchronizing Technique for an Error Correcting Digital Transmission System," (1981).

4,275,466, "Block Sync Signal Extracting Apparatus," (1981).

4,276,646, "Method and Apparatus for Detecting Errors in a Data Set," Haggard, et al., (1981).

4,276,647, "High Speed Hamming Code Circuit and Method for the Correction of Error Bursts," Thacker, et al., (1981).

4,277,844, "Method of Detecting and Correcting Errors in Digital Data Storage Systems," (1981).

4,281,355, "Digital Audio Signal Recorder," Wada, et al., (1981).

4,283,787, "Cyclic Redundancy Data Check Encoding Method and Apparatus," (1981).

4,291,406, "Error Correction on Burst Channels by Sequential Decoding," (1981).

4,292,684, "Format for Digital Tape Recorder," Kelley, et al., (1981).

4,295,218, "Error-Correcting Coding System," Tanner, (1981).

4,296,494, "Error Correction and Detection Systems," Ishikawa, et al., (1981).

4,298,981, "Decoding Shortened Cyclic Block Codes," Byford, (1981).

4,300,231, "Digital System Error Correction Arrangement," Moffitt, (1981).

4,306,305, "PCM Signal Transmitting System With Error Detecting and Correcting Capability," Doi, et al., (1981).

4,309,721, "Error Coding for Video Disc System," (1982).

4,312,068, "Parallel Generation of Serial Cyclic Redundancy Check," Goss, et al., (1982).

4,312,069, "Serial Encoding-Decoding for Cyclic Block Codes," Ahamed, (1982).

4,317,201, "Error Detecting and Correcting RAM Assembly," Sedalis, (1982).

4,317,202, "Circuit Arrangement for Correction of Data," Markwitz, (1982).

4,319,356, "Self-Correcting Memory System," Kocol, et al., (1982).

4,319,357, "Double Error Correction Using Single Error Correcting Code," Bossen, (1982).

4,320,510, "Error Data Correcting System," Kojima, (1982).

4,328,580, "Apparatus and an Improved Method for Processing of Digital Information," Stockham, Jr., et al., (1982).

4,330,860, "Error Correcting Device," Wada, et. al, (1982).

4,334,309, "Error Correcting Code System," Bannon, et. al, (1982).

4,335,458, "Memory Incorporating Error Detection and Correction," Krol (1982).

4,336,611, "Error Correction Apparatus and Method," Bernhardt, et al., (1982).

4,336,612, "Error Correction Encoding and Decoding System," Inoue, et. al, (1982).

4,337,458, "Data Encoding Method and System Employing Two-Thirds Code Rate with Full Word Look-Ahead," Cohn, et al., (1982).

4,344,171, "Effective Error Control Scheme for Satellite Communications," Lin, et al., (1982).

4,345,328 "ECC Check Bit Generation Using Through Checking Parity Bits," White, (1982).

4,355,391, "Apparatus and Method of Error Detection and/or Correction in a Data Set," Alsop IV, (1982).

4,355,392, "Burst-Error Correcting System," Doi, et al., (1982).

4,356,566, "Synchronizing Signal Detecting Apparatus," Wada, et al., (1982).

4,357,702, "Error Correcting Apparatus," Chase, et al., (1982).

4,358,848, "Dual Function ECC System with Block Check Byte," Patel, (1982).

4,359,772, "Dual Function Error Correcting System," Patel, (1982).

4,360,916, "Method and Apparatus for Providing for Two Bits-Error Detection and Correction," Kustedjo, et al., (1982).

4,360,917, "Parity Fault Locating Means," Sindelar, et al., (1982).

4,365,332, "Method and Circuitry for Correcting Errors in Recirculating Memories," Rice, (1982).

4,368,533, "Error Data Correcting System," Kojima, (1983).

4,369,510, "Soft Error Rewrite Control System," Johnson, et al., (1983).

4,375,100, "Method and Apparatus for Encoding Low Redundancy Check Words from Source Data," Tsuji, et al., (1983).

4,377,862, "Method of Error Control in Asynchronous Communications," Koford, et al., (1983).

4,377,863, "Synchronization Loss Tolerant Cyclic Error Checking Method and Apparatus," Legory, et al., (1983).

4,380,071, "Method and Apparatus for Preventing Errors in PCM Signal Processing Apparatus," Odaka, (1983).

4,380,812, "Refresh and Error Detection and Correction Technique for a Data Processing System," Ziegler II, et al., (1983).

4,382,300, "Method and Apparatus for Decoding Cyclic Codes Via Syndrome Chains, " Gupta, (1983).

4,384,353, "Method and Means for Internal Error Check in a Digital Memory," Varshney, (1983).

4,388,684, "Apparatus for Deferring Error Detection of Multibyte Parity Encoded Data Received From a Plurality of Input/Output Data Sources," Nibby, Jr., et al., (1983).

4,393,502, "Method and Apparatus for Communicating Digital Information Words by Error-Correction Encoding," Tanaka, et al., (1983).

4,394,763, "Error-Correcting System," Nagano, et al., (1983).

4,395,768, "Error Correction Device for Data Transfer System," Goethals, et al., (1983).

4,397,022 "Weighted Erasure Codec for the (24,12) Extended Golay Code," Weng, et al., (1983).

4,398,292, "Method and Apparatus for Encoding Digital with Two Error-Correcting Codes," Doi, et al., (1983).

4,402,045, "Multi-Processor Computer System," Krol, (1983).

4,402,080 "Synchronizing Device for a Time Division Multiplex System," Mueller, (1983).

4,404,673, "Error Correcting Network," Yamanouchi, (1983).

4,404,674, "Method and Apparatus for Weighted Majority Decoding of FEC Codes Using Soft Detection," Rhodes, (1983).

4,404,675, "Frame Detection and Synchronization System for High Speed Digital Transmission Systems," Karchevski, (1983).

4,404,676, "Partitioning Method and Appartus Using Data-Dependent Boundary-Marking Code Words," DeBenedictis, (1983).

4,412,329, "Parity Bit Lock-On Method and Apparatus," Yarborough, Jr., (1983).

4,413,339, "Multiple Error Detecting and Correcting System Employing Reed-Solomon Codes," Riggle, et al., (1983).

4,413,340, "Error Correctable Data Transmission Method," Odaka, et al., (1983).

4,414,667, "Forward Error Correcting Apparatus," Bennett, (1983).

4,417,339, "Fault Tolerant Error Correction Circuit," Cantarella, (1983).

4,418,410, "Error Detection and Correction Apparatus for a Logic Array," Goetze, et al., (1983).

4,425,644, "PCM Signal System," Scholz, (1984).

4,425,645, "Digital Data Transmission with Parity Bit Word Lock-On," Weaver, et al., (1984).

4,425,646, "Input Data Synchronizing Circuit," Kinoshita, et al., (1984).

4,429,390, "Digital Signal Transmitting System," Sonoda, et al., (1984).

4,429,391, "Fault and Error Detection Arrangement," Lee, (1984).

4,433,348, "Apparatus and Method for Requiring Proper Synchronization of a Digital Data Flow," Stockham, Jr., et al., (1984).

4,433,415, "PCM Signal Processor," Kojima, (1984.)

4,433,416, "PCM Signal Processor," Kojima, (1984).

4,434,487, "Disk Format for Secondary Storage System," Rubinson, et al., (1984).

4,435,807, "Orchard Error Correction System," Scott, et al., (1984).

4,441,184, "Method and Apparatus for Transmitting a Digital Signal," Sonoda, et al., (1984).

4,447,903, "Forward Error Correction Using Coding and Redundant Transmission," Sewerinson, (1984).

4,450,561, "Method and Device for Generating Check Bits Protecting a Data Word," Gotze, et al., (1984).

4,450,562, "Two Level Parity Error-Correction System," Wacyk, et al., (1984).

4,451,919, "Digital Signal Processor for Use in Recording and/or Reproducing Equipment," Wada, et al., (1984).

4,451,921, "PCM Signal Processing Circuit," Odaka, (1984).

4,453,248, "Fault Alignment Exclusion Method to Prevent Realignment of Previously Paired Memory Defects," Ryan, (1984).

4,453,250, "PCM Signal Processing Apparatus," Hoshimi, et al., (1984).

4,453,251, "Error-Correcting Memory with Low Storage Overhead and Fast Correction Mechanism," Osman, (1984).

4,454,600, "Parallel Cyclic Redundancy Checking Circuit," LeGresley, (1984).

4,454,601, "Method and Apparatus for Communication of Information and Error Checking," Helms, et al., (1984).

4,455,655, "Real Time Fault Tolerant Error Correction Mechanism," Galen, et al., (1984).

4,456,996, "Parallel/Series Error Correction Circuit," Haas, et al., (1984).

4,458,349, "Method for Storing Data Words in Fault Tolerant Memory to Recover Uncorrectable Errors," Aichelmann, Jr., et al. (1984).

4,459,696, "PCM Signal Processor with Error Detection and Correction Capability Provided by a Data Error Pointer," Kojima, (1984).

4,462,101, "Maximum Likelihood Error Correcting Technique," Yasuda, et al., (1984).

4,462,102, "Method and Apparatus for Checking the Parity of Disassociated Bit Groups," Povlick, (1984).

4,464,752, "Multiple Event Hardened Core Memory," Schroeder, et al., (1984).

4,464,753, "Two Bit Symbol SEC/DED Code," Chen, (1984).

4,464,754, "Memory System with Redundancy for Error Avoidance," Stewart, et al., (1984).

4,464,755, "Memory System with Error Detection and Correction," Stewart, et al., (1984).

4,468,769, "Error Correcting System for Correcting Two or Three Simultaneous Errors in a Code," Koga, (1984).

4,468,770, "Data Receivers Incorporating Error Code Detection and Decoding," Metcalf, et al., (1984).

4,472,805, "Memory System with Error Storage," Wacyk, et al., (1984).

4,473,902, "Error Correcting Code Processing System," Chen, (1984).

4,476,562, "Method of Error Correction," Sako, et al., (1984).

4,477,903, "Error Correction Method for the Transfer of Blocks of Data Bits, a Device for Preforming such a Method, A Decoder for Use with such a Method, and a Device Comprising such a Decoder," Schouhamer Immink, et al., (1984).

4,494,234, "On-The-Fly Multibyte Error Correcting System," Patel, (1985).

4,495,623, "Digital Data Storage in Video Format," George, et al., (1984).

4,497,058, "Method of Error Correction," Sako, et al., (1985).

4,498,174, "Parallel Cyclic Redundancy Checking Circuit," LeGresley, (1985).

4,498,175, "Error Correcting System," Nagumo, et al., (1985).

4,498,178, "Data Error Correction Circuit," Ohhashi, (1985).

4,502,141, "Circuits for Checking Bit Errors in a Received BCH Code Succession by the Use of Primitive and Non-Primitive Polynomials," Kuki, (1985).

4,504,948, "Syndrome Processing Unit for Multibyte Error Correcting Systems," Patel, (1985).

4,506,362, "Systematic Memory Error Detection and Correction Apparatus and Method," Morley, (1985).

4,509,172, "Double Error Correction - Triple Error Detection Code," Chen, (1985).

4,512,020, "Data Processing Device for Processing Multiple-Symbol Data-Words Based on a Symbol-Correcting Code and Having Multiple Operating Modes," Krol, et al., (1985).

4,519,058, "Optical Disc Player," Tsurushima, et al., (1985).

4,525,838, "Multibyte Error Correcting System Involving a Two-Level Code Structure," Patel, (1985).

4,525,840, "Method and Apparatus for Maintaining Word Synchronization After a Synchronizing Word Dropout in Reproduction of Recorded Digitally Encoded Signals," Heinz, et al., (1985).

4,527,269, "Encoder Verifier," Wood, et al., (1985).

4,538,270, "Method and Apparatus for Translating a Predetermined Hamming Code to an Expanded Class of Hamming Codes," Goodrich, Jr., et al., (1985).

4,541,091, "Code Error Detection and Correction Method and Apparatus," Nishida, et al., (1985).

4,541,092, "Method for Error Correction," Sako, et al., (1985).

4,541,093, "Method and Apparatus for Error Correction," Furuya, et al., (1985).

4,544,968, "Sector Servo Seek Control," Anderson, et al., (1985).

4,546,474, "Method of Error Correction," Sako, et al., (1985).

4,549,298, "Detecting and Correcting Errors in Digital Audio Signals," Creed, et al., (1985).

4,554,540, "Signal Format Detection Circuit for Digital Radio Paging Receiver," Mori, et al., (1985).

4,555,784, "Parity and Syndrome Generation for Error Detection and Correction in Digital Communication Systems," Wood, (1985).

4,556,977, "Decoding of BCH Double Error Correction - Triple Error Detection (DEC-TED) Codes," Olderdissen, et al., (1985).

4,559,625, "Interleavers for Digital Communications," Berlekamp, et al., (1985).

4,562,577, "Shared Encoder/Decoder Circuits for Use with Error Correction Codes of an Optical Disk System," Glover, et al., (1985).

4,564,941, "Error Detection System," Woolley, et al., (1986).

4,564,944, "Error Correcting Scheme," Arnold, et al., (1986).

4,564,945, "Error-Correction Code for Digital Data on Video Disc," Glover, et al., (1986).

4,566,105, "Coding, Detecting or Correcting Transmission Error System," Oisel, et al., (1986).

4,567,594, "Reed-Solomon Error Detecting and Correcting System Employing Pipelined Processors," Deodhar, (1986).

4,569,051, "Methods of Correcting Errors in Binary Data," Wilkinson, (1986).

4,573,171, "Sync Detect Circuit," McMahon, Jr., et al., (1986).

4,583,225, "Reed-Solomon Code Generator," Yamada, et al., (1986).

4,584,686, "Reed-Solomon Error Correction Apparatus," Fritze, (1986).

4,586,182, "Source Coded Modulation System," Gallager, (1986).

4,586,183, "Correcting Errors in Binary Data," Wilkinson, (1986).

4,589,112, "System for Multiple Error Detection with Single and Double Bit Error Correction," Karim, (1986).

4,592,054, "Decoder with Code Error Correcting Function," Namekawa, et al., (1986).

4,593,392, "Error Correction Circuit for Digital Audio Signal," Kouyama, (1986).

4,593,393, "Quasi Parallel Cyclic Redundancy Checker," Mead, et al., (1986).

4,593,394, "Method Capable of Simultaneously Decoding Two Reproduced Sequences," Tomimitsu, (1986).

4,593,395, "Error Correction Method for the Transfer of Blocks of Data Bits, a Device and Performing such a Method, A Decoder for Use with such a Method, and a Device Comprising such a Decoder," Schouhamer Immink, et al., (1986).

4,597,081, "Encoder Interface with Error Detection and Method Therefor," Tassone, (1986).

4,597,083, "Error Detection and Correction in Digital Communication Systems," Stenerson, (1986).

4,598,402, "System for Treatment of Single Bit Error in Buffer Storage Unit," Matsumoto, et al., (1986).

4,604,747, "Error Correcting and Controlling System," Onishi, et al., (1986).

4,604,750, "Pipeline Error Correction," Manton, et al., (1986).

4,604,751, "Error Logging Memory System for Avoiding Miscorrection of Triple Errors," Aichelmann, Jr., et al., (1986).

4,606,026, "Error-Correcting Method and Apparatus for the Transmission of Word-Wise Organized Data," Baggen, (1986).

4,607,367, "Correcting Errors in Binary Data," Ive, et al., (1986).

4,608,687, "Bit Steering Apparatus and Method for Correcting Errors in Stored Data, Storing the Address of the Corrected Data and Using the Address to Maintain a Correct Data Condition," Dutton, (1986).

4,608,692, "Error Correction Circuit," Nagumo, et al., (1986).

4,617,664, "Error Correction for Multiple Bit Output Chips," Aichelmann, Jr., et al., (1986).

4,623,999, "Look-up Table Encoder for Linear Block Codes," Patterson, (1986).

4,627,058, "Code Error Correction Method," Moriyama, (1986).

4,630,271, "Error Correction Method and Apparatus for Data Broadcasting System," Yamada, (1986).

4,630,272, "Encoding Method for Error Correction," Fukami, et al., (1986).

4,631,725, "Error Correcting and Detecting System," Takamura, et al., (1986).

4,633,471, "Error Detection and Correction in an Optical Storage System," Perera, et al., (1986).

4,637,023, "Digital Data Error Correction Method and Apparatus," Lounsbury, et al., (1987).

4,639,915, "High Speed Redundancy Processor," Bosse, (1987).

4,642,808, "Decoder for the Decoding of Code Words which are Blockwise Protected Against the Occurrence of a Plurality of Symbol Errors within a Block by Means of a Reed-Solomon Code, and Reading Device for Optically Readable Record Carriers," Baggen, (1987).

4,646,301, "Decoding Method and System for Doubly-Encoded Reed-Solomon Codes," Okamoto, et al., (1987).

4,646,303, "Data Error Detection and Correction Circuit," Narusawa, et al., (1987).

PERIODICALS

Abramson, N., "Cascade Decoding of Cyclic Product Codes." IEEE Trans. on Comm. Tech., Com-16 (3), 398-402 (June 1968).

Alekar, S. V., "M6800 Program Performs Cyclic Redundancy Checks." Electronics, 167 (Dec. 1979).

Bahl, L. R. and R. T. Chien, "Single- and Multiple-Burst-Correcting Properties of a Class of Cyclic Product Codes." IEEE Trans. on Info. Theory, IT-17 (5), 594-600 (Sept. 1971).

Bartee, T. C. and D. I. Schneider, "Computation with Finite Fields." Info. and Control, 6, 79-98 (1963).

Basham G. R., "New Error-Correcting Technique for Solid-State Memories Saves Hardware." Computer Design, 110-113 (Oct. 1976).

Baumert L. D. and R. J. McEliece, "Soft Decision Decoding of Block Codes." DSN Progress Report 42-47, 60-64 (July/Aug. 1978).

Beard, Jr., J., "Computing in GF(q)." Mathematics of Comp., 28 (128), 1159-1166 (Oct. 1974).

Berlekamp, E. R., "On Decoding Binary Bose-Chaudhuri-Hocquenghem Codes." IEEE Trans. on Info. Theory, IT-11 (4), 577-579 (Oct. 1965).

Berlekamp, E. R., "The Enumeration of Information Symbols in BCH Codes." The Bell Sys. Tech. J., 1861-1880 (Oct. 1967).

Berlekamp, E. R., "Factoring Polynomials Over Finite Fields." The Bell Sys. Tech. J., 1853-1859 (Oct. 1967).

Berlekamp, E. R., "Factoring Polynomials Over Large Finite Fields." Mathematics of Comp., 24 (111), 713-735 (July 1970).

Berlekamp, E. R., "Algebraic Codes for Improving the Reliability of Tape Storage." National Computer Conference, 497-499 (1975).

Berlekamp, E. R., "The Technology of Error-Correcting Codes." Proceedings of the IEEE, 68 (5), 564-593 (May 1980).

Berlekamp, E. R. and J. L. Ramsey, "Readable Erasures Improve the Performance of Reed-Solomon Codes." IEEE Trans. on Info. Theory, IT-24 (5), 632-633 (Sept. 1978).

Berlekamp, E. R., et. al., "On the Solution of Algebraic Equations Over Finite Fields." Info. and Control, 10, 553-564 (1967).

Blum, R., "More on Checksums." Dr. Dobb's J., (69), 44-45 (July 1982).

Bossen, D. C., "b-Adjacent Error Correction." IBM J. Res. Develop., 402-408 (July 1970).

Bossen, D. C. and M. Y. Hsiao, "A System Solution to the Memory Soft Error Problem." IBM J. Res. Develop., 24 (3), 390-397 (May 1980).

Bossen, D. C. and S. S. Yau, "Redundant Residue Polynomial Codes." Info. and Control, 13, 597-618 (1968).

Boudreau, P. E. and R. F. Steen, "Cyclic Redundancy Checking by Program." Fall Joint Computer Conference, 9-15 (1971).

Brown, D. T. and F. F. Sellers, Jr., "Error Correction for IBM 800-Bit-Per-Inch Magnetic Tape." IBM J. Res. Develop., 384-389 (July 1970).

Bulthuis, K., et. al., "Ten Billion Bits on a Disk." IEEE Spectrum, 18-33 (Aug. 1979).

Burton, H. O., "Some Asymptotically Optimal Burst-Correcting Codes and Their Relation to Single-Error-Correcting Reed-Solomon Codes." IEEE Trans. on Info. Theory, IT-17 (1), 92-95 (Jan. 1971).

Burton, H. O. "Inversionless Decoding of Binary BCH Codes." IEEE Trans. on Info. Theory, IT-17 (4), 464-466 (July 1971).

Carter, W. C. and C. E. McCarthy, "Implementation of an Experimental Fault-Tolerant Memory System." IEEE Trans. on Computers, C-25 (6), 557-568 (June 1976).

Chen, C. L. and R. A. Rutledge, "Error Correcting Codes for Satellite Communication Channels." IBM J. Res. Develop., 168-175 (Mar. 1976).

Chien, R. T., "Cyclic Decoding Procedures for Bose- Chaudhuri-Hocquenghem Codes." IEEE Trans. on Info. Theory, 357-363 (Oct. 1963).

Chien, R. T., "Block-Coding Techniques for Reliable Data Transmission." IEEE Trans. on Comm. Tech., Com-19 (5), 743-751 (Oct. 1971).

Chien, R. T., "Memory Error Control: Beyond Parity." IEEE Spectrum, 18-23 (July 1973).

Chien, R. T. and B. D. Cunningham, "Hybrid Methods for Finding Roots of a Polynomial-With Application to BCH Decoding." IEEE Trans. on Info. Theory, 329-335 (Mar. 1969).

Chien, R. T., et. al., "Correction of Two Erasure Bursts." IEEE Trans. on Info. Theory, 186-187 (Jan. 1969).

Comer, E., "Hamming's Error Corrections." Interface Age, 142-143 (Feb. 1978).

Davida, G. I. and J. W. Cowles, "A New Error-Locating Polynomial for Decoding of BCH Codes." IEEE Trans. on Info. Theory, 235-236 (Mar. 1975).

Delsarte, P., "On Subfield Subcodes of Modified Reed-Solomon Codes." IEEE Trans. on Info. Theory, 575-576 (Sept. 1975).

Doi, T. T., et. al., "A Long-Play Digital Audio Disk System." Journal of the Audio Eng. Soc., 27 (12), 975-981 (Dec. 1979).

Duc. N. Q., "On the Lin-Weldon Majority-Logic Decoding Algorithm for Product Codes." IEEE Trans. on Info. Theory, 581-583 (July 1973).

Duc, N. Q. and L. V. Skattebol, "Further Results on Majority-Logic Decoding of Product Codes." IEEE Trans. on Info. Theory, 308-310 (Mar. 1972).

Forney, Jr., G. D., "On Decoding BCH Codes." IEEE Trans. on Info. Theory, IT-11 (4), 549-557 (Oct. 1965).

Forney, Jr., G. D., "Coding and Its Application in Space Communications." IEEE Spectrum, 47-58 (June 1970).

Forney, Jr., G. D., "Burst-Correcting Codes for the Classic Bursty Channel." IEEE Trans. on Comm. Tech., Com-19 (5), 772-781 (Oct. 1971).

Gorog, E., "Some New Classes of Cyclic Codes Used for Burst-Error Correction." IBM J., 102-111 (Apr. 1963).

Greenberger, H., "An Iterative Algorithm for Decoding Block Codes Transmitted Over a Memoryless Channel." DSN Progress Report 42-47, 51-59 (July/Aug. 1978).

Greenberger, H. J., "An Efficient Soft Decision Decoding Algorithm for Block Codes." DSN Progress Report 42-50, 106-109 (Jan./Feb. 1979).

Gustavson, F. G., "Analysis of the Berlekamp-Massey Linear Feedback Shift-Register Synthesis Algorithm." IBM J. Res. Develop., 204-212 (May 1976).

Gustlin, D. P. and D. D. Prentice, "Dynamic Recovery Techniques Guarantee System Reliability." Fall Joint Computer Conference, 1389-1397 (1968).

Hartmann, C. R. P., "A Note on the Decoding of Double- Error-Correcting Binary BCH Codes of Primitive Length." IEEE Trans. on Info. Theory, 765-766 (Nov. 1971).

Hellman, M. E., "Error Detection in the Presence of Synchronization Loss." IEEE Trans. on Comm., 538-539 (May 1975).

Herff, A. P., "Error Detection and Correction for Mag Tape Recording." Digital Design, 16-18 (July 1978).

Hindin, H. J., "Error Detection and Correction Cleans Up Wide-Word Memory Act." Electronics, 153-162 (June 1982).

Hodges, D. A., "A Review and Projection of Semiconductor Components for Digital Storage." Proceedings of the IEEE, 63 (8), 1136-1147 (Aug. 1975).

Hong, S. J. and A. M. Patel, "A General Class of Maximal Codes for Computer Applications." IEEE Trans. on Computers, C-21 (12), 1322-1331 (Dec. 1972).

Hsiao, M. Y. and K. Y. Sih, "Serial-to-Parallel Transformation of Linear-Feedback Shift-Register Circuits." IEEE Trans. on Elec. Comp., 738-740 (Dec. 1964).

Hsu, H. T., et. al, "Error-Correcting Codes for a Compound Channel." IEEE Trans. on Info. Theory, IT-14 (1), 135-139 (Jan. 1968).

Imamura, K., "A Method for Computing Addition Tables in GF(p^n)." IEEE Trans. on Info. Theory, IT-26 (3), 367-368 (May 1980).

Iwadare, Y., "A Class of High-Speed Decodable Burst-Correcting Codes." IEEE Trans. on Info. Theory, IT-18 (6), 817-821 (Nov. 1972).

Johnson, R. C., "Three Ways of Correcting Erroneous Data." Electronics, 121-134 (May 1981).

Justesen, J., "A Class of Constructive Asymptotically Good Algebraic Codes." IEEE Trans. Info. Theory, IT-18, 652-656 (Sept. 1972).

Justesen, J., "On the Complexity of Decoding Reed-Solomon Codes." IEEE Trans. on Info. Theory, 237-238 (Mar. 1976).

Kasami, T., and S. Lin, "On the Construction of a Class of Majority- Logic Decodable Codes." IEEE Trans. on Info. Theory, IT-17 (5), 600-610 (Sept. 1971).

Kobayashi, H., "A Survey of Coding Schemes for Transmission or Recording of Digital Data." IEEE Trans. on Comm. Tech., Com-19 (6), 1087-1100 (Dec. 1971).

Koppel, R., "Ram Reliability in Large Memory Systems-Improving MTBF With ECC." Computer Design, 196-200 (Mar. 1979).

Korodey, R. and D. Raaum, "Purge Your Memory Array of Pesky Error Bits." EDN, 153-158 (May 1980).

Laws, Jr., B. A. and C. K. Rushforth, "A Cellular-Array Multiplier for GF(2^m)." IEEE Trans. on Computers, 1573-1578 (Dec. 1971).

Leung, K. S. and L. R. Welch, "Erasure Decoding in Burst-Error Channels." IEEE Trans. on Info. Theory, IT-27 (2), 160-167 (Mar. 1981).

Levine, L. and W. Meyers, "Semiconductor Memory Reliability With Error Detecting and Correcting Codes." Computer, 43-50 (Oct. 1976).

Levitt, K. N. and W. H. Kautz, "Cellular Arrays for the Parallel Implementation of Binary Error-Correcting Codes." IEEE Trans. on Info. Theory, IT-15 (5), 597-607 (Sept. 1969).

Liccardo M. A., "Polynomial Error Detecting Codes and Their Implementation." Computer Design, 53-59 (Sept. 1971).

Lignos D., "Error Detection and Correction in Mass Storage Equipment." Computer Design, 71-75 (Oct. 1972).

Lim, R. S. and J. E. Korpi, "Unicon Laser Memory: Interlaced Codes for Multiple-Burst-Error Correction." Wescon, 1-6 (1977).

Lim, R. S., "A (31,15) Reed-Solomon Code for Large Memory Systems." National Computer Conf., 205-208 (1979).

Lin, S. and E. J. Weldon, "Further Results on Cyclic Product Codes." IEEE Trans. on Info. Theory, IT-16 (4), 452-459 (July 1970).

Liu, K. Y., "Architecture for VLSI Design of Reed-Solomon Encoders." IEEE Transactions on Computers, C-31 (2), 170-175 (Feb. 1982).

Locanthi, B., et. al., "Digital Audio Technical Committee Report." J. Audio Eng. Soc., 29 (1/2), 56-78 (Jan./Feb. 1981).

Lucy, D., "Choose the Right Level of Memory-Error Protection." Electronics Design, ss37-ss39 (Feb. 1982).

Maholick, A. W. and R. B. Freeman, "A Universal Cyclic Division Circuit." Fall Joint Computer Conf., 1-8 (1971).

Mandelbaum, D., "A Method of Coding for Multiple Errors." IEEE Trans. on Info. Theory, 518-521 (May 1968).

Mandelbaum, D., "On Decoding of Reed-Solomon Codes." IEEE Trans. on Info. Theory, IT-17 (6), 707-712 (Nov. 1971).

Mandlebaum, D., "Construction of Error Correcting Codes by Interpolation." IEEE Trans. on Info. Theory, IT-25 (1), 27-35 (Jan. 1979).

Mandelbaum, D. M., "Decoding of Erasures and Errors for Certain RS Codes by Decreased Redundancy." IEEE Trans. on Info. Theory, IT-28 (2), 330-335 (Mar. 1982).

Massey, J. L., "Shift-Register Synthesis and BCH Decoding." IEEE Trans. on Info. Theory, IT-15 (1), 122-127, (Jan. 1969).

Matt, H. J. and J. L. Massey, "Determining the Burst-Correcting Limit of Cyclic Codes." IEEE Trans. on Info. Theory, IT-26 (3), 289-297 (May 1980).

Miller R. L. and L. J. Deutsch, "Conceptual Design for a Universal Reed-Solomon Decoder." IEEE Trans. on Comm., Com-29 (11), 1721-1722 (Nov. 1981).

Miller, R. L., et. al., "A Reed Solomon Decoding Program for Correcting Both Errors and Erasures." DSN Progress Report 42-53, 102-107 (July/Aug. 1979).

Miller, R. L. et. al., "An Efficient Program for Decoding the (255, 223) Reed-Solomon Code Over $GF(2^8)$ with Both Errors and Erasures, Using Transform Decoding." IEEE Proc., 127 (4), 136-142 (July 1980).

Morris, D., "ECC Chip Reduces Error Rate in Dynamic Rams." Computer Design, 137-142 (Oct. 1980).

Naga, M. A. E., "An Error Detecting and Correcting System for Optical Memory." Cal. St. Univ., Northridge, (Feb. 1982).

Oldham, I. B., et. al., "Error Detection and Correction in a Photo-Digital Storage System." IBM J. Res. Develop., 422-430 (Nov. 1968).

Patel, A. M., "A Multi-Channel CRC Register." Spring Joint Computer Conf., 11-14 (1971).

Patel, A. M., "Error Recovery Scheme for the IBM 3850 Mass Storage System." IBM J. Res. Develop., 24 (1), 32-42 (Jan. 1980).

Patel A. M. and S. J. Hong, "Optimal Rectangular Code for High Density Magnetic Tapes." IBM J. Res. Develop., 579-588 (Nov. 1974).

Peterson, W. W., "Encoding and Error-Correction Procedures for the Bose-Chaudhuri Codes." IRE Trans. on Info. Theory, 459-470 (Sept. 1960).

Peterson, W. W. and D. T. Brown, "Cyclic Codes for Error Detection." Proceedings of the IRE, 228-235 (Jan. 1961).

Plum, T., "Integrating Text and Data Processing on a Small System." Datamation, 165-175 (June 1978).

Pohlig, S. C. and M. E. Hellman, "An Improved Algorithm for Computing Logarithms Over GF(p) and Its Cryptographic Significance." IEEE Trans. on Info. Theory, IT-24 (1), 106-110 (Jan. 1978).

Poland, Jr., W. B., et. al., "Archival Performance of NASA GFSC Digital Magnetic Tape." National Computer Conf., M68-M73 (1973).

Pollard, J. M., "The Fast Fourier Transform in a Finite Field." Mathematics of Computation, 23 (114), (Apr. 1971).

Promhouse, G. and S. E. Tavares, "The Minimum Distance of All Binary Cyclic Codes of Odd Lengths from 69 to 99." IEEE Trans. on Info. Theory, IT-24 (4), 438-442 (July 1978).

Reddy, S. M., "On Decoding Iterated Codes." IEEE Trans. on Info. Theory, IT-16 (5), 624-627 (Sept. 1970).

Reddy, S. M. and J. P. Robinson, "Random Error and Burst Correction by Iterated Codes." IEEE Trans. on Info. Theory, IT-18 (1), 182-185 (Jan. 1972).

Reed, I. S. and T. K. Truong, "The Use of Finite Fields to Compute Convolutions." IEEE Trans. on Info. Theory, IT-21 (2), 208-213 (Mar. 1975).

Reed, I. S. and T. K. Truong, "Complex Integer Convolutions Over a Direct Sum of Galois Fields." IEEE Trans. on Info. Theory, IT-21 (6), 657-661 (Nov. 1975).

Reed, I. S. and T. K. Truong, "Simple Proof of the Continued Fraction Algorithm for Decoding Reed-Solomon Codes." Proc. IEEE, 125 (12), 1318-1320 (Dec. 1978).

Reed, I. S., et. al., "Simplified Algorithm for Correcting Both Errors and Erasures of Reed-Solomon Codes." Proc. IEEE, 126 (10), 961-963 (Oct. 1979).

Reed, I. S., et. al., "The Fast Decoding of Reed-Solomon Codes Using Fermat Theoretic Transforms and Continued Fractions." IEEE Trans. on Info. Theory, IT-24 (1), 100-106 (Jan. 1978).

Reed, I. S., et. al., "Further Results on Fast Transforms for Decoding Reed-Solomon Codes Over GF(2^n) for n=4,5,6,8." DSN Progress Report 42-50, 132-155 (Jan./Feb. 1979).

Reno, C. W. and R. J. Tarzaiski, "Optical Disc Recording at 50 Megabits/Second." SPIE, 177, 135-147 (1979).

Rickard, B., "Automatic Error Correction in Memory Systems." Computer Design, 179-182 (May 1976).

Ringkjob, E. T., "Achieving a Fast Data-Transfer Rate by Optimizing Existing Technology." Electronics, 86-91 (May 1975).

Sanyal, S. and K. N. Venkataraman, "Single Error Correcting Code Maximizes Memory System Efficiency." Computer Design, 175-184 (May 1978).

Sloane, N. J. A., "A Survey of Constructive Coding Theory, and a Table of Binary Codes of Highest Known Rate." Discrete Mathematics, 3, 265-294 (1972).

Sloane, N. J. A., "A Simple Description of an Error-Correcting Code for High-Density Magnetic Tape." The Bell System Tech. J., 55 (2), 157-165) (Feb. 1976).

Steen, R. F., "Error Correction for Voice Grade Data Communication Using a Communication Processor." IEEE Trans. on Comm., Com-22 (10), 1595-1606 (Oct. 1974).

Stiffler, J. J., "Comma-Free Error-Correcting Codes." IEEE Trans. on Info. Theory, 107-112 (Jan. 1965).

Stone, H. S., "Spectrum of Incorrectly Decoded Bursts for Cyclic Burst Error Codes." IEEE Trans. on Info. Theory, IT-17 (6), 742-748 (Nov. 1971).

Stone, J. J., "Multiple Burst Error Correction." Info. and Control, 4, 324-331 (1961).

Stone, J. J., "Multiple-Burst Error Correction with the Chinese Remainder Theorem." J. Soc. Indust. Appl. Math., 11 (1), 74-81 (Mar. 1963).

Sundberg, C. E. W., "Erasure and Error Decoding for Semiconductor Memories." IEEE Trans. on Computers, C-27 (8), 696-705 (Aug. 1978).

Swanson, R., "Understanding Cyclic Redundancy Codes." Computing Design, 93-99 (Nov. 1975).

Tang, D. T. and R. T. Chien, "Coding for Error Control." IBM Syst. J., (1), 48-83 (1969).

Truong, T. K. and R. L. Miller, "Fast Technique for Computing Syndromes of B.C.H. and Reed-Solomon Codes." Electronics Letters, 15 (22), 720-721 (Oct. 1979).

Ullman, J. D., "On the Capabilities of Codes to Correct Synchronization Errors." IEEE Trans. on Info. Theory, IT-13 (1), 95-105 (Jan. 1967).

Ungerboeck, G., "Channel Coding With Multilevel/Phase Signals." IEEE Trans. on Info. Theory, IT-28 (1), 55-67 (Jan. 1982).

Van Der Horst, J. A., "Complete Decoding of Triple-Error-Correcting Binary BCH Codes." IEEE Trans. on Info. Theory, IT-22 (2), 138-147 (Mar. 1976).

Wainberg, S., "Error-Erasure Decoding of Product Codes." IEEE Trans. on Info. Theory, 821-823 (Nov. 1972).

Wall, E. L., "Applying the Hamming Code to Microprocessor-Based Systems." Electronics, 103-110 (Nov. 1979).

Welch, L. R. and R. A. Scholtz, "Continued Fractions and Berlekamp's Algorithm." IEEE Trans. on Info. Theory, IT-25 (1), 19-27 (Jan. 1979).

Weldon, Jr., E. J., "Decoding Binary Block Codes on Q-ary Output Channels." IEEE Trans. on Info. Theory, IT-17 (6), 713-718 (Nov. 1971).

Weng, L. J., "Soft and Hard Decoding Performance Comparisons for BCH Codes." IEEE, 25.5.1-25.5.5 (1979).

White, G. M., "Software-Based Single-Bit I/0 Error Detection and Correction Scheme." Computer Design, 130-146 (Sept. 1978).

Whiting, J. S., "An Efficient Software Method for Implementing Polynomial Error Detection Codes." Computer Design, 73-77 (Mar. 1975).

Willett, M., "The Minimum Polynomial for a Given Solution of a Linear Recursion." Duke Math. J., 39 (1), 101-104 (Mar. 1972).

Willett, M., "The Index of an M-Sequence." Siam J. Appl. Math., 25 (1), 24-27 (July 1973).

Willett, M., "Matrix Fields Over GF(Q)." Duke Math. J., 40 (3), 701-704 (Sept. 1973).

Willett, M., "Cycle Representations for Minimal Cyclic Codes." IEEE Trans. on Info. Theory, 716-718 (Nov. 1975).

Willett, M., "On a Theorem of Kronecker." The Fibonacci Quarterly, 14 (1), 27-30 (Feb. 1976).

Willett, M., "Characteristic m-Sequences." Math. of Computation, 30 (134), 306-311 (Apr. 1976).

Willett, M., "Factoring Polynomials Over a Finite Field." Siam J. Appl. Math., 35 (2), 333-337 (Sept. 1978).

Willett, M., "Arithmetic in a Finite Field." Math. of Computation, 35 (152), 1353-1359 (Oct. 1980).

Wimble, M., "Hamming Error Correcting Code." BYTE Pub. Inc., 180-182 (Feb. 1979).

Wolf, J., "Nonbinary Random Error-Correcting Codes." IEEE Trans. on Info. Theory, 236-237 (Mar. 1970).

Wolf, J. K., et. al., "On the Probability of Undetected Error for Linear Block Codes." IEEE Trans. on Comm., Com-30 (2), 317-324 (Feb. 1982).

Wong, J., et. al., "Software Error Checking Procedures for Data Communication Protocols." Computer Design, 122-125 (Feb. 1979).

Wu, W. W., "Applications of Error-Coding Techniques to Satellite Communications." Comsat Tech. Review, 1 (1), 183-219 (Fall 1971).

Wyner, A. D., "A Note on a Class of Binary Cyclic Codes Which Correct Solid-Burst Errors." IBM J., 68-69 (Jan. 1964).

Yencharis, L., "32-Bit Correction Code Reduces Errors on Winchester Disks." Electronics Design, 46-47 (Mar. 1981).

Ziv, J., "Further Results on the Asymptotic Complexity of an Iterative Coding Scheme." IEEE Trans. on Info. Theory, IT-12 (2), 168-171 (Apr. 1966).

INDEX